DEVELOPMENTS IN **WATER SCIENCE 56**

ENVIRONMENTAL HYDRAULICS

HYDRODYNAMIC AND POLLUTANT TRANSPORT MODELLING OF LAKES AND COASTAL WATERS

OTHER TITLES AVAILABLE IN *DEVELOPMENTS IN WATER SCIENCE:*

41. D. STEPHENSON AND M.S. PETERSON
WATER RESOURCES DEVELOPMENT IN DEVELOPING COUNTRIES

43. J. ZÁRUBA
WATER HAMMER IN PIPE-LINE SYSTEMS

44. W.E. KELLY AND S. MARES (EDITORS)
APPLIED GEOPHYSICS IN HYDROGEOLOGICAL AND ENGINEERING PRACTICE

46. G.A. BRUGGEMAN
ANALYTICAL SOLUTIONS OF GEOHYDROLOGICAL PROBLEMS

47. S.M. HASSANIZADEH, R.J. SCHOLTTING, W.G. GRAY AND G.F. PINDER
COMPUTATIONAL METHODS IN WATER RESOURCES

49. P. WILDERER AND S. WUERTZ
MODERN SCIENTIFIC TOOLS IN BIOPROCESSING

50. A.S. ALSHARHAN AND W.W. WOOD
WATER RESOURCES PERSPECTIVES: EVALUATION, MANAGMENT AND POICY

51. S.K. JAIN AND V.P. SINGH
WATER RESOURCES SYSTEMS: PLANNING AND MANAGEMENT

55. C.T. MILLER, M.W. FARTHING, W.G. GRAY AND G.F. PINDER
COMPUTATIONAL METHODS IN WATER RESOURCES (Volumes 1 and 2)

DEVELOPMENTS IN **WATER SCIENCE 56**

ENVIRONMENTAL HYDRAULICS

HYDRODYNAMIC AND POLLUTANT TRANSPORT MODELLING OF LAKES AND COASTAL WATERS

I.K. TSANIS
DEPARTMENT OF CIVIL ENGINEERING
MCMASTER UNIVERSITY
HAMILTON, CANADA

Dr. JIAN WU
DEPARTMENT OF CIVIL ENGINEERING
MCMASTER UNIVERSITY
HAMILTON, CANADA

Dr. HUIHUA SHEN
DEPARTMENT OF CIVIL ENGINEERING
MCMASTER UNIVERSITY
HAMILTON, CANADA

CATERINA VALEO
DEPARTMENT OF GEOMATICS ENGINEERING
UNIVERSITY OF CALGARY
CALGARY, CANADA

ELSEVIER

Amsterdam – Boston – Heidelberg – London – New York – Oxford
Paris – San Diego – San Francisco – Singapore – Sydney – Tokyo

Elsevier
Radarweg 29, PO Box 211, 1000 AE Amsterdam, The Netherlands
The Boulevard, Langford Lane, Kidlington, Oxford OX5 1GB, UK

First edition 2007

Library of Congress Cataloguing-in-Publication Data
A catalog record for this book is available from the Library of Congress

British Library Cataloguing in Publication Data
A catalogue record for this book is available from the British Library

ISBN-13: 978-0-444-52712-7
ISBN-10: 0-444-52712-5
ISSN: 0167-5648

For information on all Elsevier publications visit our website at books.elsevier.com

Printed and bound in The Netherlands

07 08 09 10 11 10 9 8 7 6 5 4 3 2 1

CONTENTS

PREFACE

Hydrodynamic and pollutant modelling is a useful tool when addressing remedial options for polluted water bodies. There is a wealth of knowledge on the subject, including highly theoretical solutions of phenomena to black box representations of water bodies. The numerical approach toward hydrodynamic/pollutant transport modelling is addressed in this book; however, laboratory and field data are used for calibration and verification. The first chapter provides the physical background into the causes behind transport of pollutants and mixing in lakes and coastal waters. The methodology for predicting the fate and transport of pollutants is also given. The governing equations of circulation models, assumptions, boundary conditions, and discussions of turbulence closure and parameterization are given in Chapter 2. The first two chapters are an excellent introduction to the physics and mathematical theory behind circulation modelling.

Chapters 3–5 present alternatives to 3D circulation modelling as well as new advances in this field. A vertically integrated 2D model is introduced in Chapter 3, along with a detailed description of the numerical methods involved and the numerical experiments used. Two-dimensional models have proven to be adequate in situations of large-scaled flows in shallow lakes, thus reducing the amount of computational time required. Chapter 4 presents new developments in 3D circulation modelling, including a multi-layered 3D model, a computationally efficient 3D wind-induced model, a pollutant transport model; and a particle-trajectory model. This chapter closes with an integrated lake modelling system designed to provide a single tool for the hydraulic engineer. Chapter 5 develops an efficient quasi-3D circulation model and compares its performance to the 2D model, another quasi-3D model, and a fully 3D model. More accurate than 2D models, these quasi-3D models are often suitable when 3D models are not practical. Following these chapters is a chapter on pollutant transport models, which use the current field obtained by the circulation model as input. The mathematical theory and the numerical methods used to obtain solutions are also detailed. Chapter 7 is the last of the chapters devoted to theory and numerical analysis. It describes an approach to modelling near-shore areas of lakes by nesting a fine grid of the shore into a coarse grid of the entire lake. The coarse grid provides boundary conditions to the fine grid.

Chapters 8 and 9 are devoted to verification with analytical solutions and laboratory data, as well as an evaluation of model performances applied to the Great Lakes. The Great Lakes in Canada and the United States comprise the greatest fresh-water lake system in the world and are an important resource to millions of people in both nations. For this reason, there is an urgency to maintain the Great Lakes as a valuable resource. Chapter 9 focuses on examining the circulation and pollutant transport in each of the major lakes in the Great Lakes-St. Lawrence River system, as well as in the smaller bays and lakes within the system and some nearshore regions. Chapter 10 is devoted to the applications of fully 3D models to the northern Crete waterfront and Lake Biwa in Japan and two of its rivers. Chapter 11 presents a view into the future of hydrodynamic modelling and possibilities for future research. The source code for a 2D hydrodynamic model is provided in the Appendix of this book. The program is not intended to be a conventional computer package, nor should it be used as a black box. It is only intended for educational purposes and the reader is encouraged to develop and generalize it further. The authors are not responsible for any results produced by the model, or the consequences resulting from their use.

The authors would like to thank Dr. David Schwab of the NOAA-Great Lakes Environmental Research Laboratory for taking the time to review this book and for his many helpful suggestions. Thanks are also due to Mrs. Ginny Riddell, Ms. Laura Honda, Ms. Janet Nurnberg, and Mrs. Diana Maltese for typing this lengthy volume. Thanks also to Mrs. Diana Hurdowar-Castro for providing the final edits.

Chapter 1

INTRODUCTION

1.1 ENVIROMENTAL HYDRAULICS

Increasing development worldwide has created an increasing need for water of acceptable quality. Though there are numerous sources of water in the Earth's hydrosphere, the availability of water suitable for human consumption and use is limited. Lakes are one of the most valuable water resources, and have had a close association with human existence for centuries. Lakes have made enormous contributions to the development of industry, agriculture, and culture. However, these same developments have caused increasing amounts of pollutants to be introduced into these water bodies, thereby causing serious problems and concerns regarding water quality. The eutrophication of a lake, for example, is a natural phenomenon which occurs over geological time scales. However, eutrophication is remarkably accelerated by the presence of excessive amounts of nutrients such as phosphorous and nitrogen, which are found in industrial waste water and sewage. The slow flow of water leads to the accumulation of pollutants in lakes. The closure of lakes leads to long timeloss in clearing the polluted water with clean water. The engineer's ability to supply clean fresh water in ample amounts gets hindered as more and more lakes are polluted.

Remedial Action Plans (RAPs) have been developed for many ailing water bodies. The major objectives of these plans are to (a) identify the existing environmental problems; (b) delineate the physical and chemical mechanisms of various processes; (c) study the impact of any environmental engineering plans on water quality; and (d) make a vibrant centerpiece in the local community. Within these plans, hydraulic engineers are frequently asked to analyze and predict mixing in natural bodies of water, in order to assure safe and reliable supplies under water quality-controlled conditions. This requires a detailed knowledge of the hydrodynamics of transport and mixing, and of the chemistry and biology of natural water systems.

Hydrodynamics is the study of the mechanisms that give rise to transport and mixing in water bodies. The hydrodynamic properties of lakes are largely determined by lake geometry, and the flow is predominately turbulent with a horizontal length scale far exceeding the vertical length scale. Generally, mixing is accomplished by large-scale wind-driven gyres, intermediate-size eddies, and molecular viscosity and diffusion. Variations within a basin will cause variations in velocity and can lead to advective zones where currents are highly advected, mixed zones of varying velocities, or dead zones that have very small velocities and tend to occur in shallow or hydraulically rough areas.

Determining the hydrodynamic mechanisms behind circulation, and the presence of advective or dead zones, requires various factors to be examined. An important starting point is the air–water interface, where there are continual exchanges of energy and matter. Mechanical energy from the wind contributes significantly to the dynamics occurring at the air-water boundary. The frictional movement of wind blowing over water produces travelling surface waves. Wind setup, or wind effect, generates currents in the upper layers of the lake while a return flow is generated in the lower layers to restore equilibrium. Furthermore, the wind effect may give rise to rhythmic motions in the lake, referred to as a *seiche*. Internal waves may form in a stratified lake at the mid-level of the metalimnion if the wind velocity increases to such an extent that the system becomes unstable. Thermal stratification is also an important factor affecting circulation.

The inflows to a lake from tributaries can affect circulation at the entry to the lakes; particularly if the volume of influent is large in relation to the volume of water in the basin. The temperature, composition, and velocity of the inflow play a role in the distribution of the flow and in the generation of circulation patterns within a lake.

Other important factors in hydrodynamic modelling include the Coriolis effect and the barometric effect. The Coriolis effect is caused by the rotation of the earth and it is significant in large lakes. The barometric effect results from non-uniform barometric pressure and can contribute to deviations in the geodetic free water surface level.

Hydrodynamic and pollutant transport processes can be studied by field observations and laboratory experiments, and simulated by mathematical models. Hydrodynamic models are the basis of any water quality modelling system since the current distributions in both the horizontal and vertical planes of a lake are needed in physical limnological studies. Hence, the water quality model performance is largely dependent on the description of the circulation given by the hydrodynamic model.

1.2 METHODOLOGY[1]

Accurate analysis of transport and mixing processes are carried out by field observations and laboratory experiments, and simulated by mathematical models, as illustrated in Figure 1.1. The laboratory field data are used for delineation of the physical processes of transport and mixing. Specifically, laboratory experimentation is one of the most widely used approaches for understanding important mechanisms of air–water interaction, hydraulically induced and wind-induced flows, and pollutant transport and dispersion. The experiments are less costly and more easily controlled than field measurements. Though expensive, field measurements are most desirable for monitoring climatological features of current, temperature, water levels, and water quality. As a result, these studies have spatial and temporal limitations.

[1]Material in Section 1.2 is reproduced from Tsanis and Wu (1994), copyright with kind permission from Elsevier Science Ltd., The Boulevard, Langford Lane, Kidlington OX5 1GB, UK.

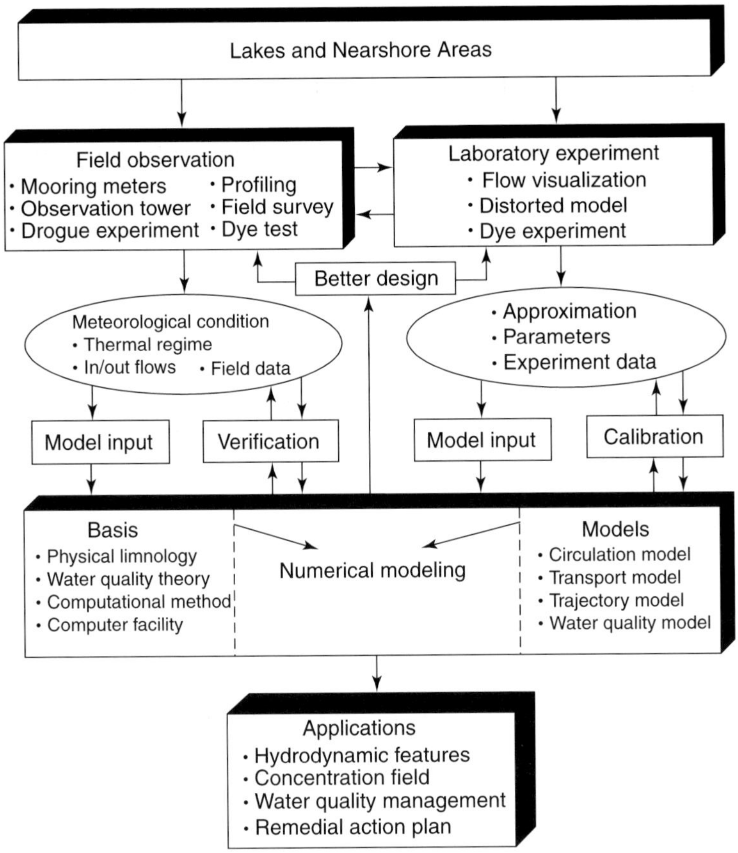

Fig. 1.1. Schematic diagram of approaches for lake research, including field observation, laboratory experiment, and numerical modelling (Reprinted from Tsanis and Wu (1994), copyright with permission from Elsevier Science Ltd., The Boulevard, Langford Lane, Kidlington OX5 1GB, UK).

1.2.1 Field Studies

Field studies are necessary to provide climatological features of physical, chemical, and biological processes, and to provide data sets for model verification and calibration. The related field measurements include wind direction and speed, solar radiation and refraction, air temperature, humidity, wind wave, water level, water currents, water temperatures, transparency, conductivity, sediment concentration, and water chemical elements. A Vertical Automatic Profiling System (VAPS) can be deployed to measure the velocity profiles of current.

1.2.2 Laboratory Experiments

Experiments in laboratories can also provide datasets for model calibration. A physical model is a scaled-down version of the prototype; on the other hand, a numerical model is a set of equations which are thought to represent the characteristics of the process being studied. Physical models have a long tradition, and are appealing because the results may be easily visualized. The models may be expensive to build but, once constructed, may be used for a variety of studies. For example, to study wind-induced flow, a series of experiments have been conducted by using conventional laboratory air–water tunnels and a volume of air in a moving bottomless box. Distorted scale models have been used to study hydraulic flow, density-driven flow, and dye transport in a basin.

1.2.3 Numerical Modelling

In the last two decades, with the development of powerful computers and computational methods, numerical modelling has become a powerful tool for engineers working to solve complex environmental problems. Before applying model results, evaluation of model performance is very important, and necessary in enhancing the product's quality and credibility. Consequently, laboratory and field data are required for model calibration and verification. Understanding of the hydraulic features in natural lakes and the feasibility study of remedial actions in the areas of environmental concern are significantly enhanced with the aid of

Table 1.1. Existing hydrodynamic/pollutant transport models for the Great Lakes.

Model	Developers	Application	Theory	Method	Modelling
Rigid-lid	Schwab (1981)	Lake Michigan,	RL	FDM	V2
(2D)	Schwab et al. (1983)	Lake Erie, Lake			V2
	Schwab et al. (1987)	St. Clair			V2,CC
	Schwab et al. (1989)				V2,TR
2D	Simons and Lam (1986)	Test Basin, Lake St. Clair	LW	FDM	V2,CC
	Simons et al. (1989)				V2,ZZ
RAND(2D)	Leendertse (1969)	Estuaries, Lake	LW	FDM	V2,TR,CC
	MOE (1989)	Ontario			V2,ZZ
3D	Ibrahim (1985)	Lake St. Clair	EK	FDM	V3,CC
3D	Simons (1972, 1974, 1983)	Lake Ontario, Hamilton	LW	FDM	V3,ZZ,NS,V3
	James and Eid (1978)	Harbour			
3D	Huang (1987)	Lake Ontario	LW	FDM	V3,ZZ
CH3D	Sheng (1982, 1984)	Lake Erie	LW	FDM	V3,ZZ,SS
LEIFS	Bedford et al. (1990)	The Great Lakes	LW	FDM	V3,ZZ

RL, rigid-lid theory; LW, long-wave theory for nearly horizontal flows; EK, Ekman flow theory; V2, two-dimensional velocities (u, v); V3, three-dimensional velocities (u, v); ZZ, water set-up; TR, particle trajectory prediction is included; CC, pollutant transport model is included; SS, sediment transport model is included; NS, capable for modelling the nearshore areas in lakes.

Source: Reprinted from Tsanis and Wu (1994), Copyright with permission from Elsevier Science Ltd., the Boulevard, Langford Lane, Kidlington OX5 1GB, UK.

numerical models. Furthermore, hydrodynamic features form the basis of any water quality modelling system.

Numerous hydrodynamic models have been developed and applied to the Great Lakes over the last two decades, as shown in Table 1.1. Excellent reviews can be found in Cheng and Smith (1990) and Schwab (1992). However, since 1990, as the computations for two-dimensional (2D) and three-dimensional (3D) circulation and pollutant transport problems became routine, the modelling task has gradually shifted from equation solving to preprocessing model inputs, post-processing model results, and data processing to aid model calibration and verification. Model products provided to users should be userfriendly and operational. These are the major features of the fifth-generation hydrodynamic models of surface waters (Abbott, 1991). The use of computer graphic visualization and animation is an effective method for displaying large volumes of model input and output. Progress has also been made in the development of the Great Lakes Forecasting System (GLFS) project (Bedford et al., 1990; Yen et al., 1992), an operational oil spill model (Spaulding et al., 1992) a pollutant transport model of Collingwood Harbour

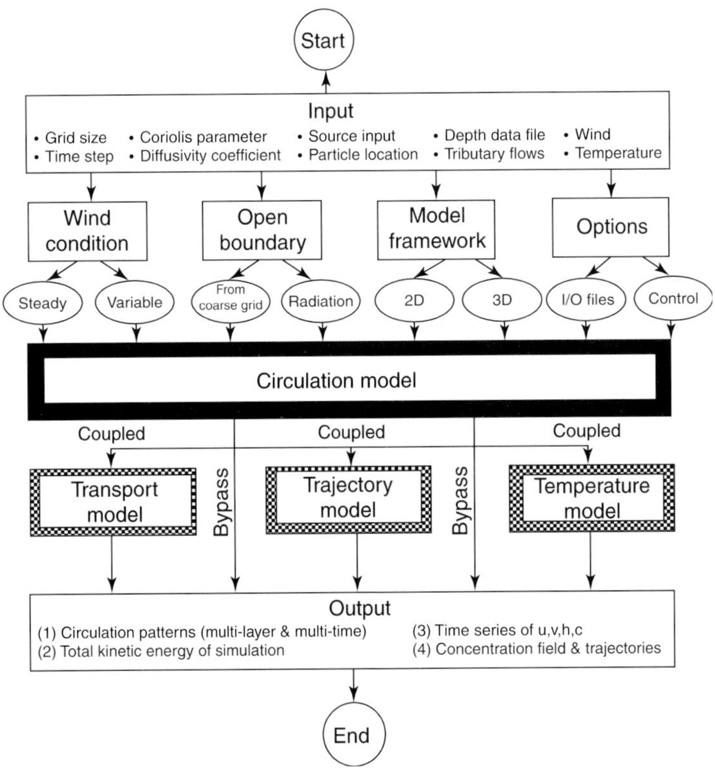

Fig. 1.2. An integrated lake modelling system (LMS) for the Great Lakes and tributaries (Reprinted from Tsanis and Wu (1994), copyright with permission from Elsevier Science Ltd., The Boulevard, Langford Lane, Kidlington OX5 1GB, UK).

in Lake Huron (MOE, 1992a), and model development at McMaster University (Tsanis et al., 1990; Tsanis and Wu, 1991, 1992; Tsanis and Shen, 1994c).

The construction of an operational modelling system is significant to model evaluation and application. All sub-models should be integrally coupled to make the system capable of solving various complicated engineering problems. This includes pre-processing (model inputs and various options), processing (simulation), and post-processing (outputs and applications). Following this procedure, an integrated Lake Modelling System (LMS) was designed for the Great Lakes and its tributaries, and is shown in Figure 1.2. The system operates under steady or variable wind conditions, using 2DH, Q3D, VHI3D, or 3D hydrodynamic models and permits the use of a nested-grid modelling system. The output from the

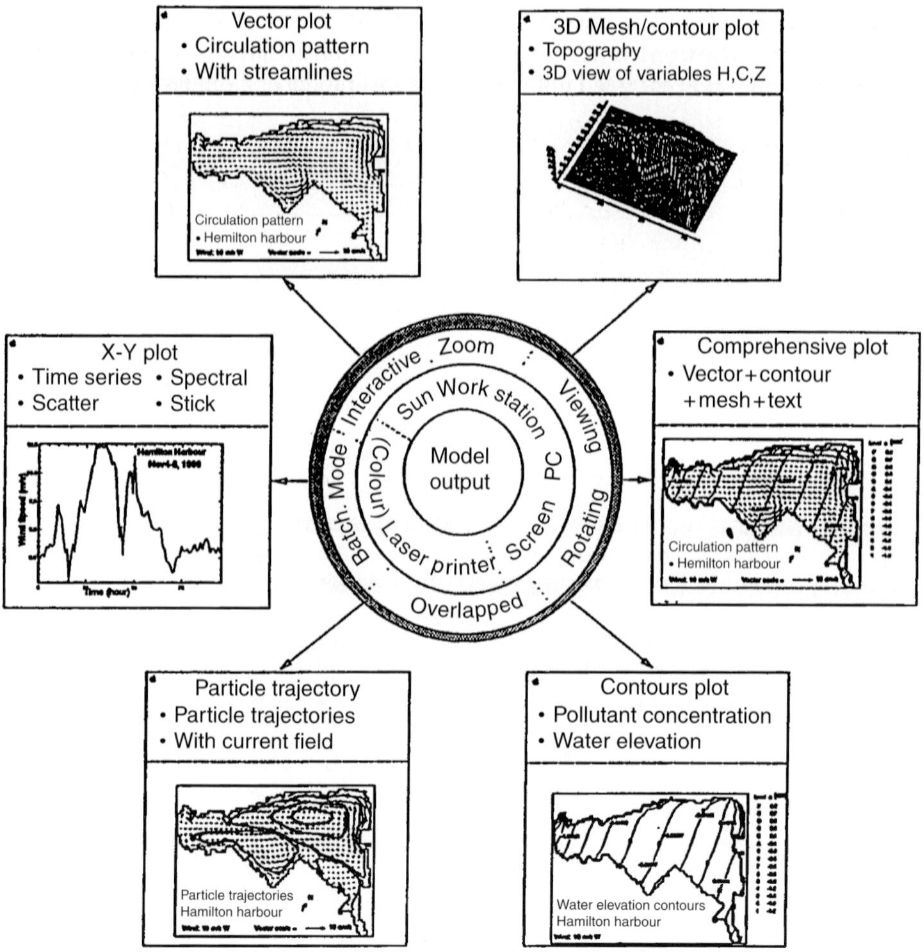

Fig. 1.3. The graphics capability for the outputs of LMS (Reprinted from Tsanis and Wu (1994), copyright with permission from Elsevier Science Ltd., The Boulevard, Langford Lane, Kidlington OX5 1GB, UK).

hydrodynamic model is then used as input to pollutant transport and trajectory prediction models (Wu, 1993). The pollutant transport and dispersion model can accept inputs from various point sources such as creeks, sewer treatment plants (STPs), and combined sewer outfalls (CSOs), while, in the trajectory model, a large number of particles can be released in space and time and their trajectories can be predicted.

Figure 1.3 shows the major graphics capability including X-Y plot—for time series of any variables such as wind, current, water elevation; vector plot—for depth-averaged or multi-layered circulation patterns; contours plot—for pollutant concentration field and spatial distribution of any variables; trajectory plot—for particle trajectories; and 3D–mesh plot—for 3D view of any variables.

Usage of zooming, rotating, and overlapping features allows a series of comprehensive plots to be made, and, as a result, many simulated features can be clearly displayed. For example, by overlapping a vector plot of circulation pattern into a flooded contour plot of pollutant concentration field or a scattered plot of particle trajectories, the effect of current on the pollutant transport or particle movement can be clearly presented and easily analyzed. A batch mode allows non-specialist users to run the modelling system and display the simulation results. Animation of these plots has also been developed for LMS, which provides a dynamic visualization of simulation results. It is also a good tool for public education on environmentally related issues.

Chapter 2

THE MATHEMATICAL THEORY OF CIRCULATION MODELS

Modelling 3D circulation is a mixture of both art and science. It requires a thorough understanding of the dynamics in order to allow trade-offs between terms of secondary importance and simplicity. Although there are various model formulations for oceans, estuaries, and lakes, the basic model equations and assumptions of these models are similar. Therefore, these models should be reviewed before developing any new model. Following the description of the model equations and the three basic approximations, the turbulence, closure scheme, and circulation theories will be presented, with the consideration of important terms in the model equations.

2.1 GOVERNING EQUATIONS

The governing system of equations for the flow and transport of scalar variables in a lake includes the conservation equations of mass and momentum, the conservative equation for a scalar variable, and the equation of state. The lake is considered large enough for Coriolis acceleration to be included, yet small enough for the acceleration to be treated as a constant. Large lake dimensions result in quite a large Reynolds number (greater than 10^8), even for a small current of 1 cm/s. Therefore, the flow is always turbulent. In Cartesian coordinates (see Fig. 2.1), the governing equations written for turbulent mean variables, include (Cheng and Smith, 1990):

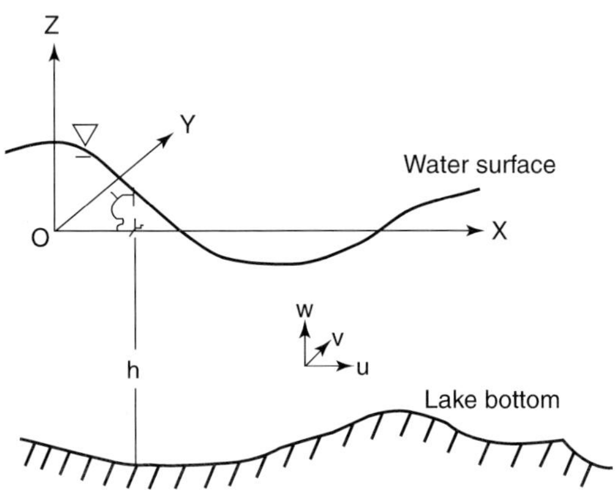

Fig. 2.1. Cartesian coordinate system for the hydrodynamic model of lakes (Wu, 1993).

Continuity equation of mass

$$\frac{D\rho}{Dt} = \frac{\partial \rho}{\partial t} + \rho\left(\frac{\partial u}{\partial x} + \frac{\partial v}{\partial y} + \frac{\partial w}{\partial z}\right) = 0 \tag{2.1}$$

Momentum equation in x-, y- and z-directions

$$\frac{\partial u}{\partial t} + \frac{\partial uu}{\partial x} + \frac{\partial uv}{\partial y} + \frac{\partial uw}{\partial z} = fv - \frac{1}{\rho}\frac{\partial p}{\partial x} - \left[\frac{\partial \overline{u'u'}}{\partial x} + \frac{\partial \overline{v'u'}}{\partial y} + \frac{\partial \overline{w'u'}}{\partial z}\right] \tag{2.2}$$

$$\frac{\partial v}{\partial t} + \frac{\partial vu}{\partial x} + \frac{\partial vv}{\partial y} + \frac{\partial vw}{\partial z} = -fu - \frac{1}{\rho}\frac{\partial p}{\partial y} - \left[\frac{\partial \overline{u'v'}}{\partial x} + \frac{\partial \overline{v'v'}}{\partial y} + \frac{\partial \overline{w'v'}}{\partial z}\right] \tag{2.3}$$

$$\frac{\partial w}{\partial t} + \frac{\partial wu}{\partial x} + \frac{\partial wv}{\partial y} + \frac{\partial ww}{\partial z} = -\frac{1}{\rho}\frac{\partial p}{\partial z} - g - \left[\frac{\partial \overline{u'w'}}{\partial x} + \frac{\partial \overline{v'w'}}{\partial y} + \frac{\partial \overline{w'w'}}{\partial z}\right] \tag{2.4}$$

$$\frac{\partial T}{\partial t} + \frac{\partial Tu}{\partial x} + \frac{\partial Tv}{\partial y} + \frac{\partial Tw}{\partial z} = -\left[\frac{\partial \overline{u'T'}}{\partial x} + \frac{\partial \overline{v'T'}}{\partial y} + \frac{\partial \overline{w'T'}}{\partial z}\right] \tag{2.5}$$

Conservation equation for a scalar variable, C_j

$$\frac{\partial C_j}{\partial t} + \frac{\partial C_j u}{\partial x} + \frac{\partial C_j v}{\partial y} + \frac{\partial C_j w}{\partial z} = -\left[\frac{\partial \overline{u'C_j'}}{\partial x} + \frac{\partial \overline{v'C_j'}}{\partial y} + \frac{\partial \overline{w'C_j'}}{\partial z}\right]$$
$$j = 0, 1, 2, ... \tag{2.6}$$

Equation of state

$$\rho = \rho(s,\ T) \tag{2.7}$$

where t is the time; x, y, z are the Cartesian coordinates; u, v, w are the velocity components in the x, y, and z directions respectively; ρ the density of water; g the acceleration due to gravity; s the salinity; T the temperature; C_j a conservative scalar variable, such as s or T; u', v', w', C_j' are the turbulent fluctuations of u, v, w, C_j; p the pressure; $f = 2\Omega \sin(\phi)$ the Coriolis coefficient; ϕ the geographic latitude of the domain; and Ω the angular rotation of the Earth. The turbulence correlations are indicated by overbars where $\overline{u'u'}$, $\overline{v'v'}$ and $\overline{w'w'}$ are the normal turbulent stresses, and $\overline{u'v'}$, $\overline{u'w'}$ and $\overline{v'w'}$ are the Reynolds turbulent stresses.

The system of equations is not closed until the turbulent stresses are defined and the appropriate initial and boundary conditions are given. Typically, a kinematic condition at the water surface has the wind stress specified as a given function. At air–water and water–sediment interfaces, fluxes of mass and conservative solutes must be specified. Necessary assumptions and approximations must be introduced to simulate a complex real-world system, and the resultant simplified governing equations are then solved by a chosen numerical method suitable for that particular formulation.

2.2 BASIC ASSUMPTIONS

There are three basic approximations used in most hydrodynamic models applied to oceans, estuaries, lakes, and impoundments.

2.2.1 *Incompressible Approximation*

For lakes or shallow estuary systems, the water can be assumed to be incompressible, and the continuity equation can be written as

$$\frac{\partial u}{\partial x}+\frac{\partial v}{\partial y}+\frac{\partial w}{\partial z}=0 \tag{2.8}$$

2.2.2 *Hydrostatic Approximation*

The characteristic length in the vertical direction is two to three orders of magnitude smaller than the characteristic horizontal dimension. Therefore, the vertical pressure gradient can be balanced only by the gravitational acceleration in the z direction, that is,

$$0=-\frac{\partial p}{\partial z}-\rho g \tag{2.9}$$

By integrating the preceding equation from z to the free surface ($z=\zeta$), where ζ is the water elevation above the mean water level, the horizontal pressure gradient can be separated into an atmospheric gradient term, a *barotropic* (water-surface gradient) term and a *baroclinic* (density gradient) term:

$$\frac{\partial p}{\partial x}=\frac{\partial p_a}{\partial x}+\rho_s g\frac{\partial \zeta}{\partial x}+\int_z^{\zeta} g\frac{\partial \rho}{\partial x}dz \tag{2.10}$$

where ρ_s is the surface water density. Atmospheric pressure (p_a) gradients when compared to wind stress (both are external forces on the lake), are relatively unimportant and are usually neglected.

2.2.3 *Boussinesq Approximation*

Density variations are of order 10^{-3}, resulting in a negligible effect on barotropic terms (ρ_s can be substituted by a constant reference density ρ_0). On the other hand, the effect of buoyancy arising from the differences in fluid density along horizontal surfaces is not negligible. As a result, the effect of density variation is only retained in the baroclinic terms. This is called the *Boussinesq Approximation*. The x-momentum equation in Eq. (2.2) with the hydrostatic and *Boussinesq Approximation* becomes

$$\frac{\partial u}{\partial t}+\frac{\partial uu}{\partial x}+\frac{\partial uv}{\partial y}+\frac{\partial uw}{\partial z}=fv-g\frac{\partial \zeta}{\partial x}-\frac{g}{\rho_0}\int_z^{\zeta}\frac{\partial \rho}{\partial x}dz - \left[\frac{\partial \overline{u'u'}}{\partial x}+\frac{\partial \overline{u'v'}}{\partial y}+\frac{\partial \overline{u'w'}}{\partial z}\right] \tag{2.11}$$

and the y-momentum equation becomes

$$\frac{\partial v}{\partial t}+\frac{\partial vu}{\partial x}+\frac{\partial vv}{\partial y}+\frac{\partial vw}{\partial z}=-fu-g\frac{\partial \zeta}{\partial y}-\frac{g}{\rho_0}\int_z^{\zeta}\frac{\partial \rho}{\partial y}dz - \left[\frac{\partial \overline{u'v'}}{\partial x}+\frac{\partial \overline{v'v'}}{\partial y}+\frac{\partial \overline{w'v'}}{\partial z}\right] \tag{2.12}$$

2.3 TURBULENCE CLOSURE

There are two major turbulence closure approaches, namely eddy viscosity-diffusivity and second-order closure. Most models have adopted the *viscosity-diffusivity approach*. In this approach, the turbulent transport terms can be written as

$$-\left[\frac{\partial \overline{u'u'}}{\partial x}+\frac{\partial \overline{v'u'}}{\partial y}+\frac{\partial \overline{w'u'}}{\partial z}\right]=\frac{\partial}{\partial x}\left(K_h\frac{\partial u}{\partial x}\right)+\frac{\partial}{\partial y}\left(K_h\frac{\partial u}{\partial y}\right)+\frac{\partial}{\partial z}\left(K_v\frac{\partial u}{\partial z}\right) \tag{2.13}$$

$$-\left[\frac{\partial \overline{u'v'}}{\partial x}+\frac{\partial \overline{v'v'}}{\partial y}+\frac{\partial \overline{w'v'}}{\partial z}\right]=\frac{\partial}{\partial x}\left(K_h\frac{\partial v}{\partial x}\right)+\frac{\partial}{\partial y}\left(K_h\frac{\partial v}{\partial y}\right)+\frac{\partial}{\partial z}\left(K_v\frac{\partial v}{\partial z}\right) \tag{2.14}$$

$$-\left[\frac{\partial \overline{u'C'}}{\partial x}+\frac{\partial \overline{v'C'}}{\partial y}+\frac{\partial \overline{w'C'}}{\partial z}\right]=\frac{\partial}{\partial x}\left(N_h\frac{\partial C}{\partial x}\right)+\frac{\partial}{\partial y}\left(N_h\frac{\partial C}{\partial y}\right)+\frac{\partial}{\partial z}\left(N_v\frac{\partial C}{\partial z}\right) \tag{2.15}$$

where K_h and N_h are the horizontal eddy viscosity and diffusivity coefficients respectively, and K_v and N_v are the vertical eddy viscosity and diffusivity coefficients respectively. Because the ratio of vertical length scale to horizontal length scale is very small for most lakes, the horizontal mixing terms are orders of magnitude smaller than the vertical mixing terms. As a result, the use of a sophisticated turbulence closure model for horizontal mixing terms is generally not warranted. Most models treat the horizontal eddy viscosity coefficients as constants, or these terms are simply neglected. After this simplification, the role of turbulent closure becomes that of determining the variation of turbulent transport in the vertical. Assuming that the eddy viscosity is constant or using an empirical equation to relate it to other parameters, a semi-empirical theory of vertical mixing has been used in most circulation models. The effect of stratification—as measured by the Richardson number, R_i—on the intensity of vertical turbulent mixing is parameterized by a number of empirical stability functions (Sheng, 1983; Blumberg, 1986):

$$K_v = K_{v0}\phi_1(R_i); \quad N_v = N_{v0}\phi_2(R_i) \tag{2.16}$$

where

$$R_i = -\frac{g}{\rho_o}\frac{\partial\rho/\partial z}{(\partial|U|/\partial z)^2} \tag{2.17}$$

in which K_{v0} and N_{v0} are the eddy viscosity coefficients in the absence of any density stratification and the ϕ_1 and ϕ_2 are the stability functions (Sheng, 1983; Blumberg, 1986). More sophisticated turbulence models may be found in the literature, such as the one-equation or $k-l_0$ model (Rodi, 1978; Koutitas and O'Connor, 1980), and the two-equation or $k-\varepsilon$ model (Mellor, 1973; Svensson, 1978; Pearce and Cooper, 1981; Blumberg and Mellor, 1987).

2.3.1 Turbulence Parameterization

The effect of stratification on the intensity of vertical turbulent mixing is parameterized by the empirical stability functions, see Eq. (2.16). One popular form of the stability functions ϕ_1 and ϕ_2 is (Blumberg, 1986):

$$\phi_1 = (1.0 + R_i)(1.0 - 0.1R_i)^{1/2} \tag{2.18}$$

$$\phi_2 = (1.0 - 0.1R_i)^{1/2} \tag{2.19}$$

$$\phi_1 = 0.0; \quad \phi_2 = 0.0; \quad for\ R_i > 10 \tag{2.20}$$

For wind-generated circulation, a parabolic distribution is sometimes used (Koutitas and O'Connor, 1980):

$$K_{v0} = z/h(1 + z/h)\sqrt{\tau_s/\rho}C_k h \tag{2.21}$$

where τ_s is the wind shear stress at the water surface. C_κ is in the range of 0.2–0.4.

Ushijima and Moriya (1988) proposed the following stability function:

$$\phi_1 = (1 + 5R_i)^{-1}, \quad \phi_2 = 1.6(1 + 6R_i)^{-1} \tag{2.22}$$

The stability functions ϕ_1 and ϕ_2 have also been proposed as (Leendertse and Liu, 1975):

$$\phi_1 = e^{-1.5R_i} \quad \phi_2 = e^{-3.0R_i} \tag{2.23}$$

It is obvious from Eq. (2.16) that, in the case of stable thermal stratification (positive high Richardson number), the vertical transport of momentum and heat are suppressed, while in the unstable case (negative Richardson number), the transport is enhanced. Finally, it should be pointed out that Eq. (2.23) is not universal and further investigation is needed.

The following empirical formula is often used to determine the horizontal eddy viscosity coefficients (Tomidokoro, 1984):

$$K_h = 10^{-3}\Delta_h^{4/3} \quad (\mathrm{m}^2/\mathrm{s}) \tag{2.24}$$

where Δ_h is the mesh interval (m) in the horizontal plane. Physically, turbulence should not depend on the computational mesh size, but turbulence motions exist on scales smaller than the mesh size and their effects are parameterized into the variables of mean flow. Therefore, the values of the eddy viscosity and diffusivity inevitably depend upon the mesh sizes adopted. In the view of numerical analysis, the eddy viscosity and diffusivity coefficients are not pure physical quantities, but complex quantities that include the effect of numerical viscosity. Therefore, they should depend on the mesh sizes in order to obtain stable solutions according to the stability conditions. For lake flows, the eddy viscosity coefficients and eddy diffusivity coefficients in the horizontal plane and in the vertical direction are usually given to be 10^2–10^{-1} m^2/s and 10^{-2}–10^{-6} m^2/s respectively.

2.3.2 Boussinesq's Eddy Viscosity[2]

The concept of eddy viscosity was first introduced by Boussinesq in 1877 (Schlichting, 1968). In analogy with laminar flow, the Reynolds stresses are represented as

$$\tau = -\rho\overline{u'w'} = \rho v_t \frac{d\bar{u}}{dz} \tag{2.25}$$

where $v_t = \mu_t/\rho$ is the kinematic eddy viscosity, which is a property of the flow. In laminar flow, the viscosity is a constant property of the fluid. In order to calculate the velocity distribution for the case of turbulent flow, it is necessary to specify an appropriate distribution of the eddy viscosity in the vertical. Thus, if the eddy viscosity μ_t is assumed constant over the depth, the velocity distribution is parabolic.

[2]Material in Sections 2.3.2 to 2.3.5 are reproduced from Tsanis (1989), copyright with permission from ASCE.

This, however, is not the case in turbulent flow. The velocity profile is different from that in laminar flow and, hence, the eddy viscosity must be a function of z. For instance, the assumption of a logarithmic velocity distribution suggests a parabolic distribution of the eddy viscosity over the depth of the flow, as demonstrated by Elder (1959) and Lau and Krishnappan (1981) in analytical and numerical studies of turbulent mixing phenomena. Similarly, Pearce and Cooper (1981) used a linearly varying eddy viscosity in their numerical model of wind-induced water currents.

2.3.3 Prandtl's Mixing Length

The mixing length hypothesis, first introduced by Prandtl in 1905, gives the relation between stress and the velocity gradient as

$$\tau = \rho l_m^2 \left|\frac{d\bar{u}}{dz}\right| \frac{d\bar{u}}{dz} \tag{2.26}$$

where l_m is the mixing length. The mixing length, like the eddy viscosity, is a property of the flow and, in first approximation, is supposed to be a purely local function. Reid (1957), in his analytical study of turbulent channel flow subject to surface wind stress, assumed a parabolic distribution of the mixing length in the vertical. This approach yields a series of solutions for the velocity distribution in terms of different shear-stress ratios $\eta = \tau_b/\tau_s$. In order to satisfy the continuity requirements for countercurrent flow with zero net volume flux, he found that the shear-stress ratio must be equal to $\eta = -0.097$, which is very close to the value of -0.10 obtained by Baines and Knapp (1965). The corresponding value for laminar countercurrent flow is $\eta = -0.50$.

Both of these so-called "zero-equation" turbulent models are based on linearly varying eddy viscosity and parabolic mixing length respectively. They provide a reasonable description of the velocity distribution in the wall region of countercurrent flows. Thus, the models are adequate when the mean velocity gradients are high. However, they fail to describe the turbulent mean velocity where the mean rate of strain is small, that is, in the core of the flow (Lauder and Spalding, 1972).

2.3.4 One-Equation, or $k-L_0$ Model of Turbulence

Prandtl and Kolmogorov suggested that the turbulent mean velocity could be determined by directly solving the different transport equations rather than by relying on phenomenological relations such as in Eqs. (2.25) and (2.26) (Pearce and Cooper, 1981). In particular, they suggested that the turbulent velocity is proportional to the time-averaged turbulent kinetic energy per unit mass, which is defined as the sum of the normal Reynolds stresses per unit mass divided by 2, viz.,

$$k = \frac{1}{2}(\overline{u^2} + \overline{v^2} + \overline{w^2}) \tag{2.27}$$

which follows from the solution of the advective transport equation. The latter equation is obtained by multiplying the momentum equation for each coordinate direction with the corresponding instantaneous velocity, followed by time averaging and, finally, summing the three equations. For large-scale turbulent motion the proper velocity scale to be used is $k^{1/2}$ because turbulent kinetic energy is contained mainly in large-scale fluctuations. After introducing this scale in the eddy viscosity concept, one finds that

$$\upsilon_t = c_t k^{1/2} L_0 \tag{2.28}$$

where L_0 is the dissipation length scale. The dissipation ε is usually modelled by the expression

$$\varepsilon = c_d \frac{k^{3/2}}{L_0} \tag{2.29}$$

where c_t and c_d are numerical constants with a value around 0.3 (Koutitas and O'Connor, 1980). The form of the equation for turbulent kinetic energy for 2D nearly horizontal flows at high Reynolds numbers is (Koutitas and O'Connor, 1980)

$$\frac{Dk}{Dt} - \frac{\partial}{\partial z}\left(\frac{\upsilon + \upsilon_t}{\sigma_k}\frac{\partial k}{\partial z}\right) - \upsilon_t\left(\frac{\partial \bar{u}}{\partial z}\right)^2 + c_d \frac{k^{3/2}}{L_0} = 0 \tag{2.30}$$

where σ_k is an effective turbulent Prandtl number. The latter is defined as the ratio between the turbulent transport coefficients for momentum and heat, respectively, and is approximately equal to unity. Careful inspection of Eq. (2.30) indicates that the mixing length hypothesis is a special case of the $k{-}L_0$ turbulent model if both the advective and diffusive transport terms are negligible. Under these circumstances the turbulent energy production k is equal to the dissipation of this energy, and the mixing length is related to the dissipation length scale, as follows (Lauder and Spalding, 1972):

$$l_m = \left(\frac{c_t^3}{c_d}\right)^{1/4} L_0 \tag{2.31}$$

The rate of change of k is balanced by the advective transport due to the mean motion, the diffusive transport due to velocity and pressure fluctuations, the production of k by interaction of Reynolds stresses and mean-velocity gradients, and the dissipation of k by viscous action into heat. Koutitas and O'Connor (1980) used the one-equation model for the modelling of wind-induced flows under the assumption of steady and uniform conditions, that is, $Dk/Dt = 0$. With the aid of a complicated expression for the dissipation length scale they eventually arrived at a solution for the eddy viscosity, which is described by the following empirical relation:

$$\upsilon_t = 0.1249 u_s^* z\left(2 - \frac{z}{h}\right) \tag{2.32}$$

where u_s^* is the surface shear velocity. This solution is, however, not satisfactorily close to the air–water interface, probably because the analysis fails to take into account the mass transport velocity of surface waves. A modified distribution of the dissipation length scale L_0 based on experimental observations solved this inconsistency, and led to an improved distribution of the eddy viscosity over the depth (Koutitas and O'Connor, 1980):

$$v_t = 0.1249 u_s^* h\left(1 - \frac{z}{h}\right)\left(5\frac{z}{h} - 1\right) \tag{2.33}$$

2.3.5 Two-Equation or $k-\varepsilon$ Model of Turbulence

The $k-\varepsilon$ model includes an expression for the eddy viscosity and two coupled differential equations for the turbulent kinetic energy k and the dissipation of this energy ε. The equation for the dissipation of k is obtained by differentiating the equation for velocity fluctuations with respect to the coordinates, and multiplying the result by twice the kinematic viscosity and the gradient of the velocity fluctuations. Svensson (1978) used the standard $k-\varepsilon$ model to calculate a wind-induced channel flow. He considered the free surface as a moving wall at which the shear stress is prescribed.

The expression for the eddy viscosity and the differential equations for the turbulent kinetic energy k and the dissipation of this energy ε, for 2D nearly horizontal flows at high Reynolds numbers are respectively (Svensson, 1978)

$$v_t = c_t \frac{k^2}{\varepsilon} \tag{2.34}$$

$$\frac{Dk}{Dt} = \frac{\partial}{\partial z}\left(\frac{v + v_t}{\sigma_k}\frac{\partial k}{\partial z}\right) + v_t\left(\frac{\partial \bar{u}}{\partial z}\right)^2 - \varepsilon \tag{2.35}$$

$$\frac{D\varepsilon}{Dt} = \frac{\partial}{\partial z}\left(\frac{v + v_t}{\sigma_\varepsilon}\frac{\partial \varepsilon}{\partial z}\right) + c_{1\varepsilon} v_t\left(\frac{\partial \bar{u}}{\partial z}\right)^2 - c_{2\varepsilon}\frac{\varepsilon^2}{k} \tag{2.36}$$

where σ_ε is the effective Schmidt number; c_t the numerical constant in eddy viscosity expression; and $c_{1\varepsilon}$ and $c_{2\varepsilon}$ are the numerical constants in the dissipation of turbulent kinetic energy equation. The turbulent Schmidt number is defined as the ratio between the turbulent transport coefficients for momentum and mass. For non-buoyant conditions the constants involved in the above equations are determined by reference to experimentally well-studied flows and have the following values: $c_t = 0.09$, $\sigma_k = 1.0$, $\sigma_\varepsilon = 1.3$, $c_{1\varepsilon} = 1.44$, $c_{2\varepsilon} = 1.9$.

The input variables used by Svensson (1978) for the wind-induced channel flow problem were the surface shear stress equal to 0.092 N/m^2 and the channel depth equal to 0.30 m, used by Baines and Knapp (1965). According to Svensson's calculations, the ratio of surface velocity to the surface shear velocity is about 22, and the turbulent mean velocity is zero at the normalized depth, $z/h = 0.65$. The maximum eddy viscosity, rendered non-dimensional by the surface shear velocity u_s^*

and the depth h, is of the order 0.08 and occurs at $z/h = 0.55$. The actual observed (Baines and Knapp, 1965) values are

$$\frac{u_s}{u_s^*} = 16.2 \text{ and } \bar{u} = 0 \text{ at } \frac{z}{h} = 0.76 \tag{2.37}$$

Mean velocity profiles based on the above numerical techniques are presented in non-dimensional form in Figure 2.2. The calculated, or pre-defined eddy viscosities used in the various numerical models, normalized by the surface shear stress and the channel depth, are depicted in Figure 2.3. Finally, the experimental and numerical mean velocity profiles referred to in the foregoing papers and others are compiled in Figure 2.4 (Forssblad, 1947; Tsanis, 1986; Masch, 1963; Goossens et al., 1982; Baines and Knapp, 1965; Koutitas and O'Connor, 1980; Tsuruya et al., 1985).

All of the numerical models require a priori knowledge of the surface shear stress u_s^* and the depth h. The assumed distribution of eddy viscosity used in the linearly varying eddy viscosity model is based on observations of open channel flow (Pearce and Cooper, 1981). In the mixing length model, the vertical distribution of the mixing length is defined empirically, using as a guide the values of the length that successfully described other shear flow experiments. For example, Prandtl (Schlichting, 1968) proposed that the mixing length l_m is proportional to the distance from the wall. The eddy viscosity in this case, based on the mixing length hypothesis, is defined as

$$v_t = l_m^2 \left| \frac{d\bar{u}}{dz} \right| \tag{2.38}$$

and is calculated from the mean velocity gradient and the mixing length. The value of the eddy viscosity, based on the mixing length, is zero at the position of zero

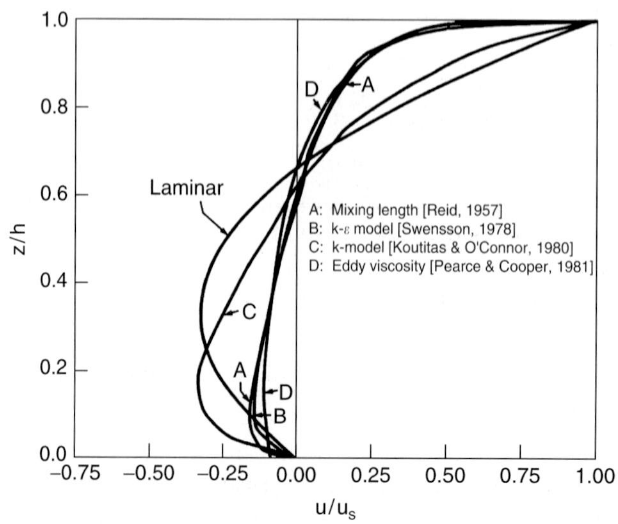

Fig. 2.2. Existing analytically predicted mean velocity distributions (Reprinted from Tsanis (1989), copyright with permission from ASCE).

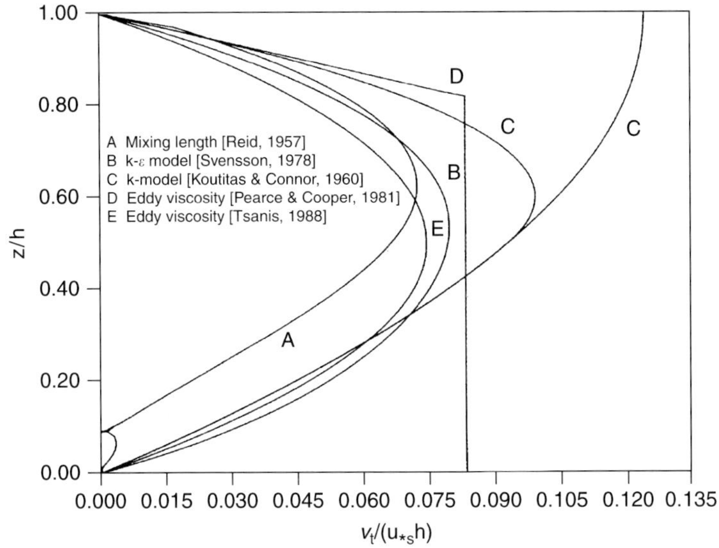

Fig. 2.3. Various distributions of eddy viscosity used in analytical models of wind-induced flows (Reprinted from Tsanis (1989), copyright with permission from ASCE).

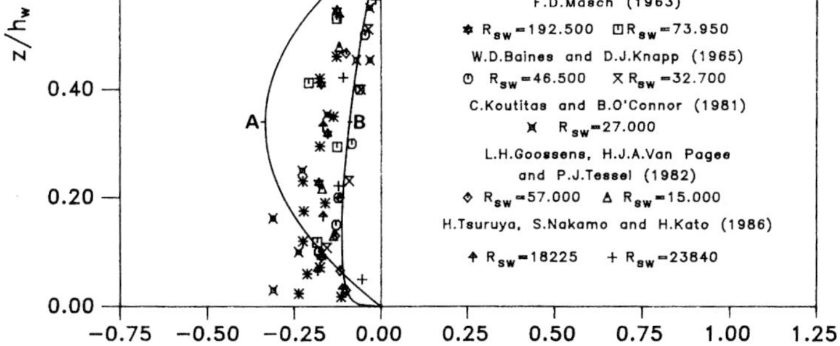

Fig. 2.4. Extant experimental mean velocity profiles and analytical predictions (Reprinted from Tsanis (1989), copyright with permission from ASCE).

mean velocity gradient (see Figs. 2.2 and 2.3), which is physically unsound (Lauder and Spalding, 1972). To demonstrate this, consider the flow downstream from a step. Near the re-attachment point, where the mean velocity gradients are low, adoption of the mixing length results in a low value of eddy viscosity. This is in contrast with experimental findings and can be explained by the fact that the mixing length does not take into account the processes of diffusion and advection of turbulence.

The $k-L_0$ model overcomes this problem. Here the vertical distribution of the dissipation length L_s is pre-defined based on empirical relations similar to those used for the mixing length. The eddy viscosity is calculated by solving the differential equation for the turbulent kinetic energy. The diffusion effects, represented by the second term on the left-hand side of Eq. (2.30), act as a spreading and smoothing mechanism. Thus, discontinuities in the solution are avoided, as can be seen in Figure 2.3.

For the reason just mentioned, the $k-L_0$ model is preferred to the mixing length model in problems related to countercurrent flows. The eddy viscosity based on the modified $k-L_0$ model has been successfully used in calculating the water circulation of the Gulf of Thessaloniki in the Aegean Sea of Greece (Ganoulis et al., 1980). The eddy viscosity based on $k-\varepsilon$ is calculated by solving the two coupled differential equations for the turbulent kinetic energy and its dissipation. This model is more complex than the other ones but it does not require a priori knowledge of the distributions of mixing or dissipation lengths. The eddy viscosities based on the aforementioned models are mutually consistent in their overall magnitudes, and are in agreement with those found in open channel flows (Pearce and Cooper, 1981).

Nevertheless, the analytically predicted velocity profiles compiled in Figure 2.2 differ significantly from each other. The reason for this is the fact that the constants used in these models are based on experimental findings of varying quality and theory. Furthermore, there is always some scatter in the experimentally measured mean velocities. Reasons for this include the 3D nature of conventional laboratory flows, the possibility that the flows are not fully developed, and the difficulties in measuring velocities close to a wavy surface. As a consequence, an accurate determination of the surface shear stress in combined wind-water tunnels is practically impossible.

The experimentally determined mean velocity profiles shown in Figure 2.4 were obtained for different Reynolds numbers, and under different characteristics of surface roughness. From Figure 2.5, it is apparent that the experimentally examined flows are in the region of transitional roughness in which the value of the Von Kármán constant κ is not necessarily equal to 0.4. A value of $\kappa = 0.4$ was used in almost all of the numerical models. All of this contributes to the discrepancies between the numerical and experimental mean velocity profiles.

The principal disadvantage of experiments conducted in conventional laboratory systems is that the flows therein are essentially 3D ($h/d = 1/1$ to $1/3$) in both air and water sides. Hence, the use of these experimental results in the development of 2D analytical models is questionable. For example, Pearce and Cooper (1981) with their linearly varying eddy viscosity, and Svensson (1978) with his $k-\varepsilon$ model, used the experimental results from Baines and Knapp (1965) to justify their predictions. Similarly, Reid (1957), with his mixing length model, used extant experimental results to justify his results. Finally, the experimental mean velocity profiles

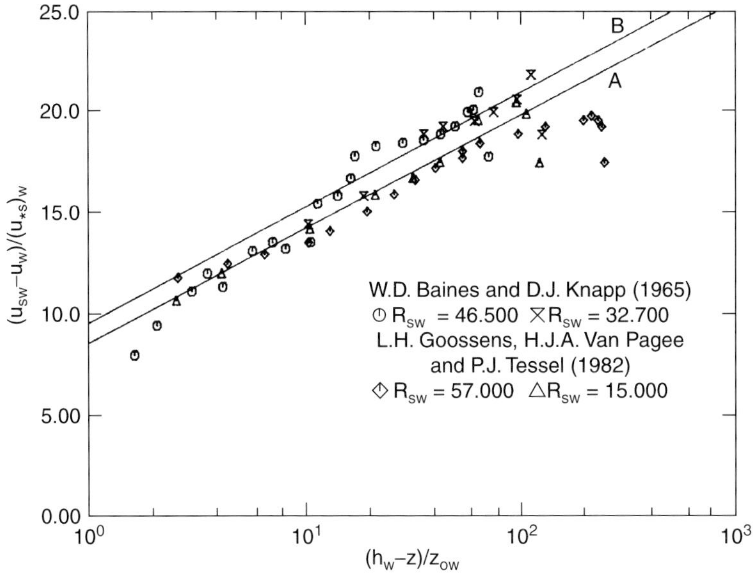

Fig. 2.5. "Rough" inner law of velocity distribution of extant experimental data relative to interface (Reprinted from Tsanis (1989), copyright with permission from ASCE).

determined by Koutitas and O'Connor (1980) were used for the development of the $k-L_0$ model.

2.3.6 Large Eddy Simulation

Turbulence models are usually differentiated by whether they filter (or average) the Navier–Stokes equations temporally or spatially. The former type includes the so-called zero-equation model, one-equation model, and two-equation model based on the Reynolds equations. The latter type is indicated by the large eddy simulation (LES) model.

As is widely known, turbulence is eddy motion that has a wide spectrum of eddy sizes. Large eddies, which correspond to frequency fluctuations, are determined by boundary conditions or external forces and their scales are of the same order as the mean flows; thus, they play the dominant role in the transport of mass, momentum, and energy. Small eddies, corresponding to high-frequency fluctuations, depend on viscous forces and contribute very little to the flow and transport of mass. The LES model divides the flows into large eddies and small eddies; the flows due to the large eddies are calculated in terms of the governing equations, and the small eddies are modelled by the use of the eddy viscosity model.

By the use of the Gaussian filter function,

$$G_i(x_i - x_i') = \sqrt{\frac{\gamma}{\pi}}\frac{1}{\Delta x_i}\exp\left[-\frac{\gamma(x_i - x_i')^2}{\Delta_i^2}\right] \quad (i = 1, 2, 3) \tag{2.39}$$

a hydraulic quantity is divided into two parts, one of which is the large-scale component $\bar{f}$ and another the residual component f'

$$\bar{f}(x_i) = \int \prod_{i=1}^{3} G_i(x_i - x_i') \, f(x_i - x_i')dx_i' \, dx_2' \, dx_3' \tag{2.40}$$

$$f'(x_i) = f(x_i) - \bar{f}(x_i) \tag{2.41}$$

where Δ_i is the filter width in the x_i-axis and γ is the filter constant, which is normally given to be 0.6 (Leonard, 1988).

The governing equations of the LES model can be derived by making volume integrals after multiplying by the Gaussian filter function. For convenience of description, the governing equations are expressed in compact notation.

Continuity equation

$$\frac{\partial \bar{u}_i}{\partial x_i} = 0 \quad (i = 1, 2, 3) \tag{2.42}$$

Momentum equation

$$\begin{aligned} \frac{\partial \bar{u}_i}{\partial t} + \frac{\partial}{\partial x_j}\bar{u}_j\bar{u}_i &= 2\varepsilon_{ijk}\bar{u}_j\omega_k - \frac{\partial \Omega}{\partial x_i} - \frac{1}{\rho}\frac{\partial \bar{p}}{\partial x_i} \\ &+ \frac{\partial}{\partial x_j}\left(\mu_m \frac{\partial \bar{u}_i}{\partial x_j} + R_{ij} - L_{ij}\right) \quad (i,j,k = 1,2,3) \end{aligned} \tag{2.43}$$

Temperature equation

$$\frac{\partial \bar{T}}{\partial t} + \frac{\partial}{\partial x_j}\bar{u}_j\bar{T} = \frac{\partial}{\partial x_j}\left(\mu_T \frac{\partial \bar{T}}{\partial x_j} + R_{jT} - L_{jT}\right) \quad (j = 1,2,3) \tag{2.44}$$

where R is the Subgrid Scale (SGS) component and L is the Leonard term:

$$\begin{aligned} R_{ij} &= -\left(\overline{u_i'\bar{u}_j} + \overline{\bar{u}'_i u_j'} + \overline{u_i' u_j'}\right) \\ L_{ij} &= \left(\overline{\bar{u}_j\bar{u}_j} - \bar{u}_i\bar{u}_j\right) \\ R_{iT} &= -\left(\overline{u_i'\bar{T}} + \overline{\bar{u}'_i T} + \overline{u_i' T}\right) \\ L_{iT} &= \left(\overline{\bar{u}_i\bar{T}} - \bar{u}_i\bar{T}\right) \end{aligned} \tag{2.45}$$

By making the Taylor expansion of the equation

$$\overline{\bar{u}_i\bar{u}_j} = \int_{-\infty}^{\infty} G_i(x_i - x_i')\bar{u}_i\bar{u}_j dx' \tag{2.46}$$

around $\bar{u}_i\bar{u}_j$, the next formula is derived:

$$\overline{\bar{u}_i\bar{u}_j} = \bar{u}_i\bar{u}_j + \frac{\Delta_j^2\partial^2(\bar{u}_i\bar{u}_j)}{4\gamma\partial x_j\partial x_i} + \ldots \tag{2.47}$$

Therefore, L_{ij} is represented in the following:

$$\frac{\partial L_{ij}}{\partial x_j} = \frac{\partial}{\partial x_j}\left(\frac{\Delta_j^2\partial^2(\bar{u}_i\bar{u}_j)}{4\gamma\partial x_j\partial x_i}\right) \tag{2.48}$$

It is known from Eq. (2.47) that L_{ij} has the accuracy of the same order as the truncation error of the difference schemes. Therefore, if the difference schemes of second-order accuracy are employed, it is not necessary to take into account the L_{ij} terms. It becomes necessary to consider the effects of L_{ij} terms when higher-order accuracy schemes are used. In this section, L_{ij} terms are ignored (Ivetic, 1988).

On the other hand, by the use of Smagorinsky's eddy viscosity model (1963) (Deardorff, 1970), SGS shear stress R_{ij} can be formulated as follows:

$$R_{ij} = v_t\left(\frac{\partial u_i}{\partial x_j} + \frac{\partial u_j}{\partial x_i}\right) - \frac{1}{3}\delta_{ij}\overline{u_l'u_l'} \tag{2.49}$$

where v_t is the SGS eddy viscosity coefficient and δ_{ij} is the Kronecker delta.

Making a dimensional analysis with the aid of $-5/3$ power law of Kolmogorov spectrum, v_t can be presented in the following (Leonard, 1988):

$$v_t = c^{4/3}\varepsilon^{4/3}\Delta^{4/3} \tag{2.50}$$

where c is the dimensionless constant, ε the energy dissipation rate, and Δ the magnitude of spatial scale. These are to be described later.

If the energy dissipation rate ε is specified by

$$\bar{\varepsilon} = v_t\frac{\partial \bar{u}_i}{\partial x_j}\left(\frac{\partial \bar{u}_i}{\partial x_j} + \frac{\partial \bar{u}_j}{\partial x_i}\right) \tag{2.51}$$

v_t can be expressed by

$$v_t = (c\Delta)^2\left\{\frac{\partial \bar{u}_i}{\partial x_j}\left(\frac{\partial \bar{u}_i}{\partial x_j} + \frac{\partial \bar{u}_j}{\partial x_i}\right)\right\}^{1/2} \tag{2.52}$$

after eliminating ε in Eqs. (2.50) and (2.51).

The SGS eddy viscosity coefficient v_t is generally calculated by Eq. (2.52). Because of the extreme difference of the scales between the horizontal plane and the vertical direction, v_t is also divided into horizontal and vertical components. The right-hand side of Eq. (2.52) is normalized by introducing the representative lengths of L in the horizontal plane, H in the vertical direction, and representative velocity U in the horizontal plane (the dimensionless quantities are indicated by the symbol +):

$$v_t = (c\Delta)^2 \frac{U^2}{L^2}\left[2\left(\frac{\partial u^+}{\partial x^2}\right)^2 + 2\left(\frac{\partial v^+}{\partial y^2}\right)^2 + 2\left(\frac{\partial w^+}{\partial z^+}\right)^2 + \left(\frac{\partial u^+}{\partial y^+} + \frac{\partial v^+}{\partial x^+}\right)^2 \right.$$
$$\left. + \frac{L^2}{H^2}\left\{\left(\frac{\partial u^+}{\partial z^+} + \frac{H^2}{L^2}\frac{\partial w^+}{\partial x^+}\right)^2 + \left(\frac{\partial v^+}{\partial z^+} + \frac{H^2}{L^2}\frac{\partial w^+}{\partial y^+}\right)^2\right\}\right]^{1/2} \tag{2.53}$$

In lakes, usually $L \gg H$, therefore v_t can be approximated by the following formula:

$$v_t = (c\Delta)^2\left[\left(\frac{\partial u}{\partial z}\right)^2 + \left(\frac{\partial v}{\partial z}\right)^2\right]^{1/2} \tag{2.54}$$

Obviously, the constants c and Δ have to be given in order to calculate the SGS eddy viscosity coefficient v_t^v from the preceding equation. $c = 0.05$–0.30 is usually used (Deardorff, 1970). The larger the c used, the more excessively the small energy is evaluated, and thus the eddy viscosity coefficients become larger and the attenuation of flow becomes faster. For this book, c is assumed to be 0.1. On the other hand, Δ is given by the formula

$$\Delta_z = \min(2\Delta_z, z_a) \tag{2.55}$$

where Δ_z is the depth-wise mesh size and z_a is the distance from the point under consideration to the bottom of the lake or the water surface.

If the horizontal momentum equations are normalized,

$$v_t^h \frac{U}{L^2} = v_t \frac{U}{H^2} \tag{2.56}$$

is obtained, and the horizontal SGS eddy coefficient v_t^h is calculated by the equation

$$v_t^h = \left(\frac{\Delta_h}{\Delta z}\right)^2 \overline{v_t^v} = \left(\frac{\Delta_h}{\partial z}\right)^2 \frac{1}{H}\sum_{\text{surf}}^{\text{btm}} v_t^v \Delta z \tag{2.57}$$

where $\overline{v_t^v}$ is the depth averaged value of v_t Δ_h, given by

$$\Delta_h = \sqrt{\Delta x \Delta y} \tag{2.58}$$

2.4 BOUNDARY AND INITIAL CONDITIONS

2.4.1 Boundary Conditions at the Surface ($z = \zeta$)

The kinematic boundary condition is imposed at the free water surface:

$$w = \frac{\partial \zeta}{\partial t} + u\frac{\partial \zeta}{\partial x} + v\frac{\partial \zeta}{\partial y} \tag{2.59}$$

The momentum and heat across the air–water interface are the primary causes of circulation. The momentum flux, described as wind shear stress, can be evaluated by the wind velocity:

$$(\tau_{sx}, \tau_{sy}) = \rho_a C_D \sqrt{W_x^2 + W_y^2}(W_x,\ W_y) \tag{2.60}$$

where $W = (W_x,\ W_y)$ is the wind velocity measured at 10 m above the still water surface; ρ_a the air density; and C_D a surface wind drag coefficient, which is typically of order $0[10^{-3}]$. According to Wu (1969), the drag coefficient for neutral stratification is given by

$$C_D = (0.8 + 0.065W) \times 10^{-3} \tag{2.61}$$

In a formula similar to the one used for the wind-wave model by Donelan et al. (1974), Eid (1981) proposed relating the drag coefficient to the wind speed at distance z (usually 10 m), as well as wave height H_s and wave velocity c:

$$C_D = \left[\frac{k}{\ln(z/\beta H_s)}\right]^2 \left(1.0 - \frac{c}{W}\right)^2 \tag{2.62}$$

After evaluation, using wind and wave data in Hamilton Harbour (Eid, 1981), a formula similar to Eq. (2.61) was presented:

$$C_D = (0.75 + 0.075W) \times 10^{-3} \tag{2.63}$$

Usually, the heat-exchange rate at the surface is taken as zero during the short-term simulation. However, this assumption may not be justified. In general, there are two approaches for determining heat-exchange flux. The first approach is a simple but efficient one, which is similar to the momentum flux formula, that is, an empirical function with environmental parameters (Large and Pond, 1981; Smith and Anderson, 1984; Wu, 1992). Ignoring the molecular transfer term, the heat flux across air–water interface can be expressed as (Wu, 1992):

$$H_0 = \rho_a C_P \overline{w'T'} = \rho_a C_p C_T U_Z (T_0 - T_Z) \tag{2.64}$$

where H_0 is the sensible heat flux; T' and w' are the temperature and the vertical velocity fluctuations respectively; T_Z the potential air temperature at the elevation Z

above the mean water surface; T_0 the potential air temperature at the surface, that is, water temperature at the surface; C_P the specific heat of air at a constant pressure; C_T the coefficient of sensible heat transfer and is also called the Stanton number S_T. Based on the direct measurement of $\overline{w'T'}$ from field studies, an empirical formula of the heat transfer coefficient—varying with wind, atmospheric stability conditions, and temperature difference—is

$$C_T = (0.72 + 0.0175W\ \Delta T) \times 10^{-3},\ W < 8\ \mathrm{m/s} \tag{2.65}$$

$$C_T = (1.000 + 0.0015W\ \Delta T) \times 10^{-3},\ W > 8\ \mathrm{m/s} \tag{2.66}$$

$\Delta T = T_0 - T_{10}$ is the temperature difference between the water surface and the temperature at 10 m above the water surface, and is expressed in degrees Celsius (Wu, 1992).

The second approach is based on the energy budget calculation: the net heat exchange is the difference between absorbed solar radiation and heat losses due to evaporation, conduction, and reverse radiation (Ryan, 1971; Shen, 1991).

The heat exchange is specified in the next formula and used as the boundary condition of the density deficit equation:

$$K_v \frac{\partial \Delta\rho}{\partial z} = \frac{\alpha Q_0}{c_p} \tag{2.67}$$

where α is the volume expansion rate of water, Q_0 the heat absorbed at the water surface per unit time and unit area, and c_p the specific heat of water.

The heat exchanges include solar radiation and heat loss. If the solar radiation is defined as Q_{s0}, then the solar radiation entering into the lake surface Q_s is expressed as follows:

$$Q_s = (1 - a_l)Q_{s0} \tag{2.68}$$

where a_l is the reflection rate of the water surface. Within Q_s, βQ_s is absorbed by the water surface, and $(1-\beta)Q_s$ enters into the body of water, where β denotes the absorption rate. The heat Q_z from the water surface to the water depth $\zeta - z$ is

$$Q_z = (1 - \beta)Q_s \exp[-\kappa(\zeta - z)] \tag{2.69}$$

where κ is the extinction coefficient.

On the other hand, the heat loss from the water surface Q_L (kcal/m^2/day) includes three parts: the evaporation loss Q_e, the conduction loss Q_c, and the reverse radiation loss Q_{ra}:

$$Q_L = Q_e + Q_c + Q_{ra} \tag{2.70}$$

Q_e+Q_c and Q_{ra} are expressed by the following Rohwer and Swinbank's empirical formulas (Donelan, 1980):

$$Q_e + Q_c = (0.000308 + 0.000185 W_s)\rho(e_s - \phi e_a) \times \left\{ L_v + c_p T_x + \frac{269.1(T_s - T_a)}{e_s - \phi e_a} \right\} \quad (2.71)$$

$$Q_{ra} = 0.97k\,\{T_W^4 - 0.937 \times 10^{-5}\ T_A^6(1.0 + 0.17C^2)\} \quad (2.72)$$

where W_s is the wind velocity 15 cm over the water surface; e_s the saturation steam pressure (mmHg) corresponding to the temperature of the water surface; e_a the saturation vapor pressure (mmHg) corresponding to the air temperature; ϕ the relative humidity (%); L the potential heat of evaporation (kcal/kg); T_s the temperature of the water surface (°C); T_a the air temperature; k the Stefan-Boltzmann constant ($k = 1.171 \times 10^{-6}$ kcal/m^2/day/°K); T_W the temperature of the water surface (°K); T_A the air temperature (°K); and C the cloudiness (%).

The heat exchanges at the water surface and the productive term q of density deficit, because of heat, are expressed by the following formulas:

$$Q_0 = \beta Q_s - Q_L \quad (2.73)$$

$$q = -\left(\frac{dQ_z}{dz}\right)\alpha/c_p \quad (2.74)$$

Temperature T (°C) and saturation vapor pressure e (mmHg) are connected by the empirical formula

$$e = 0.418T^2 - 0.6216T + 13.0 \quad (2.75)$$

The third approach is based on the surface heat flux calculation from the wind and temperature profiles, as described in ALFS (Schwab and Bedford, 1994).

2.4.2 Boundary Conditions at the Bottom

The kinematic boundary condition at the bottom ($z = h - z_b$) is

$$w_b = u_b \frac{\partial z_b}{\partial x} + v_b \frac{\partial z_b}{\partial y} \quad (2.76)$$

where subscript b refers to bottom, and h is the water depth. For the shear stress condition, there are two kinds of boundary conditions.

1. No-slip condition, that is, the velocity at the bottom is equal to zero:

$$u = v = w = 0 \quad (2.77)$$

Then, the bottom shear stress is evaluated from the velocity and vertical eddy viscosity near the bottom directly in the model.

2. A quadratic law similar to that at the surface,

$$(\tau_{bx}, \tau_{by}) = r\sqrt{u_b^2 + v_b^2}(u_b, v_b) \tag{2.78}$$

where (τ_{bx}, τ_{by}) are the shear stresses at the bottom in the x and y directions respectively; r the bottom friction coefficient; and (u_b, v_b) are the bottom velocity components (usually at the layer close to the bottom) in the x and y directions, respectively. There is no heat allowed to be transferred through the bottom, or at least, the flux is assumed to be negligible.

2.4.3 Lateral Boundary Conditions

Along the shoreline, where river inflow or outflow may occur, the variables generally take a priori known value:

$$u = u_r(x, y, z, t) \tag{2.79}$$

$$v = v_r(x, y, z, t) \tag{2.80}$$

$$w = 0.0 \tag{2.81}$$

$$T = T_r(x, y, z, t) \tag{2.82}$$

While, along the land shoreline, no normal fluxes of momentum and heat are allowed:

$$\vec{V}_n = 0; \quad N_h \frac{\partial T}{\partial n} = 0 \tag{2.83}$$

where n is the outward-directed normal.

2.4.4 Open Boundary Conditions

At the open boundary, that is, computational or artificial boundary, the surface elevation or mass flux is required, as is the flux of energy into the system. In the nested-grid model system, the velocity, elevation, and temperature may be interpolated from the coarse-grid model. Otherwise, free radiation of open boundary condition should be used (Miller and Thorpe, 1981). In the free radiation condition, no perturbation is incoming through the boundary. Outgoing signals are allowed to cross without back reflection. The condition applied is

$$hU_n = \zeta\sqrt{gh} \tag{2.84}$$

where U_n is the depth-averaged normal velocity, because only the depth-averaged velocity is provided in this open boundary condition. In order to provide 3D velocities, a profile at the open boundary must be adopted; one example is the second-order parabola. This profile is valid for a laminar type flow and is an approximation of the turbulent flow.

2.4.5 *Initial Conditions*

The initial distributions of velocities and temperature must be specified for the simulation. The velocities are taken to be zero, unless the initial velocity field is given. The wind field input can be specified in a variety of ways. For example, a step-growth wind is used in the simulation for a smooth start purpose, and a warm-up period is introduced for the variable wind process. In the case of thermal stratification, the initial temperature field is needed.

2.5 CIRCULATION THEORY

2.5.1 *Long-Wave Theory*

The velocity profile develops slowly with time and uniformly in stages, having the same form as in a river channel:

$$u = \frac{u_s^*}{\kappa} \ln \frac{h+z}{z_0} \tag{2.85}$$

The current intensity is almost uniform over the depth, that is, the same direction over the whole depth, with steep gradients developing only near the bottom. This flow structure is similar to a hydraulically induced flow, such as a river flow.

2.5.2 *Shear-Induced Countercurrent Flow Theory*

Inherently, wind-induced currents in natural lakes, estuaries, and embayments are shear-induced type of flows. The vertical mixing mechanism can be examined using steady plane Couette laminar flow (for slow wind-induced current), and turbulent flow. In these flows, the current close to the surface is in the direction of the applied shear stress, while, away from the surface at the bottom, a return current develops. Required by continuity, the integrated flow transport from the surface to the bottom is zero, that is, zero net volume flux.

Laminar flow has a parabolic velocity distribution and a linear shear stress distribution. The bottom to surface shear stress ratio equals -0.5, and the zero velocity occurs at $z/h = 0$ and $2/3$, where z is measured upward from the lake bottom. A constant viscosity over the depth leads to a parabolic velocity profile, as the exact solution of the Navier–Stokes equation, which is close to the experimentally determined profile (Tsanis, 1986). Analogous to the laminar flow theory, a constant vertical eddy viscosity, evaluated by the applied wind shear stress

τ_s, was used in many circulation models. For example, Koutitas and O'Connor (1980) used

$$K_v = \lambda h \sqrt{\tau_s / \rho} \tag{2.86}$$

where λ is a constant. Taking the mean value of a parabolic distribution of eddy viscosity, it was suggested that $\lambda = 0.066$.

In the case of turbulent shear-induced countercurrent flow, both viscous stresses (due to the molecular motion), and Reynolds stresses (due to turbulence), contribute to the total shear stress. Close to the water surface and bottom boundaries, the viscous stresses are predominant, while in the rest of the flow field, the Reynolds stresses greatly exceed the viscous contribution. The thicknesses of these viscous sublayers are very small compared to the depth. The velocity changes linearly with the distance from the boundary, while steep velocity gradients exist in the sub-layers. The velocity distribution then depends on the distribution of the turbulent shear stress with z, which is provided by introducing appropriate "closure" hypotheses.

2.6 MODEL CONSIDERATIONS

2.6.1 *Baroclinic Terms*

Baroclinic terms represent the effect of thermal stratification. Great simplification can be achieved if these terms are assumed to be small, and are therefore neglected. If the density can be justifiably assumed to be constant, then the transport equation of temperature, or salinity, can be uncoupled from the continuity and momentum equations. Because thermal stratification is usually present in the summer period, one such strategy is based on the following consideration: assuming that in the simulation period the wind did not significantly affect the stratification, and the determination of the state of stratification is not a primary modelling objective, then the density field can be treated diagnostically. The baroclinic terms in the momentum equation are not solved, but evaluated, by assuming that the temperature (or salinity) field is known as a priori from measurements or from independent studies. However, this method cannot be applied to the case of rapid change of weather conditions, or medium- and long-term simulations. In a review paper by the ASCE Task Committee, Wang et al. (1990) pointed out that the simulation of partially stratified estuaries or lakes is one of the most challenging areas in recent years. A completed field data base was required to provide the input initial density field and verify the simulation.

2.6.2 *Advection Terms*

Non-linear advection terms are often the source of difficulties in numerical solutions, as they give rise to non-linear instabilities. As a result, in some model formulations, the advection terms are neglected (Simons, 1972; Eid, 1981). This implication needs to be carefully justified for different applications (Blaisdell et al., 1991; Wu and Tsanis, 1991).

2.6.3 Bottom Shear Stress Terms

The bottom stress formulation, which is a specification of the rate of momentum loss at the water–sediment interface, assumes a variety of forms in different models. The bottom stress is treated as proportional to a quadratic form of velocity in most 2D models (Leendertse, 1967; Simons, 1972; Koutitas and Gousidou-Koutita, 1986), and enters into the 3D circulation model through the bottom boundary conditions (Sheng, 1983). The proportional constant or friction coefficient can be assigned as a constant, or calculated from the roughness or Manning's coefficient and water depth, or evaluated using the results from the turbulent boundary layer theory. Other models assume the bottom stress to be linearly proportional to velocity. A no-slip condition assuming a zero velocity at lake-bottom interface is also adopted by many 3D circulation models.

2.6.4 Rigid-Lid Approximation

In the early lake circulation models, a rigid-lid approximation was used (Cheng et al., 1976; Schwab, 1981). In this approximation, the free surface deformations are ignored, that is, $\partial\zeta/\partial t \approx 0$. As a result, along with the neglect of non-linear advection terms, a simplified model produced by solving the stream function was formed for steady circulation. However, in most cases, such an approximation is not justified.

2.7 FINITE DIFFERENCE METHOD

Traditional finite-difference models use rectangular Cartesian grids, which resolve the geometry with "stair-steps." This grid became inefficient at representing features such as coastal boundary layer dynamics, flow in channels, embayments, and around islands, and flows near engineering structures. To resolve the smallest spatial scales of interest for an engineering problem, and at the same time maintain reasonable computational cost, a transformation has been applied by many investigators to allow increased user control of grid placement. These transformations include the change of variables from Cartesian coordinates to orthogonal curvilinear coordinates (Blumberg and Herring, 1987) or boundary fitted conditions (Swanson, 1986; Sheng, 1990). Successful applications were mainly found for large bays, estuaries, and wide rivers where many tributaries exist, and where geometry is complicated. Examples of this include Chesapeake Bay (Johnson et al., 1990), the James River (Sheng et al., 1990), and Potomac Estuary (Hinz et al., 1990). However, due to the limited field data available for model calibration, the use of curvilinear or boundary fitted coordinates in lake models is limited. These transformations would provide a better shoreline description by using a large number of elements that, in turn, increase the required computational time. An alternative approach, which may also be more efficient, is the use of a nested-grid modelling technique in which a high-resolution fine grid is used for the local nearshore areas of interest, and the velocity and elevation along the artificial open boundary in the fine grid are provided from the overall coarse grid for the whole lake (Murthy et al., 1985; Tsanis and Wu, 1992, 1993).

Two types of vertical grids are generally used in the finite difference models of lakes. In the first type, as adopted by Leendertse and Liu (1975), Simons (1972), and Eid (1981), the vertical domain of the water body is separated by layers of constant depth. In general, water depth varies over a lake basin, shallow in nearshore areas and deep in the central basin. Thus, this grid is insufficient to resolve large variations of depth over the domain. The second type of grid is a vertically stretched grid, or σ-coordinate transformation, which leads to a smooth representation of the topography, and, in addition, gives the same order of resolution for the shallow and deeper parts of the water body. This transformation will introduce additional terms which require numerical treatment. In some cases, their physical significance is not apparent. However, most of the additional terms are contained in the horizontal diffusivity terms. Since the horizontal diffusion is generally small compared to the vertical diffusion and horizontal advection, only the leading terms in general need to be retained (Sheng, 1983). Tests on the different formulations of vertical and horizontal diffusivities, and pressure gradient force over steep topography in the σ-coordinate system, were conducted by Miller and Blumberg (1985) and Haney (1991) respectively. Although there are increased calculations and uncertainties using the σ-transformation, the ability to deal with a large variation of water depth offsets other inherent disadvantages. A further refinement of this transformation is to increase the vertical resolution either near the surface or near both the surface and the bottom (Swanson, 1986; Davis, 1991).

On the selection of time step, if the explicit time integration method is used, the time step will be governed by the speed of propagation of the free surface wave (Koutitas, 1988). Also, a condition will be imposed by the vertical diffusion terms which are related to the vertical eddy viscosity and the water depth. In lakes and their nearshore areas, shallow water depth and high vertical eddy viscosity results in a small time step because of the stability requirement. It is necessary to treat the diffusion terms implicitly, at least semi-implicitly, such as with the Crank–Nicholson implicit scheme (Koutitas and Gousidou-Koutita, 1986) or with the three time-level Dufort–Frankel scheme (Davies, 1985).

Chapter 3

THE VERTICALLY INTEGRATED TWO-DIMENSIONAL MODELS

For well-mixed shallow waters near coastlines and in shallow lakes, the water movements in the horizontal plane usually predominate, and the vertical velocity and density variations are very small. Therefore, adopting a vertically integrated two-dimensional model to simulate large-scaled flows in shallow waters is adequate. In this chapter, the governing equations of the vertically integrated two-dimensional model are derived. Several numerical methods have been employed to simulate the prominent flows in shallow waters, and comparisons between the methods are made.

3.1 GOVERNING EQUATIONS

In order to obtain the governing equations of the vertically integrated two-dimensional model for a well-mixed shallow lake, the following assumptions are adopted.

(a) Lake water is completely homogeneous, thereby uncoupling the temperature equation from the governing equations.

(b) The vertical accelerations are negligible in comparison to the horizontal accelerations.

(c) Compared to bottom friction, eddy viscosity terms in horizontal direction are small enough to be ignored.

In combination with the assumption of constant water density, the approximation of hydrostatic pressure can be represented as

$$p = g\int_{z}^{\zeta} \rho \; dz = \rho g(\zeta - z) \tag{3.1}$$

Moreover, by differentiating the preceding equation with respect to x and y respectively, the next two formulas are obtained:

$$\frac{\partial p}{\partial x} = \rho g \frac{\partial \zeta}{\partial x} \tag{3.2}$$

$$\frac{\partial p}{\partial y} = \rho g \frac{\partial \zeta}{\partial y} \tag{3.3}$$

By combining all the assumptions and by integrating the continuity equation and Reynolds equations vertically with the aid of Leibnitz's rule, the following governing equations can be obtained:

Continuity equation

$$\frac{\partial \zeta}{\partial t} + \frac{\partial (h+\zeta)U}{\partial x} + \frac{\partial (h+\zeta)V}{\partial y} = 0 \tag{3.4}$$

Momentum equations

x-direction

$$\frac{\partial U}{\partial t} + U\frac{\partial U}{\partial x} + V\frac{\partial U}{\partial y} = fV - g\frac{\partial \zeta}{\partial x} + \frac{\tau_{sx} - \tau_{bx}}{\rho(h+\zeta)} \tag{3.5}$$

y-direction

$$\frac{\partial V}{\partial t} + U\frac{\partial V}{\partial x} + V\frac{\partial V}{\partial y} = -fU - g\frac{\partial \zeta}{\partial y} + \frac{\tau_{sy} - \tau_{by}}{\rho(h+\zeta)} \tag{3.6}$$

where the vertically averaged velocity components U and V are defined as

$$U = \frac{1}{h+\zeta}\int_{-h}^{\zeta} u \, dz \tag{3.7}$$

$$V = \frac{1}{h+\zeta}\int_{-h}^{\zeta} v \, dz \tag{3.8}$$

where h is the depth from the reference plane to the bottom.

This model excludes the eddy viscosity terms in the horizontal direction because they are normally small, and the effects of small-scale velocity fluctuations are aggregated into the shear stress terms.

3.2 NUMERICAL METHODS

Many kinds of difference schemes for the vertically integrated two-dimensional model have been proposed (Cheng, 1976), and they can be divided into two types: implicit schemes and explicit schemes. They are differentiated by what time level is used for each discretized variable. In an explicit scheme, the unknown variables at a future time step $t+\Delta t$ are expressed by a number of variables at the previous time steps. If the initial and boundary conditions are provided, the calculation can proceed in a sequential way to get the values of all variables at any time step. In an implicit scheme, an unknown variable at time step $t+\Delta t$ depends not only on the values of variables at previous time steps, but also on the variables in other locations at time step $t+\Delta t$. Therefore, linear simultaneous algebraic equations need to be solved. In this chapter, an explicit or implicit scheme is indicated when the pressure terms (or water stage terms) in the momentum equations are treated explicitly or implicitly. Formulas of explicit schemes are usually simple, but they have to satisfy the following Courant–Friedrichs–Lewy condition, resulting from Neumann's stability analysis, in order to get a stable solution:

$$C_r = \Delta t\sqrt{gH_{\max}} \quad \sqrt{1/\Delta x^2 + 1/\Delta y^2} \leq 1 \tag{3.9}$$

$H_{\max}$ is the maximum water depth, and Δx and Δy are the spatial increments of the mesh in the x- and y-directions, respectively. The small time steps in explicit schemes usually lead to large amounts of Central Processing Unit (CPU) time, which can be unacceptable for some problems. On the other hand, implicit schemes are unconditionally stable and permit a larger time step. The difference formulas, however, are complicated and iterative calculations are required to solve the simultaneous algebraic equations, which can also consume large amounts of CPU time. In order to overcome this problem, the so-called Alternating Direction Implicit (ADI) method and the operator-splitting method have been developed. The implicit solutions can then be obtained without any iterative calculations.

In this section, an explicit scheme, the ADI method and the operator-splitting method are adopted, based on a space-staggered grid system. A space-staggered grid system is a system in which hydraulic variables are defined at different places in such a way that the numerical oscillation due to pressure terms can be avoided (Kurihara and Holloway, 1967). In the staggered grid system, which is shown in Figure 3.1, i and j are the running indices which represent the horizontal points $x = i\Delta x$ and $y = j\Delta y$. The velocity component U is defined at half-integer i and integer j; V is defined at integer i and half-integer j; and the water stage ζ is described at both integers i and j.

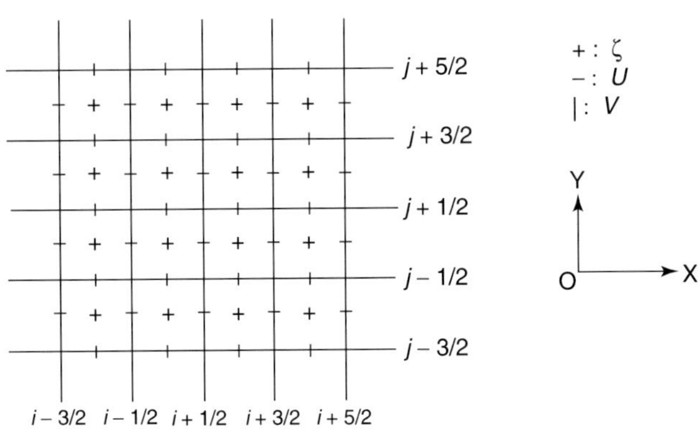

Fig. 3.1. Definition points of variables in the space-staggered grid system (Shen, 1991).

3.2.1 Explicit Method

There are various kinds of explicit methods which use different temporal difference schemes. In this study, an explicit method based on the leap-frog scheme is used. In order to satisfy the Vasiliev stability condition (Inoue, 1986), the velocities U and V in the bottom friction terms are treated as partially implicit. The governing equations are discretized into the following equations, according to the explicit method.

Momentum equation in the x-direction

$$\begin{aligned}\frac{U^{n+2}_{i+1/2,j}-U^{n}_{i+1/2,j}}{2\Delta t} = & -U^{n+1}_{i+1/2,j}\frac{U^{n+1}_{i+3/2,j}-U^{n+1}_{i-1/2,j}}{2\Delta x} \\ & -V^{n+1}_{i+1/2,j}\frac{U^{n+1}_{i+1/2,j+1}-U^{n+1}_{i+1/2,j-1}}{2\Delta y} \\ & +fV^{n+1}_{i+1/2,j}-g\frac{\zeta^{n+1}_{i+1,j}-\zeta^{n+1}_{i,j}}{\Delta x} \\ & -gn^2\frac{U^{n+2}_{i+1/2,j}\sqrt{U^{n2}_{i+1/2,j}+V^{n2}_{i+1/2,j}}}{(h+\zeta^{n+1})^{4/3}_{i+1/2,j}} \\ & +\frac{\tau_{sx}}{\rho(h+\zeta^{n+1})_{i+1/2,j}}\end{aligned} \tag{3.10}$$

Momentum equation in the y-direction

$$\begin{aligned}\frac{V^{n+2}_{i,j+1/2}-V^{n}_{i,j+1/2}}{2\Delta t} = & -U^{n+1}_{i,j+1/2}\frac{V^{n+1}_{i+1,j+1/2}-V^{n+1}_{i-1,j+1/2}}{2\Delta x} \\ & -V^{n+1}_{i,j+1/2}\frac{V^{n+1}_{i,j+3/2}-V^{n+1}_{i,j-1/2}}{2\Delta y} \\ & -fU^{n+1}_{i,j+1/2}-g\frac{\zeta^{n+1}_{i,j+1}-\zeta^{n+1}_{i,j}}{\Delta y} \\ & -gn^2\frac{V^{n+2}_{i,j+1/2}\sqrt{U^{n2}_{i,j+1/2}+V^{n2}_{i,j+1/2}}}{(h+\zeta^{n+1})^{4/3}_{i,j+1/2}} \\ & +\frac{\tau_{sy}}{\rho(h+\zeta^{n+1})_{i,j+1/2}}\end{aligned} \tag{3.11}$$

in which

$$U_{i,j+1/2}=(U_{i-1/2,j}+U_{i+1/2,j}+U_{i-1/2,j+1}+U_{i+1/2,j+1})/4 \tag{3.12}$$

$$V_{i+1/2,j} = (V_{i,j-1/2} + V_{i,j+1/2} + V_{i+1,j-1/2} + V_{i+1,j+1/2})/4 \tag{3.13}$$

Continuity equation

$$\frac{\zeta_{i,j}^{n+3} - \zeta_{i,j}^{n+1}}{2\Delta t} = -\frac{(h+\zeta^{n+1})_{i+1/2,j}U_{i+1/2,j}^{n+2} - (h+\zeta^{n+1})_{i-1/2,j}U_{i-1/2,j}^{n+2}}{\Delta x} - \frac{(h+\zeta^{n+1})_{i,j+1/2}V_{i,j+1/2}^{n+2} - (h+\zeta^{n+1})_{i,j-1/2}V_{i,j-1/2}^{n+2}}{\Delta y} \tag{3.14}$$

where

$$\zeta_{i+1/2,j} = (\zeta_{i,j} + \zeta_{i+1,j})/2 \tag{3.15}$$

$$\zeta_{i,j+1/2} = (\zeta_{i,j} + \zeta_{i,j+1})/2 \tag{3.16}$$

Stable solutions can be obtained by this method if Eq. (3.9) is satisfied. The following procedure is used:

1. Calculate $U_{i+1/2,j}^{n+2}$ for all grid points using Eq. (3.10).
2. Calculate $V_{i,j+1/2}^{n+2}$ for all grid points from Eq. (3.11).
3. Substitute U^{n+2} and V^{n+2} into Eq. (3.14) to obtain $\zeta_{i,j}^{n+3}$ for all grid points.

The values of U, V and ζ for every grid point at any time step can be obtained by repeating the foregoing procedure.

3.2.2 ADI Method

The ADI method is an implicit method widely used in numerical analysis (Murakami et al., 1985). Different from ordinary implicit methods, the ADI method splits one time step Δt into two fractional steps. In the first fractional step, the x-direction components in the continuity equation and the momentum equation in the x-direction are expressed in implicit form, and the y-direction components are discretized in explicit form. The resulting linear simultaneous algebraic equations are solved by the so-called double-sweep method. In the second fractional step, the y-direction components in the continuity equation and the momentum equation in the y-direction are expressed in implicit form, and the x-direction components in explicit form.

The governing Eqs. (3.4)–(3.6) are discretized into the following difference formulas by means of the ADI method:

First fractional step $(t \to t + \Delta t/2)$
Continuity equation

$$\frac{\zeta_{i,j}^{n+1/2} - \zeta_{i,j}^{n}}{\Delta t/2} = -\frac{(h+\zeta^n)_{i+1/2,j}U_{i+1/2,j}^{n+1/2} - (h+\zeta^n)_{i-1/2,j}U_{i-1/2,j}^{n+1/2}}{\Delta x} - \frac{(h+\zeta^n)_{i,j+1/2}V_{i,j+1/2}^{n} - (h+\zeta^n)_{i,j-1/2}V_{i,j+1/2}^{n}}{\Delta y} \tag{3.17}$$

x-wise momentum equation

$$\begin{aligned}\frac{U_{i+1/2,j}^{n+1/2} - U_{i+1/2,j}^{n}}{\Delta t/2} = &- U_{i+1/2,j}^{n+1/2}\frac{U_{i+3/2,j}^{n} - U_{i-1/2,j}^{n}}{2\Delta x} \\ &- V_{i+1/2,j}^{n}\frac{U_{i+1/2,j+1}^{n} - U_{i+1/2,j-1}^{n}}{2\Delta y} \\ &+ fV_{i+1/2,j}^{n+1} - g\frac{\zeta_{i+1,j}^{n+1/2} - \zeta_{i,j}^{n+1/2}}{\Delta x} \\ &- gn^2\frac{U_{i+1/2,j}^{n+1/2}\sqrt{U_{i+1/2,j}^{n2} + V_{i+1/2,j}^{n2}}}{(h+\zeta^n)_{i+1/2,j}^{4/3}} \\ &+ \frac{\tau_{sx}}{\rho(h+\zeta^n)_{i+1/2,j}}\end{aligned} \tag{3.18}$$

y-wise momentum equation

$$\begin{aligned}\frac{V_{i,j+1/2}^{n+1/2} - V_{i,j+1/2}^{n}}{\Delta t/2} = &- U_{i,j+1/2}^{n+1/2}\frac{V_{i+1,j+1/2}^{n} - V_{i-1,j+1/2}^{n}}{2\Delta x} \\ &- V_{i,j+1/2}^{n+1/2}\frac{V_{i,j+3/2}^{n} - V_{i,j-1/2}^{n}}{2\Delta y} \\ &- fU_{i,j+1/2}^{n+1/2} - g\frac{\zeta_{i,j+1}^{n+1/2} - \zeta_{i,j}^{n+1/2}}{\Delta y} \\ &- gn^2\frac{V_{i,j+1/2}^{n+1/2}\sqrt{U_{i,j+1/2}^{n+1/22} + V_{i,j+1/2}^{n2}}}{(h+\zeta^{n+1/2})_{i,j+1/2}^{4/3}} \\ &+ \frac{\tau_{sy}}{\rho(h+\zeta^{n+1/2})_{i,j+1/2}}\end{aligned} \tag{3.19}$$

By rearranging Eqs. (3.17) and (3.18), the following simultaneous algebraic equations are obtained:

$$A_{i-1/2,j}U_{i-1/2,j}^{n+1/2} + B_{i,j}\zeta_{i,j}^{n+1/2} + C_{i+1/2,j}U_{i+1/2,j}^{n+1/2} = D_{i,j} \tag{3.20}$$

$$A'_{i,j}\zeta_{i,j}^{n+1/2} + B'_{i+1/2,j}U_{i+1/2,j}^{n+1/2} + C'_{i+1/2,j}\zeta_{i+1,j}^{n+1/2} = D'_{i+1/2,j} \tag{3.21}$$

where

$$\begin{aligned} A_{i-1/2,j} &= -\frac{\Delta t}{2\Delta x}(h+\zeta^n)_{i+1/2,j} \\ B_{i,j} &= 1 \\ C_{i+1/2,j} &= \frac{\Delta t}{2\Delta x}(h+\zeta^n)_{i+1/2,j} \\ D_{i,j} &= \zeta_{i,j}^n - \frac{\Delta t}{2\Delta y}\left[(h+\zeta^n)_{i,j+1/2}V_{i,j+1/2}^n - (h+\zeta^n)_{i,j-1/2}V_{i,j-1/2}^n\right] \\ A'_{i,j} &= -g\frac{\Delta t}{2\Delta x} \end{aligned} \tag{3.22a}$$

$$\begin{aligned} B'_{i+1/2,j} &= 1 + \frac{\Delta t}{4\Delta x}(U_{i+3/2,j}^n - U_{i-1/2,j}^n) + \frac{\Delta t}{2}gn^2\frac{\sqrt{U_{i+1/2,j}^{n2} + V_{i+1/2,j}^{n2}}}{(h+\zeta_{i+1/2,j}^n)^{4/3}} \\ C'_{i+1,j} &= g\frac{\Delta t}{2\Delta x} \\ D'_{i+1/2,j} &= U_{i+1/2,j}^n - \frac{\Delta t}{4\Delta y}V_{i+1/2,j}^n(U_{i+1/2,j+1}^n - U_{i+1/2,j-1}^n) \\ &+ \frac{\Delta t}{2}fV_{i+1/2,j}^n + \frac{\Delta t}{2}\frac{\tau_{sx}}{\rho(h+\zeta^n)_{i+1/2,j}} \end{aligned} \tag{3.22b}$$

Because the coefficient matrices of Eqs. (3.20) and (3.21) are tri-diagonal matrices, $U^{n+1/2}$ and $\zeta^{n+1/2}$ can be obtained simply by the double-sweep method, without any iterative calculation. Then, $V^{n+1/2}$ is calculated in Eq. (3.19) by using the values of $U^{n+1/2}$ and $\zeta^{n+1/2}$.

Double-sweep method
If the following formulas are defined,

$$E_{i,j} = C_{i+1/2,j}/B_{i,j} \tag{3.23}$$

$$F_{i,j} = \left(D_{i,j} - A_{i-1/2,j}U_{i-1/2,j}^{n+1/2}\right)/B_{i,j} \tag{3.24}$$

Eq. (3.20) can be represented as

$$\zeta_{i,j}^{n+1/2} = -E_{i,j}U_{i+1/2,j}^{n+1/2} + F_{i,j} \tag{3.25}$$

Moreover, by substituting Eq. (3.25) into Eq. (3.21), the following equation results:

$$U_{i+1/2,j}^{n+1/2} = -E'_{i,j}\zeta_{i+1,j}^{n+1/2} + F'_{i,j} \tag{3.26}$$

where

$$E'_{i,j} = \frac{C'_{i+1/2,j}}{B'_{i+1/2,j} - A'_{i,j}E'_{i,j}} \tag{3.27}$$

$$F'_{i,j} = \frac{D'_{i+1/2,j} - A'_{i,j}F'_{i,j}}{B'_{i+1/2,j} - A'_{i,j}E'_{i,j}} \tag{3.28}$$

By replacing i with $i+1$ in Eq. (3.20) and substituting Eq. (3.21) into Eq. (3.20), the following equation is obtained:

$$\zeta^{n+1/2}_{i+1,j} = -E_{i+1,j}U^{n+1/2}_{i+3/2,j}F_{i+1,j} \tag{3.29}$$

where

$$E_{i+1,j} = \frac{C_{i+3/2,j}}{B_{i+1,j} - A_{i+1/2,j}E_{i,j}} \tag{3.30}$$

$$F_{i+1,j} = \frac{D_{i+1,j} - A_{i+1/2,j}F'_{i,j}}{B_{i+1,j} - A_{i+1/2,j}E'_{i,j}} \tag{3.31}$$

Similarly, by replacing i with $i+1$ in Eq. (3.21) and substituting Eq. (3.29) into Eq. (3.21), the following equation results:

$$U^{n+1/2}_{i+3/2,j} = -E'_{i+1,j}\zeta^{n+1/2}_{i+2,j} + F'_{i+1,j} \tag{3.32}$$

where

$$E'_{i+1,j} = \frac{C'_{i+2,j}}{B'_{i+3/2,j} - A'_{i+1,j}E_{i+1,j}} \tag{3.33}$$

$$F'_{i+1,j} = \frac{D'_{i+3/2,j} - A'_{i+1/2,j}F_{i+1,j}}{B'_{i+3/2,j} - A'_{i+1,j}E_{i+1,j}} \tag{3.34}$$

Comparisons of Eq. (3.32) with Eq. (3.26), and Eq. (3.29) with Eq. (3.25) show that $\zeta^{n+1/2}$ and $U^{n+1/2}$ can be calculated by the following recurrence formulas:

$$\zeta^{n+1/2}_{i,j} = -E_{i,j}U^{n+1/2}_{i+1/2,j} + F_{i,j} \tag{3.35}$$

$$U^{n+1/2}_{i+1/2,j} = -E'_{i-1,j}\zeta^{n+1/2}_{i,j} + F'_{i-1,j} \tag{3.36}$$

where

$$E_{i,j} = \frac{C_{i+1/2,j}}{B_{i,j} - A_{i-1/2,j}E'_{i-1,j}} \tag{3.37}$$

$$F_{i,j} = \frac{D_{i,j} - A_{i-1/2,j}F'_{i-1,j}}{B_{i,j} - A_{i-1/2,j}E'_{i-1,j}} \tag{3.38}$$

$$E'_{i,j} = \frac{C'_{i+1,j}}{B'_{i+1/2,j} - A'_{i,j}E_{i,j}} \tag{3.39}$$

$$F'_{i,j} = \frac{D'_{i+1/2,j} - A'_{i,j}F_{i,j}}{B'_{i+1/2,j} - A'_{i,j}E'_{i,j}} \tag{3.40}$$

Firstly, the values of $E_{i,j}$, $F_{i,j}$, $E'_{i,j}$ and $F'_{i,j}$ are calculated for increasing values of i. Then $\zeta_{i,j}^{n+1/2}$ and $U_{i-1/2,j}^{n+1/2}$ are calculated by using Eqs. (3.35) and (3.36) alternately for decreasing values of i. Finally $V^{n+1/2}$ is calculated in Eq. (3.19) after all of $U^{n+1/2}$ and $\zeta^{n+1/2}$ are calculated.

In the second fractional step ($t+\Delta t/2 \rightarrow t\Delta t$), V^{n+1}, ζ^{n+1} and U^{n+1} can be calculated by a similar procedure to the first fractional step, only replacing variable x with variable y.

Because the ADI method treats the pressure terms with an implicit scheme, it is unconditionally stable and not governed by the Courant-Friedrichs-Lewy condition, and thus a large time step, Δt, can be chosen. However, since the accuracy of the method depends on both the mesh size and the time step size, if the Courant number C_r is too large, the accuracy of the calculation will drop.

3.2.3 Operator-Splitting Method

The operator-splitting method has been applied to a wide variety of numerical problems (Yanenko, 1971; He and Lin, 1984). This method splits a complicated partial differential equation into several simple differential equations by introducing one or more temporary variables, and then adopting the most suitable schemes for them. The operator-splitting method is expected to apply to flows with large Courant numbers, and thus the continuity equation and momentum equations are discretized in the following way.

In the first step ($t \rightarrow t+\Delta t$), the governing equations are split into six equations, in terms of the x- and y-directions, by dividing the time step into two fractional steps. The component equations in the x-direction are treated in the first fractional step and the component equations in the y-direction are dealt with in the second fractional step.

First fractional step

$$\frac{\zeta^*_{i,j}-\zeta^n_{i,j}}{\Delta t}+\frac{1}{2}\left[\frac{(h+\zeta^n)_{i+1/2,j}U^n_{i+1/2,j}-(h+\zeta^n)_{i-1/2,j}U^n_{i-1/2,j}}{\Delta x}\right.$$
$$\left.+\frac{(h+\zeta^n)_{i+1/2,j}U^*_{i+1/2,j}-(h+\zeta^n)_{i-1/2,j}U^*_{i-1/2,j}}{\Delta x}\right]=0 \tag{3.41}$$

$$\frac{U^*_{i+1/2,j}-U^n_{i+1/2,j}}{\Delta t}+\frac{1}{2}(U^*_{i+1/2,j}+U^n_{i+1/2,j})\frac{U^n_{i+3/2,j}-U^n_{i-1/2,j}}{2\Delta x}$$
$$=fV^n_{i+1/2,j}-\frac{g}{2}\left(\frac{\zeta^n_{i+1,j}-\zeta^n_{i,j}}{\Delta x}+\frac{\zeta^*_{i+1,j}-\zeta^*_{i,j}}{\Delta x}\right)$$
$$-\frac{gn^2U^*_{i+1/2,j}\sqrt{U^{n2}_{i+1/2,j}+V^{n2}_{i+1/2,j}}}{(h+\zeta^n)^{4/3}_{i+1/2,j}}$$
$$+\frac{\tau_{sx}}{\rho(h+\zeta^n)_{i+1/2,j}} \tag{3.42}$$

$$\frac{V^*_{i,j+1/2}-V^n_{i,j+1/2}}{\Delta t}-U^*_{i,j+1/2}\frac{V^n_{i+1,j+1/2}-U^n_{i-1,j+1/2}}{2\Delta x}$$
$$-\left|U^*_{i,j+1/2}\right|\frac{V^n_{i+1,j+1/2}-2V^n_{i,j+1/2}+V^n_{i-1,j+1/2}}{2\Delta x}=0 \tag{3.43}$$

where the superscript (*) denotes the temporary variables.

Second fractional step

$$\frac{\zeta^{n+1}_{i,j}-\zeta^*_{i,j}}{\Delta t}+\frac{1}{2}\left[\frac{(h+\zeta^*)_{i,j+1/2}V^*_{i,j+1/2}-(h+\zeta^*)_{i,j-1/2}V^*_{i,j-1/2}}{\Delta y}\right.$$
$$\left.+\frac{(h+\zeta^*)_{i,j+1/2}V^{n+1}_{i,j+1/2}-(h+\zeta^*)_{i,j-1/2}V^{n+1}_{i,j-1/2}}{\Delta y}\right]=0 \tag{3.44}$$

$$\frac{V^{n+1}_{i,j+1/2}-V^*_{i,j+1/2}}{\Delta t}+\frac{1}{2}(V^*_{i,j+1/2}+V^{n+1}_{i,j+1/2})\frac{V^*_{i,j+3/2}-V^*_{i,j-1/2}}{2\Delta y}=-fU^*_{i,j+1/2}$$
$$-\frac{g}{2}\left(\frac{\zeta^*_{i,j+1}-\zeta^*_{i,j}}{\Delta y}+\frac{\zeta^{n+1}_{i,j+1}-\zeta^{n+1}_{i,j}}{\Delta y}\right)+\frac{\tau_{sy}}{\rho(h+\zeta^n_{i,j+1/2})}$$
$$-\frac{gn^2V^{n+1}_{i,j+1/2}\sqrt{U^{*2}_{i,j+1/2}+V^{*2}_{i,j+1/2}}}{(h+\zeta^n_{i,j+1/2})^{4/3}} \tag{3.45}$$

$$\frac{U^{n+1}_{i+1/2,j}-U^{*}_{i+1/2,j}}{\Delta t}-V^{n+1}_{i+1/2,j}\frac{U^{*}_{i+1/2,j+1}-U^{*}_{i+1/2,j}}{2\Delta y}$$
$$-\left|V^{n+1}_{i+1/2,j}\right|\frac{U^{*}_{i+1/2,j+1}-2U^{*}_{i+1/2,j}+U^{*}_{i+1/2,j-1}}{2\Delta y}=0 \qquad (3.46)$$

By using the double-sweep method, ζ^*, U^* are calculated in Eqs. (3.41) and (3.42), then V^* is obtained from Eq. (3.43). Similarly, ζ^{n+1}, V^{n+1} are calculated in Eqs. (3.44) and (3.45), and U^{n+1} is obtained from Eq. (3.46).

In the second step ($t+\Delta t \rightarrow t+2\Delta t$), the component equations in the y-direction are treated in the first fractional step and the component equations in the x-direction are dealt with in the second fractional step.

First fractional step

$$\frac{\zeta^{**}_{i,j}-\zeta^{n+1}_{i,j}}{\Delta t}+\frac{1}{2}\left[\frac{(h+\zeta^{n+1})_{i,j+1/2}V^{n+1}_{i,j+1/2}-(h+\zeta^{n+1})_{i,j-1/2}V^{n+1}_{i,j-1/2}}{\Delta y}\right.$$
$$\left.+\frac{(h+\zeta^{n+1})_{i,j+1/2}V^{**}_{i,j+1/2}-(h+\zeta^{n+1})_{i,j-1/2}V^{**}_{i,j-1/2}}{\Delta y}\right]=0 \qquad (3.47)$$

$$\frac{V^{**}_{i,j+1/2}-V^{n+1}_{i,j+1/2}}{\Delta t}+\frac{1}{2}(V^{**}_{i,j+1/2}+V^{n+1}_{i,j+1/2})\frac{V^{n+1}_{i,j+3/2}-V^{n+1}_{i,j-1/2}}{2\Delta y}=-fU^{n+1}_{i,j+1/2}$$
$$-\frac{g}{2}\left(\frac{\zeta^{**}_{i,j+1}-\zeta^{**}_{i,j}}{\Delta y}+\frac{\zeta^{n+1}_{i,j+1}-\zeta^{n+1}_{i,j}}{\Delta y}\right)+\frac{\tau_{sy}}{\rho(h+\zeta^{n+1}_{i,j+1/2})}$$
$$-\frac{gn^2V^{**}_{i,j+1/2}\sqrt{U^{n+12}_{i,j+1/2}+V^{n+12}_{i,j+1/2}}}{(h+\zeta^{n+1}_{i,j+1/2})^{4/3}} \qquad (3.48)$$

$$\frac{U^{**}_{i+1/2,j}-U^{n+1}_{i+1/2,j}}{\Delta t}-V^{**}_{i+1/2,j}\frac{U^{n+1}_{i+1/2,j+1}-U^{n+1}_{i-1/2,j}}{2\Delta y}$$
$$-\left|V^{**}_{i+1/2,j}\right|\frac{U^{n+2}_{i+1/2,j+1}-2U^{n+1}_{i+1/2,j}+U^{n+1}_{i+1/2,j-1}}{2\Delta y}=0 \qquad (3.49)$$

Second fractional step

$$\frac{\zeta^{n+2}_{i,j}-\zeta^{**}_{i,j}}{\Delta t}+\frac{1}{2}\left[\frac{(h+\zeta^{**})_{i+1/2,j}U^{**}_{i+1/2,j}-(h+\zeta^{**})_{i-1/2,j}U^{**}_{i-1/2,j}}{\Delta x}\right.$$
$$\left.+\frac{(h+\zeta^{**})_{i+1/2,j}U^{n+2}_{i+1/2,j}-(h+\zeta^{**})_{i-1/2,j}U^{n+2}_{i-1/2,j}}{\Delta x}\right]=0 \qquad (3.50)$$

$$\frac{U^{n+2}_{i+1/2,j} - U^{**}_{i+1/2,j}}{\Delta t} + \frac{1}{2}(U^{**}_{i+1/2,j} + U^{n+2}_{i+1/2,j})\frac{U^{**}_{i+3/2,j} - U^{**}_{i+1/2,j}}{2\Delta x}$$
$$= -fV^{**}_{i+1/2,j} - \frac{g}{2}\left(\frac{\zeta^{**}_{i+1,j} - \zeta^{**}_{i,j}}{\Delta x} + \frac{\zeta^{n+2}_{i+1,j} - \zeta^{n+2}_{i,j}}{\Delta x}\right) + \frac{\tau_{sx}}{\rho(h + \zeta^{**}_{i+1/2,j})}$$
$$- \frac{gn^2 U^{n+2}_{i+1/2,j}\sqrt{U^{**2}_{i+1/2,j} + V^{**2}_{i+1/2,j}}}{(h + \zeta^{**}_{i+1/2,j})^{4/3}} \tag{3.51}$$

$$\frac{V^{n+2}_{i,j+1/2} - V^{**}_{i,j+1/2}}{\Delta t} - U^{n+2}_{i,j+1/2}\frac{V^{**}_{i+1,j+1/2} - V^{**}_{i-1,j+1/2}}{2\Delta x}$$
$$-\left|U^{n+2}_{i,j+1/2}\right|\frac{V^{**}_{i+1,j+1/2} - 2V^{**}_{i,j+1/2} + V^{**}_{i-1,j+1/2}}{2\Delta x} = 0 \tag{3.52}$$

where the superscript (**) denotes the temporary variables.

By using the same computational procedure as in the first step, U^{n+2}, ζ_i^{n+2} and V^{n+2} are obtained from the preceding equations. The time step proceeds every $2\Delta t$ for one cycle of calculation. As with the ADI method, the operator-splitting method is unconditionally stable and not limited by the Courant–Friedrichs–Lewy stability condition.

3.3 NUMERICAL EXPERIMENTS

For a particular application, it is not always necessary for a model to reproduce the smallest scales of motion. There are various flows of different scales in shallow waters, and it is nearly impossible, and not necessary, to calculate all of them. However, it is extremely important to specify the spatial and temporal scales over which the simulated results can be meaningfully interpreted. In other words, it is necessary to determine the smallest scale flow which can be correctly simulated according to the models employed and the specific interests of the problem. Because the specific interests corresponding to the vertically integrated two-dimensional models are the total mass transport, the plane distribution of vertically averaged velocity and the variations of water stage due to wind, the various flows of large scales in the horizontal plane which are closely connected to these components are selected. Because large-scaled flows in shallow lakes and coastal waters are always coupled to each other, due to the random wind, complicated basin bathymetry, and the rotation of the Earth, selecting some significant phenomena from these complicated flows and treating them separately is considered to be an efficient and effective approach.

3.3.1 *Computational Conditions*

For the convenience of discussion, a simplified, enclosed model basin, whose configuration and depth distribution are shown in Figure 3.2 is employed. The

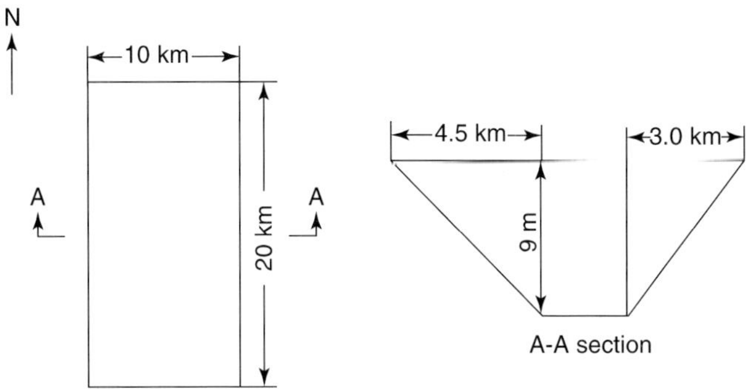

Fig. 3.2. Sketch of a simplified water basin (Shen, 1991).

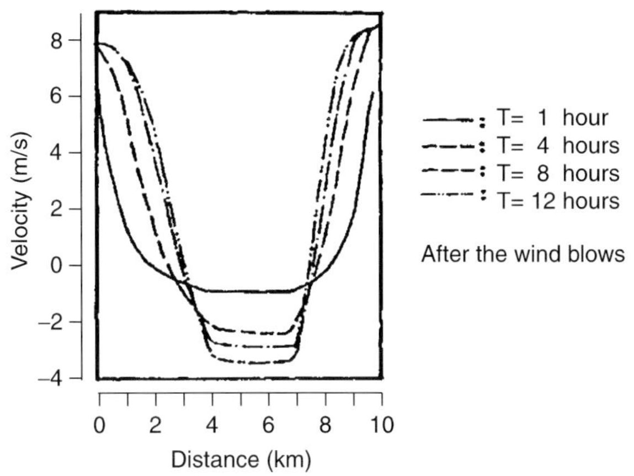

Fig. 3.3. Transverse distribution of the velocity component V in the A-A cross section (Shen, 1991).

water basin is divided by the mesh intervals $\Delta x = \Delta y = 500\,\text{m}$, and a southern wind at the velocity of 5 m/s is assumed to blow over the water surface. The other parameters used in the computation are given as follows: Manning's coefficient, $n = 0.023$; friction coefficient at the water surface, $C_D = 1.3 \times 10^{-3}$; air density, $\rho_a = 1.2\,\text{kg/m}^3$; water density, $\rho_0 = 1000\,\text{kg/m}^3$; and, the Coriolis parameter, $f = 8.34 \times 10^{-5}$. The time steps are chosen as 10, 90 and 120 s for the explicit, ADI and operator splitting methods respectively (Iwasa et al., 1988).

3.3.2 Horizontal Circulation

Figure 3.3 illustrates the transverse distribution of the velocity component V in the A-A cross section. Figure 3.4 shows the velocity vector distributions in the horizontal plane at different times. As the south wind blows, the strong flows in the leeward direction occur in the east and west shores, and the weak flows in the windward direction occur in the center area of the basin. These flows, which form a

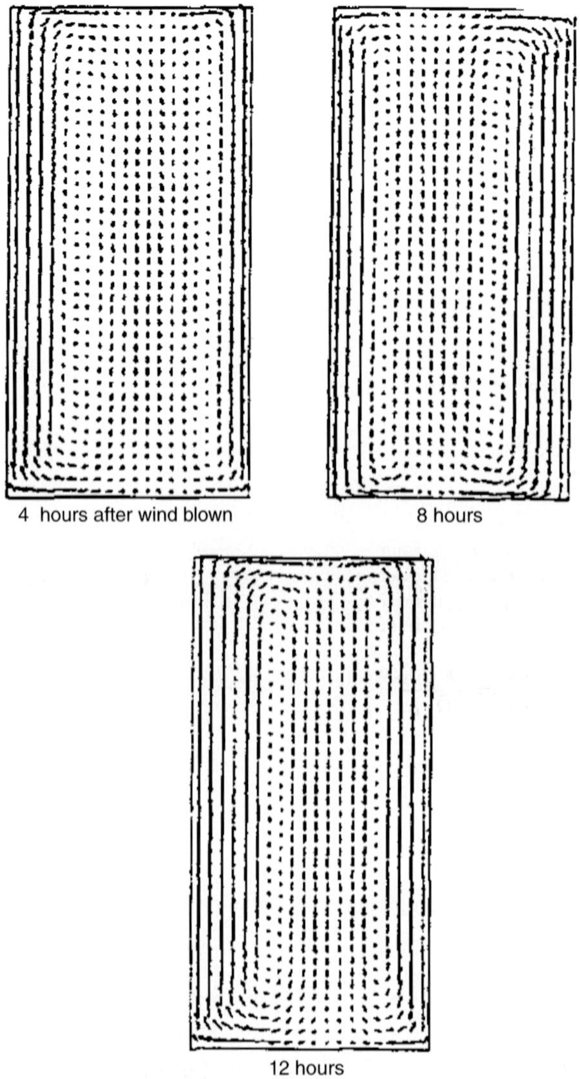

Fig. 3.4. Velocity vectors in horizontal plane at different times (Shen, 1991).

horizontal circulation, become stronger with blowing time and finally attain a steady state. The flows in the shallow regions arrive at steady state faster than in the central region. This circulation can be explained as follows.

Ignoring the Earth's rotation and convective terms, the following equation can be obtained from Eq. (3.6):

$$\frac{\partial HV}{\partial t} = -gH\frac{\partial \zeta}{\partial y} + \frac{\tau_{sy} - \tau_{by}}{\rho} \tag{3.53}$$

where $H = h+\zeta$. Integrating Eq. (3.53) with respect to the transverse direction (x-direction), and using the relation

$$\int_0^b HV\,dx - 0 \tag{3.54}$$

yields

$$-g\bar{H}\frac{\partial \zeta}{\partial y} + \frac{\tau_{sy} - \bar{\tau}_{by}}{\rho} = 0 \tag{3.55}$$

where b is the width in transverse direction, $\bar{H}$ is the average depth in the transverse direction,

$$\bar{H} = \frac{1}{b}\int_0^b H\,dx = 0 \tag{3.56}$$

and $\bar{\tau}_{by}$ is the average value of shear stress at the bottom. Since $\bar{\tau}_{by}$ is small enough to be ignored, at the onset of wind blowing, Eq. (3.55) becomes

$$\frac{\partial \zeta}{\partial y} = \frac{1}{g\bar{H}}\frac{\tau_{sy}}{\rho} \tag{3.57}$$

Substituting Eq. (3.57) into Eq. (3.53) we get

$$\frac{\partial HV}{\partial t} = \left(1 - \frac{H}{\bar{H}}\right)\frac{\tau_{sy}}{\rho} \tag{3.58}$$

$$V = \frac{t}{H}\left(1 - \frac{H}{\bar{H}}\right)\frac{\tau_{sy}}{\rho} \tag{3.59}$$

Equation (3.57) implies that the slope of the water surface due to the wind is proportional to the wind shear stress and is independent of the blowing time. In addition, it is obvious that if the water depth H is uniform, Eq. (3.59) results in $V = 0$, which implies that no horizontal circulation will occur and the wind shear stress will balance the set up of the water surface. On the other hand, when H is not uniform, velocity V is positive (the same direction as the wind) in the shore region, where H is small, and is opposite to the wind direction in the central region, where H is large. The absolute value of velocity in the central region is smaller than that of the shore region. Moreover, velocity V is proportional to the blowing time as well as to the wind shear stress.

The physical mechanism of these results can be considered as follows: in shallow regions, the total pressure gradient along the wind direction is so small that the major portion of wind shear stress is transformed to the momentum of water. On

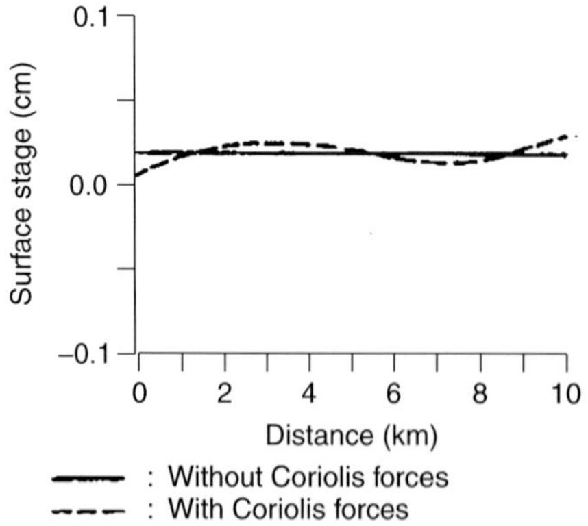

Fig. 3.5. Influences of the Coriolis force (Shen, 1991).

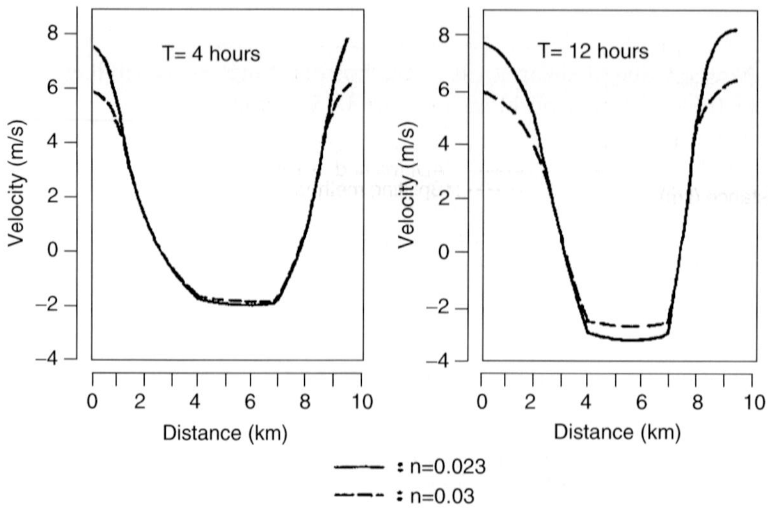

Fig. 3.6. The influence of bottom friction (Shen, 1991).

the other hand, the pressure gradient is large in deep regions, and a considerable part of wind shear stress is converted to the pressure gradient, so that the momentum of water becomes small.

When the Earth's rotation is considered, the water surface gets lower at the west shore and higher at the east shore, due to the geostrophic balance in the Northern Hemisphere. This phenomenon can be reproduced by the simulation as shown in Figure 3.5.

Figure 3.6 shows the influences of bottom friction. This figure shows that the bottom friction becomes larger as wind blows, and the velocities in shallow regions vary more with Manning's coefficient than in deeper areas. If the wind continues to blow, the influences of bottom friction also become large in the central areas.

3.3.3 Comparison of Numerical Methods

The explicit, ADI and operator-splitting method were applied under the same computational conditions. By comparing the results of the three methods as shown in Figure 3.7, it is obvious that the results are very similar to each other.

The ADI method permits a larger time step than the explicit method and has second-order accuracy temporally. This method is unconditionally stable like the implicit method from Neumann's stability analysis. It is possible to simulate

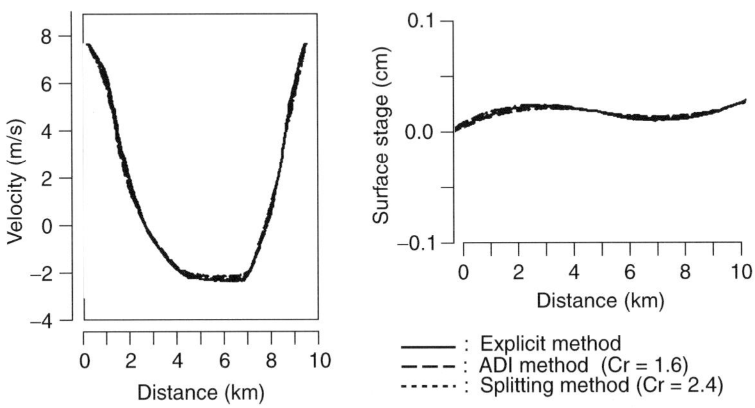

Fig. 3.7. Comparison of the three methods (Shen, 1991).

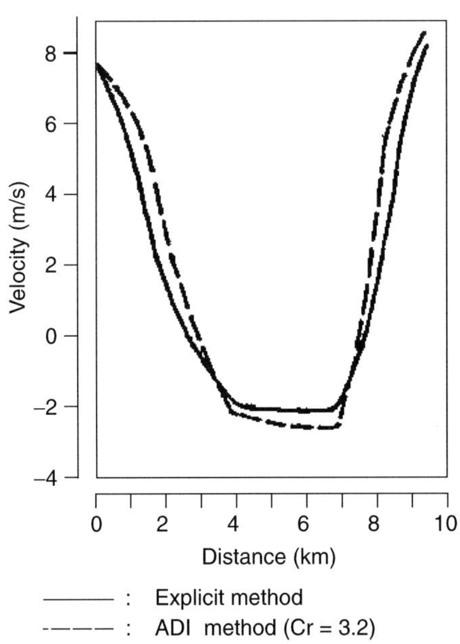

Fig. 3.8. Comparison of the ADI method with the explicit method (Shen, 1991).

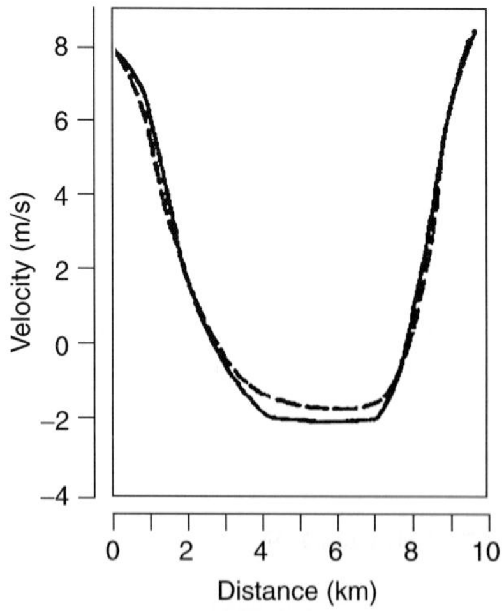

—— : Explicit method
– – – : Splitting method (Cr = 4.8)

Fig. 3.9. Comparison of the operator-splitting method with the explicit method (Shen, 1991).

two-dimensional flows by a one-dimensional numerical computational method because the coefficient matrices are tri-diagonal and can be solved by the double-sweep method. For flows with large Courant numbers, however, this method brings out numerical attenuation and yields a poor reproduction of wave propagation. The ADI method can give very similar results to the explicit method for flows with Courant numbers less than 3–5 (see Fig. 3.8).

The operator-splitting method has almost the same advantages as the ADI method, and its computational formulas are simpler than the ADI method. In addition, this method is available for slightly larger Courant numbers than the ADI method (which is about 4–6; see Fig. 3.9). As the time step of the operator-splitting method proceeds every $2\Delta t = 180\,\text{s}$ in one calculation cycle, the operator-splitting method consumes less CPU time than the ADI method.

Chapter 4

THREE-DIMENSIONAL CIRCULATION MODELS

This chapter presents various 3D circulation models, including an external-internal splitting 3D model, a vertical-horizontal splitting 3D model. and a control-volume 3D model.

4.1 A VERTICAL–HORIZONTAL SPLITTING 3D MODEL

4.1.1 Model Equations in the σ-Coordinate System

The σ-coordinate transformation, shown in Figure 4.1, leads to a smooth representation of topography. The same order of resolution for the shallow and deep parts of the water body will be applied to the model equations. The model equations are transformed from (x, y, z, t) to (x', y', σ', t') coordinates (Swanson, 1986), where

$$x' = x,\ y' = y,\ \sigma = \frac{z - \zeta(x, y, t)}{h(x, y) + \zeta(x, y, t)},\ t' = t \tag{4.1}$$

The differential operator transformation can be derived from the following relationships based on the chain rule:

$$\frac{\partial}{\partial x} = \frac{\partial}{\partial x'} - \frac{1}{H}\left[(1 + \sigma)\frac{\partial \zeta}{\partial x'} + \sigma \frac{\partial h}{\partial x'}\right]\frac{\partial}{\partial \sigma} \tag{4.2}$$

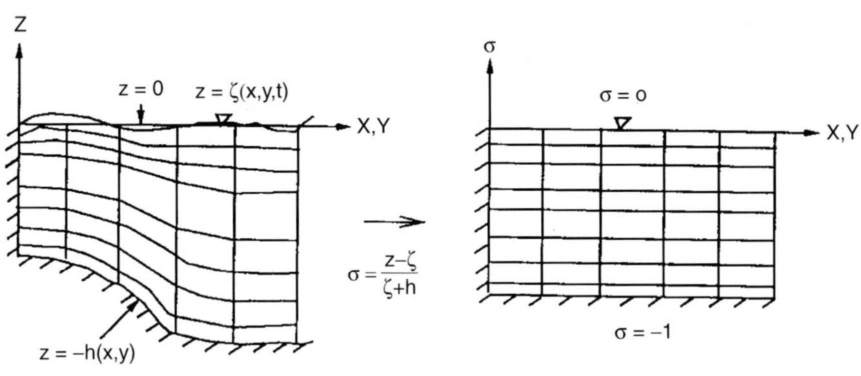

Fig. 4.1. σ-coordinates transformation in the vertical plane, from layer discretization in the Cartesian coordinates (left) to the one in σ coordinates (right) (Wu, 1993).

$$\frac{\partial}{\partial y} = \frac{\partial}{\partial y'} - \frac{1}{H}\left[(1+\sigma)\frac{\partial \zeta}{\partial y'} + \sigma\frac{\partial h}{\partial y'}\right]\frac{\partial}{\partial \sigma} \tag{4.3}$$

$$\frac{\partial}{\partial z} = \frac{1}{H}\frac{\partial}{\partial \sigma} \tag{4.4}$$

$$\frac{\partial}{\partial t} = \frac{\partial}{\partial t'} - \frac{1}{H}\left[(1+\sigma)\frac{\partial \zeta}{\partial t}\right]\frac{\partial}{\partial \sigma} \tag{4.5}$$

Therefore, the 3D hydrodynamic model equations in σ coordinates are now written as (all primes will be dropped for notational convenience):

$$\frac{\partial uH}{\partial t} + \frac{\partial uuH}{\partial x} + \frac{\partial uvH}{\partial y} + \frac{\partial u\omega H}{\partial \sigma} = fvH - gH\frac{\partial \zeta}{\partial x} + B_x + \tau_u + \frac{\partial}{\partial \sigma}\left(\frac{K_v}{H}\frac{\partial u}{\partial \sigma}\right) \tag{4.6}$$

$$\frac{\partial vH}{\partial t} + \frac{\partial vuH}{\partial x} + \frac{\partial vvH}{\partial y} + \frac{\partial v\omega H}{\partial \sigma} = -fuH - gH\frac{\partial \zeta}{\partial y} + B_y + \tau_v + \frac{\partial}{\partial \sigma}\left(\frac{K_v}{H}\frac{\partial v}{\partial \sigma}\right) \tag{4.7}$$

$$\frac{\partial p}{\partial \sigma} = -\frac{1}{H}\rho g \tag{4.8}$$

$$\frac{\partial \zeta}{\partial t} + \frac{\partial Hu}{\partial x} + \frac{\partial Hv}{\partial y} + \frac{\partial H\omega}{\partial \sigma} = 0 \tag{4.9}$$

$$\frac{\partial TH}{\partial t} + \frac{\partial uTH}{\partial x} + \frac{\partial vTH}{\partial y} + \frac{\partial \omega TH}{\partial \sigma} = \tau_T + \frac{\partial}{\partial \sigma}\left(\frac{N_v}{H}\frac{\partial T}{\partial \sigma}\right) \tag{4.10}$$

ω is the vertical velocity component in σ coordinates:

$$\omega = \frac{w}{H} - \frac{u}{H}\left(\sigma\frac{\partial H}{\partial x} + \frac{\partial \zeta}{\partial x}\right) - \frac{v}{H}\left(\sigma\frac{\partial H}{\partial y} + \frac{\partial \zeta}{\partial y}\right) - \frac{1}{H}\left(\sigma\frac{\partial H}{\partial t} + \frac{\partial \zeta}{\partial t}\right) \tag{4.11}$$

τ_u, τ_v, and τ_T are the horizontal eddy diffusivity terms in σ coordinates:

$$\tau_u = \frac{\partial H\tau_x}{\partial x} - \frac{\partial}{\partial \sigma}\left[\left(\sigma\frac{\partial H}{\partial x} + \frac{\partial \zeta}{\partial x}\right)\tau_x\right] + \frac{\partial H\tau_y}{\partial y} - \frac{\partial}{\partial \sigma}\left[\left(\sigma\frac{\partial H}{\partial y} + \frac{\partial \xi}{\partial y}\right)\tau_y\right] \tag{4.12}$$

where

$$\tau_u = K_h\left[\frac{\partial u}{\partial x} - \frac{\partial u}{\partial \sigma}\left(\frac{\sigma}{H}\frac{\partial H}{\partial x} + \frac{1}{H}\frac{\partial \zeta}{\partial x}\right)\right] \tag{4.13}$$

$$\tau_y = K_h\left[\frac{\partial u}{\partial y} - \frac{\partial u}{\partial \sigma}\left(\frac{\sigma}{H}\frac{\partial H}{\partial y} + \frac{1}{H}\frac{\partial \zeta}{\partial y}\right)\right] \tag{4.14}$$

Similar formula can be derived for τ_v and τ_T.

Atmospheric pressure gradients are neglected and the baroclinic density gradients B_x and B_y, because of the temperature difference, are:

$$B_x = -\frac{gH^2}{\rho_0}\frac{\partial}{\partial x}\int_\sigma^0 \rho d\sigma + \frac{gH}{\rho_0}\frac{gH}{\partial x}\int_\sigma^0 \sigma\frac{\partial \rho}{\partial \sigma}d\sigma \tag{4.15}$$

$$B_y = -\frac{gH^2}{\rho_0}\frac{\partial}{\partial y}\int_\sigma^0 \rho d\sigma + \frac{gH}{\rho_0}\frac{gH}{\partial y}\int_\sigma^0 \sigma\frac{\partial \rho}{\partial \sigma}d\sigma \tag{4.16}$$

All the boundary conditions can easily be transformed to σ coordinates using the appropriate operator transformation.

4.1.2 Fractional Step Method for the 3D Model[3]

To the fractional step method, a consistency analysis has been carried out for the linear equations. Although the analysis has not yet been made for the full non-linear Navier–Stokes equations, it has been used successfully in a number of hydrodynamic models and advection–diffusion transport models. The key point is that the equation to be solved is broken into a series of intermediate steps, such as advection and diffusion. The most suitable scheme can be employed for a specific step according to the mathematical and physical features of each step equation. Splitting momentum equations into two steps, that is, an advection step and a propagation + diffusion + source step, Eqs. (4.6) and (4.7), gives (Wu and Tsanis, 1995a):

$$\frac{\partial uH^*}{\partial t} = -\frac{\partial uuH}{\partial x} - \frac{\partial uvH}{\partial y} - \frac{\partial u\omega H}{\partial \sigma} + fvH - gH\frac{\partial \zeta}{\partial x} + B_x + \tau_u \tag{4.17}$$

$$\frac{\partial vH^*}{\partial t} = -\frac{\partial vuH}{\partial x} - \frac{\partial vvH}{\partial y} - \frac{\partial v\omega H}{\partial \sigma} - fHu - gH\frac{\partial \zeta}{\partial y} + B_y + \tau_v \tag{4.18}$$

[3]Material in Sections 4.1.2 to 4.1.6 are reproduced from Wu and Tsanis (1995a), Copyright, with permission from ASCE.

$$\frac{\partial uH}{\partial t} = \frac{\partial}{\partial \sigma}\left(\frac{K_v}{H}\frac{\partial u^*}{\partial \sigma}\right) \tag{4.19}$$

$$\frac{\partial vH}{\partial t} = \frac{\partial}{\partial \sigma}\left(\frac{K_v}{H}\frac{\partial v^*}{\partial \sigma}\right) \tag{4.20}$$

4.1.3 Grid Discretization and Finite Difference Representation

As shown in Figure 4.2, a staggered finite difference grid system known as the Arakawa C-grid is used. Horizontally, ζ and T are computed at the centre of the mesh with u at the left and right faces and v at the front and back vertical faces of the mesh. This arrangement is convenient for derivatives and boundary conditions. Vertically, the horizontal velocity components are calculated at the level surface; therefore, the boundary conditions at the bottom, that is, no-slip conditions $u = v = 0$, can be accurately satisfied. The vertical velocity w is calculated at the middle of each layer. A 3D view of discretization is also shown in the figure. In order to increase resolution, both in the free surface and bottom layers, a mesh of varying thickness $\Delta\sigma$ can be used. The finite difference representation of a variable can be written as

$$\phi(x, y, \sigma, t) = \phi(i\Delta x, j\Delta y, k\Delta\sigma, n\Delta t) = \phi_{ijk}^{n} \tag{4.21}$$

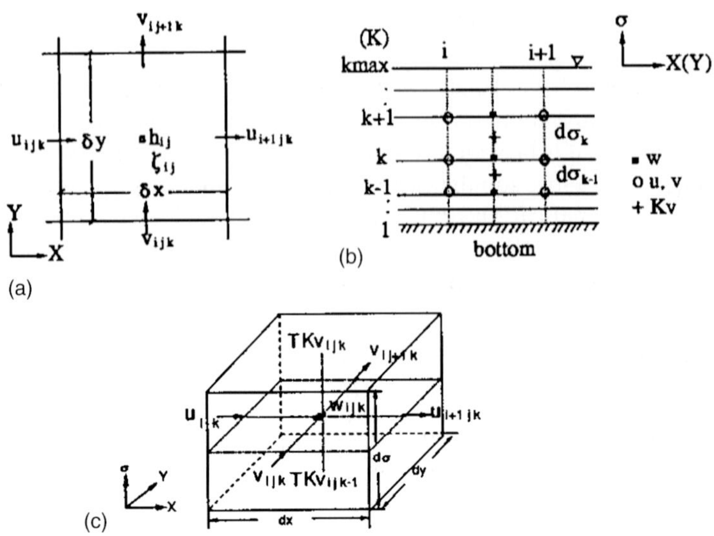

Fig. 4.2. Variable arrangements in a staggered finite difference grid system for the 3D hydrodynamic model. (a) View of a horizontal grid; (b) View of a vertical mesh; and (c) A 3D view (Wu and Tsanis, 1995a, Copyright, pp. 388–395, with permission from ASCE).

where i is the 1, 2, ..., $i_{\max}$; j the 1, 2, ..., $j_{\max}$; k the 1, 2, ..., $k_{\max}$; $i_{\max}, j_{\max}$ and $k_{\max}$ are the maximum grid number in the x, y, and σ directions respectively; n the time level of integration.

$$\delta_t \phi - (\phi_{ijk}^{n+1} - \phi_{ijk}^{n})/\Delta t \tag{4.22}$$

Let the following expressions

$$\delta_x \phi = (\phi_{i+1jk} - \phi_{i-1jk})/(2\Delta x) \tag{4.23}$$

$$\delta_x^2 \phi = (\phi_{i+1jk} - 2\phi_{ijk} + \phi_{i-1jk})/\Delta x^2 \tag{4.24}$$

$$\bar{\phi}^x = (\phi_{i+1jk} + \phi_{i-1jk})/2 \tag{4.25}$$

represent the first derivative for time and the x direction, the second derivative in the x direction, and the average value of ϕ in the x direction respectively: similar expressions are straightforward in the y and z directions. In addition, one can define

$$\delta_\sigma^2 \phi_{ijk}^n = \left(K_{v\ ijk}^n \frac{\phi_{ijk+1}^n - \phi_{ijk}^n}{\Delta\sigma_k} - K_{v\ ijk-1}^n \frac{\phi_{ijk}^n - \phi_{ijk-1}^n}{\Delta\sigma_{k-1}} \right) / [(\Delta\sigma_k + \sigma_{k-1})/2] \tag{4.26}$$

as the finite difference of the vertical diffusivity term, where ϕ is a horizontal velocity component.

4.1.4 Numerical Scheme for the Advection Step

An explicit scheme is applied for the advection step (Wu and Tsanis, 1995a):

$$\delta_t uH = -\delta_x uuH - \delta_y \bar{v}^y uH - \delta_\sigma \bar{\omega}^x uH + \overline{fHv}^{xy} - gH\delta_x\zeta + \bar{B}_x^{x\sigma} + \tau_u \tag{4.27}$$

$$\delta_t vH = -\delta_x \bar{u}^x vH - \delta_y vvH - \delta_\sigma \bar{\omega}^y vH - \overline{fHu}^{xy} - gH\delta_y\zeta + \bar{B}_y^{y\sigma} + \tau_v \tag{4.28}$$

$$\delta_t \zeta = -\delta_x(HU) - \delta_y(HV) \tag{4.29}$$

The baroclinic terms at layer k can be expressed as

$$B_x = -\frac{g}{\rho_0}\delta_y \left(H^2 \sum_{m=k}^{k_{\max}-1} \rho_m \Delta\sigma_m \right) + \frac{gH}{\rho_0}\frac{\partial H}{\partial x} \sum_{m=k}^{k_{\max}-1} (\rho_{m+1} - \rho_m)\sigma_m \tag{4.30}$$

$$B_y = -\frac{g}{\rho_0}\delta_y \left(H^2 \sum_{m=k}^{k_{\max}-1} \rho_m \Delta\sigma_m \right) + \frac{gH}{\rho_0}\frac{\partial H}{\partial y} \sum_{m=k}^{k_{\max}-1} (\rho_{m+1} - \rho_m)\sigma_m \tag{4.31}$$

4.1.5 Implicit Scheme for the Vertical Eddy Diffusion Step

As pointed out in Chapter 3, the explicit scheme is constrained by the time step, due to the stability requirement; therefore, the implicit scheme has to be used. A Crank–Nickolson semi-implicit scheme is used here. Equations for the vertical eddy viscosity step become (Wu and Tsanis, 1995a):

$$Hu_{ijk}^{n+1} = Hu_{ijk}^{*n} + \frac{\Delta t}{\bar{H}}\left(\delta_\sigma^2 u_{ijk}^{*n} + \delta_\sigma^2 u_{ijk}^{n+1}\right)/2 \tag{4.32}$$

$$Hv_{ijk}^{n+1} = Hv_{ijk}^{*n} + \frac{\Delta t}{\bar{H}}(\delta_\sigma^2 v_{ijk}^{*n} + \delta_\sigma^2 v_{ijk}^{n+1})/2 \tag{4.33}$$

By re-arranging Eqs. (4.32) and (4.33), the following equations can be obtained:

$$au_{ijk+1}^{n+1} + bu_{ijk}^{n+1} + cu_{ijk-1}^{n+1} = f \tag{4.34}$$

$$av_{ijk+1}^{n+1} + bv_{ijk}^{n+1} + cv_{ijk-1}^{n+1} = g \tag{4.35}$$

where the coefficients a, b, c, f, g are:

$$b = H + \frac{\Delta t}{\bar{H}}\left(\frac{k_{vijk}^n}{\Delta\sigma_k} + \frac{k_{vijk-1}^n}{\Delta\sigma_{k-1}}\right)\frac{1}{\Delta\sigma_k + \Delta\sigma_{k-1}} \tag{4.36}$$

$$a = -\frac{\Delta t}{\bar{H}}\left(\frac{K_{vijk}^n}{\Delta\sigma_k}\right)\frac{1}{\Delta\sigma_k + \Delta\sigma_{k-1}} \tag{4.37}$$

$$c = -\frac{\Delta t}{\bar{H}}\left(\frac{K_{vijk-1}^n}{\Delta\sigma_{k-1}}\right)\frac{1}{\Delta\sigma_k + \Delta\sigma_{k-1}} \tag{4.38}$$

$$f = Hu_{ijk}^n + \frac{\Delta t}{\bar{H}}\delta_\sigma^2 u_{ijk}^{*n}/2 \tag{4.39}$$

$$g = Hv_{ijk}^n + \frac{\Delta t}{\bar{H}}\delta_\sigma^2 v_{ijk}^{*n}/2 \tag{4.40}$$

It should be pointed out that the vertical eddy viscosity coefficients in a, b, and c, at the new time level, are approximated by the ones, at the previous time level. Then a tri-diagonal set of equations for the velocity components u and v can be obtained and solved, using the Thomas algorithm (Wu and Tsanis, 1995a). The depth-averaged velocities can be obtained from the calculated multi-layered velocities:

$$U = \sum_{k=1}^{k=k_{\max}-1} \bar{u}^k \Delta\sigma_k \tag{4.41}$$

$$V = \sum_{k=1}^{k=k_{\max}-1} \bar{v}^k \Delta\sigma_k \tag{4.42}$$

The vertical velocity component in σ coordinates can be calculated from

$$\omega_{ijk+1}^{n+1} = \omega_{ijk}^{n+1} - \frac{\Delta\sigma_k}{H}\left(\delta_t\zeta + \delta_x(\bar{H}^x\bar{u}^x) + \delta_y(\bar{H}^y\bar{v}^y)\right) \tag{4.43}$$

4.1.6 Numerical Scheme for the Temperature Equation

The propagation of surface gravity waves is excluded in the temperature equation. Therefore, the constraints on the time step are related only to advection and diffusion, especially the vertical diffusion terms. Similar to the hydrodynamic model scheme, the fractional step method is employed. For the first step, an explicit scheme is used (Wu and Tsanis, 1995a):

$$\delta_t TH = -\delta_x \bar{u}^x TH - \delta_y \bar{v}^y TH - \omega\delta_\sigma T + \tau_T \tag{4.44}$$

A similar implicit scheme for the momentum is applied to the vertical diffusivity term:

$$HT_{ijk}^{n+1} = HT_{ijk}^{*n} + \frac{\Delta+}{\bar{H}}(\delta_\sigma^2 T_{ijk}^{*n} + \delta_\sigma^2 T_{ijk}^{n+1})/2 \tag{4.45}$$

Figure 4.3 summarizes the order of operations for the multi-layered 3D circulation model.

4.2 AN EXTERNAL INTERNAL 3D MODEL (PRINCETON MODEL)

Blumberg and Mellor (1987) developed a 3D circulation model for coastal waters known as the "Princeton Model." Its unique feature is an imbedded turbulence closure sub-model. Model development begins by noting that the equations governing the dynamics of coastal circulation contain propagation of fast-moving

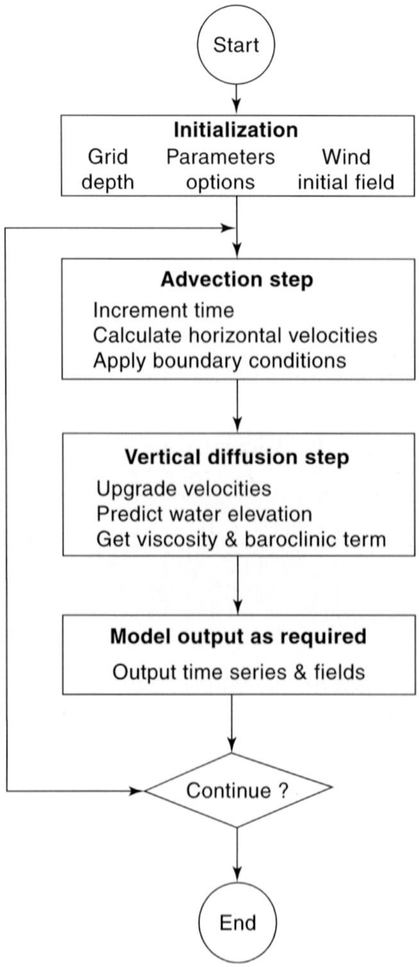

Fig. 4.3. Order of operations for the multi-layered 3D model (Wu, 1993).

external gravity waves and slow-moving internal gravity waves. Blumberg and Mellor (1987) separate out vertically integrated equations (external mode) from the vertical structure equations (internal mode) using a mode splitting technique which permits the calculation of the free surface elevation, with little sacrifice in computational time, by solving the volume transport separately from the vertical velocity shear.

The volume transport, external mode equations are obtained by integrating the internal mode equations over the depth. The equation for the surface elevation is

$$\frac{\partial \zeta}{\partial t} + \frac{\partial \bar{u} H}{\partial x} + \frac{\partial \bar{v} H}{\partial y} = 0 \tag{4.46}$$

and the momentum equations are (after vertical integration):

$$\frac{\partial \bar{u}H}{\partial t} + \frac{\partial \bar{u}^2 H}{\partial x} + \frac{\partial \bar{u}\bar{v}H}{\partial y} - f\bar{v}H + gH\frac{\partial \zeta}{\partial x} - H\bar{F}_x = -\overline{wu}(0)$$
$$+ \overline{wu}(-1) - \frac{\partial \overline{Hu'^2}}{\partial y} - \frac{\partial \overline{Hu'v'}}{\partial y} - \frac{gH^2}{\rho_o}\frac{\partial}{\partial x}\int_{-1}^{0}\int_{\sigma}^{0} \rho \, d\sigma' \, d\sigma$$
$$+ \frac{gH}{\rho_o}\frac{\partial H}{\partial x}\int_{-1}^{0}\int_{\sigma}^{0} \sigma' \frac{\partial \rho}{\partial \sigma'} d\sigma' d\sigma \tag{4.47}$$

$$\frac{\partial \bar{v}H}{\partial t} + \frac{\partial \bar{u}\bar{v}H}{\partial x} + \frac{\partial \bar{v}^2 H}{\partial y} + f\bar{u}H + gH\frac{\partial \zeta}{\partial y} - H\bar{F}_y = -\overline{wu}(0)$$
$$+ \overline{wv}(-1) - \frac{\partial \overline{Hu'v'}}{\partial x} - \frac{\partial \overline{Hv'^2}}{\partial y} - \frac{gH^2}{\rho_o}\frac{\partial}{\partial y}\int_{-1}^{0}\int_{\sigma}^{0} \rho d\sigma' d\sigma$$
$$+ \frac{gH}{\rho_o}\frac{\partial H}{\partial y}\int_{-1}^{0}\int_{\sigma}^{0} \sigma' \frac{\partial \rho}{\partial \sigma'} d\sigma' d\sigma \tag{4.48}$$

The vertically integrated velocities are:

$$(\bar{u}, \bar{v}) \equiv \int_{-1}^{0} (u, v) d\sigma \tag{4.49}$$

The wind stress components are $-\overline{wu}(0)$ and $-\overline{wu}(0)$, and the bottom stress components are $-\overline{wu}(1)$ and $-\overline{wu}(1)$. The terms in Eqs. (4.47) and (4.48), involving u'^2, $u'v'$, and v'^2, represent vertical averages of the cross-products of the velocity departures from the vertically integrated (average) velocity, and are often referred to as the dispersion terms. Hence,

$$\overline{u'^2}, \overline{v'^2}, \overline{u'v'}) = \int_{-1}^{0} (u'^2, v'^2, u'v') d\sigma \tag{4.50}$$

where $(u', v') = (u - \bar{u}, v - \bar{v})$. The quantities $\bar{F}_x$ and $\bar{F}_y$ are vertical integrals of the horizontal momentum diffusion and are defined as

$$H\bar{F}_x = \frac{\partial}{\partial x}\left(2K_h \frac{\partial \bar{u}H}{\partial x}\right) + \frac{\partial}{\partial y} K_h \left(\frac{\partial \bar{u}H}{\partial y} + \frac{\partial \bar{u}H}{\partial x}\right) \tag{4.51}$$

$$H\bar{F}_y = \frac{\partial}{\partial y}\left(2K_h \frac{\partial \bar{v}H}{\partial y}\right) + \frac{\partial}{\partial x} K_h \left(\frac{\partial \bar{v}H}{\partial y} + \frac{\partial \bar{v}H}{\partial x}\right) \tag{4.52}$$

For the fast-moving external gravity waves, which are governed by the shallow water wave Eqs. (4.46), (4.47), and (4.48), a short-time step must be used to get stable solutions. On the other hand, for the slow-moving internal gravity waves, a longer time step can be used to decreasetotal computational time. Ther efore, the computational procedure is as follows:

(1) Solve the Eqs. (4.46), (4.47), and (4.48) for the external mode by fixing the terms on the right-hand side of the Eqs. (4.47) and (4.48) in time.
(2) Solve the internal mode equations by inserting the $\partial\zeta/\partial x$ and $\partial\zeta/\partial y$ obtained from the external mode after a large number of time steps, usually of the order of 100.
(3) Update the right-hand side of Eqs. (4.47) and (4.48) after the vertical structure has been determined.
(4) Start another external mode by using the new advective and diffusive terms in Eqs. (4.47) and (4.48) supplied by the internal mode.

Because the truncation errors for external and internal modes are different, there may be a slow tendency for the internal mode velocities to differ from the external mode velocities. In order to minimize the accumulated mismatch, it is suggested the vertical mean of the internal velocity be replaced by the external mode velocity at every time step (Blumberg and Mellor, 1987).

4.3 AN OPERATOR-SPLITTING 3D MODEL

The operator-splitting method discussed in this section splits the momentum equations into several simple equations in accordance with the characteristics of the terms involved. This is done by dividing one time step into some fractional steps. For lake flows, the most important terms in the momentum equations are the pressure terms. The terms which often give rise to instability in calculations are the pressure terms and advective terms. Therefore, it is reasonable to deal with the pressure and advective fractional steps. The horizontal eddy viscosity terms, though usually contributing little to flows, play a great role in the stability of a calculation. Therefore, they are dealt with in the same fractional step as the advective terms. Moreover, since the velocity gradient in the vertical direction is much larger than that in the horizontal plane, and, the wind shear stresses and bottom friction stresses act on the water surface and bottom as the boundary conditions, the vertical eddy viscosity term is treated in another fractional step, along with the Coriolis term. A time step is divided into three fractional steps. According to these fractional steps, momentum equations are split into three parts and the best suitable scheme is applied for each part (Shen, 1988):

$$\frac{\partial X}{\partial t} = \frac{X^{n+1} - X^{**}}{\Delta t} + \frac{X^{**} - X^{*}}{\Delta t} + \frac{X^{*} - X^{n}}{\Delta t} \tag{4.53}$$

where X expresses any dependent variable; $(n+1)$ and n are the quantities at time $(n+1)\ \Delta t$ and $n\Delta t$ respectively; superscripts * and ** indicate the intermediate quantities. In the following, only the x-wise momentum equation and temperature equation are illustrated.

4.3.1 First Fractional Step

All the advective terms and horizontal eddy viscosity terms of the momentum equations, and all the advective terms and eddy diffusive terms of the temperature equation are calculated in this fractional step. The Adams–Bashforth scheme is adopted for the temporal difference.

Momentum equation

$$\frac{u^{*}-u^{n}}{\Delta t}\Delta V=\frac{3}{2}F^{n}-\frac{1}{2}F^{n-1} \tag{4.54}$$

where

$$F^{n}=\int_{\Delta v}\left\{K_{h}\frac{\partial^{2}u^{n}}{\partial x^{2}}+K_{h}\frac{\partial^{2}u^{n}}{\partial y^{2}}-\left(\frac{\partial(uu)^{n}}{\partial x}+\frac{\partial(vu)^{n}}{\partial y}+\frac{\partial(wu)^{n}}{\partial z}\right)\right\}dV$$

Temperature equation

$$\frac{T^{n+1}-T}{\Delta t}\Delta V=\frac{3}{2}M^{n}-\frac{1}{2}M^{n-1} \tag{4.55}$$

where

$$M^{n}=\int_{\Delta V}\left\{\left(\frac{\partial(uT)^{n}}{\partial x}+\frac{\partial(vT)^{n}}{\partial Y}+\frac{\partial(WT)^{n}}{\partial z}\right)\right.$$
$$\left.+N_{h}\frac{\partial^{2}T^{n}}{\partial x^{2}}+N_{h}\frac{\partial^{2}T^{n}}{\partial y^{2}}+\frac{\partial}{\partial z}\left(N_{v}\frac{\partial T^{n}}{\partial z}\right)\right\}dV \tag{4.56}$$

The Doner-cell scheme, with second-order accuracy, and the center difference scheme are applied to advective terms and eddy viscosity terms respectively, and thus the above equations are discretized into the following difference equations:

Momentum equation

$$\frac{u_{i,j+1/2}^{k+1/2,*}-u_{i,j+1/2}^{k+1/2,n}}{\Delta t}\Delta V_{i,j+1/2}^{k+1/2,n}=\frac{3}{2}F_{i,j+1/2}^{k+1/2,n}-\frac{1}{2}F_{i,j+1/2}^{k+1/2,n-1} \tag{4.57}$$

Temperature equation

$$\frac{T_{i+1/2,j+1/2}^{k+1/2,n+1}-T_{i+1/2,j+1/2}^{k+1/2,n}}{\Delta t}\Delta V_{i+1/2,j+1/2}^{k+1/2,n}=\frac{3}{2}M_{i+1/2,j+1/2}^{k+1/2,n}-\frac{1}{2}M_{i+1/2,j+1/2}^{k+1/2,n-1} \tag{4.58}$$

where

$$
\begin{aligned}
F^{k+1/2,n}_{i,j+1/2} = K_h \Bigg[\frac{1}{\Delta x} \Bigg\{ & \left(u^{k+1/2,n}_{i+1,j+1/2} - u^{k+1/2,n}_{i,j+1/2} \right) \Delta y \Delta z^{k+1/2,n}_{i+1/2,j+1/2} \\
& - \left(u^{k+1/2,n}_{i,j+1/2} - u^{k+1/2,n}_{i-1,j+1/2} \right) \Delta y \Delta z^{k+1/2,n}_{i-1/2,j+1/2} \\
& + \frac{1}{\Delta y} \Bigg\{ \left(u^{k+1/2,n}_{i+1/2,j+3/2} - u^{k+1/2,n}_{i+1/2,j+1/2} \right) \Delta x \Delta z^{k+1/2,n}_{i+1/2,j+1} /2 \\
& + \left(u^{k+1/2,n}_{i-1/2,j+3/2} - u^{k+1/2,n}_{i-1/2,j+1/2} \right) \Delta x \Delta z^{k+1/2,n}_{i-1/2,j+1} /2 \\
& - \left(u^{k+1/2,n}_{i+1/2,j+1/2} - u^{k+1/2,n}_{i+1,j-1/2} \right) \Delta x \Delta z^{k+1/2,n}_{i+1/2,j} /2 \Bigg] \\
& - \left(u^{k+1/2,n}_{i+1/2,j+1/2}, u^{k+1/2,n}_{m,j+1/2}, \theta_h \right)^{m=i+1}_{m=i} \Delta y \Delta z^{k+1/2,n}_{i+1/2,j+1/2} \\
& + \left(u^{k+1/2,n}_{i-1/2,j+1/2}, u^{k+1/2,n}_{m,j+1/2}, \theta_h \right)^{m=i}_{m=i-1} \Delta y \Delta z^{k+1/2,n}_{i-1/2,j+1/2} \\
& - \left(v^{k+1/2,n}_{i-1/2,j+1}, u^{k+1/2,n}_{i-1/2,m}, \theta_h \right)^{m=j+3/2}_{m=j+1/2} \Delta x \Delta z^{k+1/2,n}_{i-1/2,j+1} /2 \\
& - \left(v^{k+1/2,n}_{i+1/2,j+1}, u^{k+1/2,n}_{i+1/2m}, \theta_h \right)^{m=j+3/2}_{m=j+1/2} \Delta x \Delta z^{k+1/2,n}_{i+1/2,j+1} /2 \\
& + \left(v^{k+1/2,n}_{i-1/2,j}, u^{k+1/2,n}_{i+1/2,m}, \theta_h \right)^{m=j+1/2}_{m=j-1/2} \Delta x \Delta z^{k+1/2,n}_{i-1/2,j} /2 \\
& + \left(v^{k+1/2,n}_{i+1/2,j}, u^{k+1/2,n}_{i+1/2,m}, \theta_h \right)^{m=j+1/2}_{m=j-1/2} \Delta x \Delta z^{k+1/2,n}_{i+1/2,j} /2 \\
& - \left(w^{k,n}_{i+1/2,j+1/2}, u^{m,n}_{i+1/2,j+1/2}, \theta_v \right)^{m=k-1/2}_{m=k+1/2} \Delta x \Delta y /2 \\
& - \left(w^{k,n}_{i-1/2,j+1/2}, u^{m,n}_{i-1/2,j+1/2}, \theta_v \right)^{m=k-1/2}_{m=k+1/2} \Delta x \Delta y /2 \\
& - \left(w^{k+1,n}_{i+1/2,j+1/2}, u^{m,n}_{i-1/2,j+1/2}, \theta_v \right)^{k+1/2}_{m=k+3} \Delta x \Delta y /2 \\
& - \left(w^{k+1,n}_{i-1/2,j+1/2}, u^{m,n}_{i-1/2,j+1/2}, \theta_v \right)^{m=k+1/2}_{m=k+3/2} \Delta x \Delta y /2
\end{aligned}
\tag{4.59}
$$

Temperature equation

$$
\begin{aligned}
M^{k+1/2,n}_{i+1/2,j+1/2} = N_h \Bigg[\frac{1}{\Delta x} \Bigg\{ & \left(T^{k+1/2}_{i+3/2,j+1/2} - T^{k+1/2,n}_{i+1/2,j+1/2} \right) \Delta y \Delta z^{k+1/2,n}_{i+1,j+1/2} \\
& - \left(T^{k+1/2,n}_{i+1/2,j+1/2} - T^{k+1/2,n}_{i-1/2,j+1/2} \right) \Delta y \Delta z^{k+1/2,n}_{i,j+1/2} \Bigg\} \\
& + \frac{1}{\Delta y} \Bigg\{ \Big(T^{k+1/2,n}_{i+1/2,j+3/2} - \Delta x \Delta z^{k+1/2,n}_{i+1/2,j+1} \\
& - T^{k+1/2,n}_{i+1/2,j+1/2} - T^{k+1/2,n}_{i+1/2,j-1/2} \Big) \Delta x \Delta z^{k+1/2,n}_{i+1/2,j} \Bigg\} \Bigg]
\end{aligned}
$$

$$
\begin{aligned}
&+ N_v \left\{ \frac{1}{\Delta z^{k,n}_{i+1/2,j+1/2}} \left(T^{k-1/2,n}_{i+1/2,j+1/2} - T^{k+1/2,n}_{i+1/2,j+1/2} \right) \right. \\
&\left. - \frac{1}{\Delta z^{k+1,n}_{i+1/2,j+1/2}} \left(T^{k+1/2,n}_{i+1/2,j+1/2} - T^{k+3/2,n}_{i+1/2,j+1/2} \right) \right\} \Delta x \Delta y \\
&- \left(u^{k+1/2,n}_{i+1/2,j+1/2}, T^{k+1/2,n}_{m,j+1/2}, \theta_h \right)^{m=i+3/2}_{m=1+1/2} \Delta y \Delta z^{k+1/2,n}_{i+1,j+1/2} \\
&+ \left(u^{k+1/2,n}_{i,j+1/2}, T^{k+1/2,n}_{m,j+1/2}, \theta_h \right)^{m=i+3/2}_{m=i+1/2} \Delta y \Delta z^{k+1/2,n}_{i,j+1/2} \\
&- \left(v^{ku+1/2,n}_{i+1/2,j+1}, T^{k+1/2,n}_{i+1/2,m}, \theta_h \right)^{j+3/2}_{m=j+1/2} \Delta x \Delta z^{k+1/2,n}_{i+1/2,j+1} \\
&+ \left(v^{k+1/2,n}_{i+1/2,j}, T^{k+1/2,n}_{i+1/2,j+1/2,}, \theta_h \right)^{m=j+1/2}_{m=j-1/2} \Delta x \Delta z^{k+1/2,n}_{i+1/2,j} \\
&- \left(w^{k,n}_{i+1/2,j+1/2}, T^{m,n}_{i+1/2,j+1/2}, \theta_v \right)^{m=k-1/2}_{m=k+1/2} \Delta x \Delta y \\
&+ \left(w^{k+1,n}_{i+1/2,j+1/2}, T^{m,n}_{i+1/2,j+1/2}, \theta_v \right)^{m=k+1/2}_{m=k+3/2} \Delta x \Delta y
\end{aligned}
\tag{4.60}
$$

4.3.2 Second Fractional Step

In this fractional step, the vertical eddy viscosity terms and Coriolis terms are included. As mentioned before, the vertical eddy viscosity terms greatly influence lake flows, and the boundary conditions at the free surface and lake bottom are aggregated into these terms. Therefore, the Crank–Nicholson scheme is used in this step; that is,

$$
\frac{u^{**} - u^{*}}{\Delta t} \Delta V = \int_{\Delta V} \left\{ \frac{1}{2} \left[\frac{\partial}{\partial z} \left(K_v \frac{\partial u^{**}}{\partial z} \right) + \frac{\partial}{\partial 3} \left(K_v \frac{\partial u^{*}}{\partial z} \right) \right] + f v^{*} \right\} dV \tag{4.61}
$$

The Crank–Nicholson scheme is unconditionally stable and has second-order accuracy in space. Two tri-diagonal matrices are deduced from the preceding two equations and can be easily solved by the double-sweep method. Discretizing Eq. (4.61), the following equation is obtained:

$$
\begin{aligned}
\frac{u^{k+1/2,**}_{i,j+1/2} - u^{k+1/2,*}_{i,j+1/2}}{\Delta t} \Delta V^{k+1/2,n}_{kj+1/2} = \frac{1}{2} \left\{ K^{k+1,n}_{vi,j+1/2} \frac{u^{k+3/2,**}_{i,j+1/2} - u^{k+1/2,**}_{i,j+1/2}}{\Delta z^{k+1,n}_{i,j+1/2}} \right. \\
- K^{k,n}_{vi,j+1/2} \frac{u^{k+1/2,**}_{i,j+1/2} - u^{k-1/2,**}_{i,j+1/2}}{\Delta z^{k,n}_{i,j+1/2}}
\end{aligned}
$$

$$
\begin{aligned}
&+ K^{k+1,n}_{vi,j+1/2}\frac{u^{k+3/2,*}_{i,j+1/2} - u^{k+1/2,*}_{i,j+1/2}}{\Delta z^{k+1,n}_{i,j+1/2}} \\
&\left.-K^{k,n}_{vi,j+1/2}\frac{u^{k+1/2,*}_{i,j+1/2} - u^{k-1/2,*}_{i,j+1/2}}{\Delta z^{k,n}_{i,j+1/2}}\right\}\Delta x \Delta y \\
&+ f v^{k+1/2,*}_{i,j+1/2}\Delta V^{k+1/2,n}_{i,j+1/2}
\end{aligned}
\tag{4.62}
$$

where

$$
v^{k+1/2,*}_{i,j+1/2} = \left(v^{k+1/2,*}_{i+1/2,j} + v^{k+1/2,*}_{i+1/2,j+1} + v^{k+1/2,*}_{i-1/2,j} + v^{k+1/2,*}_{i-1/2,j+1}\right)/4 \tag{4.63}
$$

By rearranging the preceding equation, the following Poisson equation is obtained:

$$
-A_{k+1/2}\,u^{k-1/2,**}_{i,j+1/2} + B_{k+1/2}\,u^{k+1/2,**}_{k,j+1/2} - C_{k+1/2}\,u^{k+3/2,**}_{i,j+1/2} = D_{k+1/2} \tag{4.64}
$$

where

$$
\begin{aligned}
A_{k+1/2} &= \frac{K^{k,n}_{vi,j+1/2}\Delta x \Delta y \Delta t}{2\Delta z^{k,n}_{i,j+1/2}} \\
B_{k+1/2} &= \Delta x \Delta y \Delta z^{k+1/2,n}_{i,j+1/2} + \frac{1}{2}\left(\frac{K^{k+1,n}_{vi,j+1/2}}{\Delta z^{k+1,n}_{i,j+1/2}} + \frac{K^{k,n}_{vi,j+1/2}}{\Delta z^{k,n}_{i,j+1/2}}\right)\Delta x \Delta y \Delta t \\
C_{i+1/2} &= \frac{K^{k+1,n}_{vi,j+1/2}\Delta x \Delta y \Delta t}{2\Delta z^{k+1,n}_{i,j+1/2}} \\
D_{k+1/2} &= \Delta x \Delta y \Delta z^{k+1/2,n}_{i,j+1/2}\,u^{k+1/2,*}_{i,j+1/2} \\
&\quad + \frac{1}{2}\left(K^{k+1/2,n}_{vi,j+1/2}\frac{u^{k+3/2,*}_{i,j+1/2} - u^{k+1/2,*}_{i,j+1/2}}{\Delta z^{k+1,n}_{i,j+1/2}} - K^{k,n}_{vi,j+1/2}\frac{u^{k+1/2,*}_{i,j+1/2} - u^{k-1/2,*}_{i,j+1/2}}{\Delta z^{k,n}_{i,j+1/2}}\right) \\
&\quad + \frac{f}{4}\left(v^{k+1/2,*}_{i+1/2,j} + v^{k+1/2,*}_{i+1/2,j+1} + v^{k+1/2,*}_{i-1/2,j} + v^{k+1/2,*}_{i-1/2,j+1}\right)\Delta V^{k+1/2,n}_{i,j+1/2}\Delta t
\end{aligned}
\tag{4.65}
$$

Because of Richtmyer and Morton's sufficient condition, which makes the round-off error small (Roache, 1972),

$$
\begin{aligned}
A_{k+1/2} &> 0 \\
B_{k+1/2} &> 0 \\
C_{k+1/2} &> 0 \\
B_{k+1/2} &> 0\ A_{k+1/2} + C_{k+1/2}
\end{aligned}
\tag{4.66}
$$

Equation (4.66) is satisfied and Eq. (4.64) can be solved by the double-sweep method.

Eq. (4.64) can be expressed as:

$$u_{i,j+1/2}^{k+1/2,**} = E_{k+1/2}\, u_{i,j+1/2}^{k-1/2,**} + F_{k+1/2} \tag{4.67}$$

where

$$E_{k+1/2} = \frac{A_{k-1/2}}{B_{k+1/2} - C_{k+1/2}E_{k-1/2}} \tag{4.68}$$

$$F_{k+1/2} = \frac{D_{k+1/2} + C_{k+1/2}F_{k-1/2}}{B_{k+1/2} - C_{k+1/2}E_{k-1/2}} \tag{4.69}$$

In order to solve Eq. (4.67), $E_{k+1/2}$ and $F_{k+1/2}$ are first calculated from the water surface to the bottom, using Eqs. (4.68) and (4.69) (it is given that $E_{k+1/2} = F_{k+1/2} = 0$ at the water surface). Then, $u_{i,j+1/2}^{k+1/2,**}$ is calculated from the bottom to the water surface with the aid of the equation $u_{i,j+1/2}^{k+1,**} = 0$ at the bottom.

4.3.3 Third Fractional Step

In this fractional step, the continuity equation and the pressure terms of momentum equations are calculated; that is,

$$\frac{\partial u}{\partial x} + \frac{\partial v}{\partial y} + \frac{\partial w}{\partial z} = 0 \tag{4.70}$$

$$\frac{u^{n+1} - u^{**}}{\Delta t} = -\frac{1}{\rho_0}\frac{\partial p}{\partial x} \tag{4.71}$$

$$\frac{v^{n+1} - v^{**}}{\partial t} = -\frac{1}{\rho_0}\frac{\partial p}{\partial x} \tag{4.72}$$

$$0 = -g - \frac{1}{\rho}\frac{\partial p}{\partial z} \tag{4.73}$$

Because the water stage ζ is very small, the hydrostatic pressure distribution can be linearized as

$$p = \rho_0\varsigma - g\int_z^0 \Delta\rho dz - g\Delta\rho_{sur}\varsigma = \rho_0 g\zeta\left(1 - \frac{\Delta\rho_{sur}}{\rho_0}\right) - g\int_z^0 \Delta\rho dz \tag{4.74}$$

Because the barotropic mode has a much larger propagation speed than that of the baroclinic mode, the former is treated implicitly, in order to get a larger time step, and the latter is calculated explicitly.

For the continuity equation, integrated from the bottom to the water surface,

$$\frac{\partial \zeta}{\partial t} = -\int_{-h}^{\zeta}\left(\frac{\partial u}{\partial x} + \frac{\partial v}{\partial y}\right)dz \tag{4.75}$$

If the Crank–Nicholson scheme is used, the following equation can be obtained:

$$\frac{\zeta^{n+1} - \zeta^{n}}{\Delta t} = -\frac{1}{2}\int_{-h}^{\zeta}\left(\frac{\Delta u^{n+1}}{\Delta x} + \frac{\Delta u^{n}}{\Delta x} + \frac{\Delta v^{n+1}}{\Delta y} + \frac{\Delta v^{n}}{\Delta y}\right)dz \tag{4.76}$$

By differentiating Eqs. (4.71) and (4.72) about x and y respectively, the following equations are obtained:

$$\frac{\partial u^{n+1}}{\partial x} = \frac{\partial u^{**}}{\partial x} - \frac{\Delta t}{\rho_0}\int_{-h}^{\zeta}\frac{\partial^2 p}{\partial x^2} \tag{4.77}$$

$$\frac{\partial v^{n+1}}{\partial y} = \frac{\partial v^{**}}{\partial y} - \frac{\Delta t}{\rho_0}\int_{-h}^{\zeta}\frac{\partial^2 p}{\partial y^2} \tag{4.78}$$

Substituting Eqs. (4.77) and (4.78) into Eq. (4.76), the following equation yields

$$\frac{\zeta^{n+1} - \zeta^{n}}{\Delta t} = -\frac{1}{2}\int_{-h}^{\zeta}\left(\frac{\partial u^{**}}{\partial x} + \frac{\partial v^{**}}{\partial y} + \frac{\partial u^{n}}{\partial x} + \frac{\partial v^{n}}{\partial y}\right)dz + \frac{\Delta t}{2\rho_0}\int_{-h}^{\zeta}\left(\frac{\partial^2}{\partial x^2} + \frac{\partial^2 p}{\partial y^2}\right)dz \tag{4.79}$$

where the ζ at time $n\Delta t$ is used for the upper limit of the integration. By the use of Eq. (4.74),

$$\int_{-h}^{\zeta}\left(\frac{\partial^2 p}{\partial x^2} + \frac{\partial^2 p}{\partial y^2}\right)dz = \int_{-h}^{\zeta}\rho_0 g\left(1 - \frac{\Delta\rho_{sur}^{n+1}}{\rho_0}\right)\left(\frac{\partial^2 \zeta^{n+1}}{\partial x^2} + \frac{\partial^2 \zeta^{n+1}}{\partial y^2}\right)dz - g\int_{-h}^{\zeta}\left(\left(\frac{\partial^2}{\partial x^2} + \frac{\partial^2}{\partial y^2}\right)\int_{dz}^{0}\Delta\rho^{n+1}d\right)dz \tag{4.80}$$

yields and Eq. (4.79) can be presented as

$$\frac{\zeta_{n+1} - \zeta^{n}}{\Delta t} - \frac{\Delta t}{2}g\left(1 - \frac{\Delta\rho_{sur}^{n+1}}{\rho_0}\right)\int_{-h}^{\zeta}\left(\frac{\partial^2 \zeta^{n+1}}{\partial x^2} + \frac{\partial^2 \zeta^{n+1}}{\partial y^2}\right)dz - \frac{1}{2}\int_{-h}^{\zeta}\left(\frac{\partial u^{**}}{\partial x} + \frac{\partial v^{**}}{\partial y} + \frac{\partial u^{n}}{\partial x} + \frac{\partial v^{n}}{\partial y}\right)dz - \Delta p_{\inf}^{n+1} \tag{4.81}$$

where

$$\Delta p_{inf}^{n+1} = \frac{g\Delta t}{2\rho_0} \int_{-h}^{\zeta} \left\{ \left(\frac{\partial^2 d}{\partial x^2} + \frac{\partial^2}{\partial y^2} \right) \int_{z}^{0} \Delta\rho^{n+1} d \right\} dz \tag{4.82}$$

Because $\partial^2\zeta^{n+1}/\partial x^2$ and $\partial^2\zeta^{n+1}/\partial y^2$ are independent of the water depth z, the following formula holds:

$$\int_{-h}^{\zeta} \left(\frac{\partial^2 \zeta^{n+1}}{\partial x^2} + \frac{\partial^2 \zeta^{n+1}}{\partial y^2} \right) dz = \left(\frac{\partial^2 \zeta^{n+1}}{\partial x^2} + \frac{\partial^2 \zeta^{n+1}}{\partial y^2} \right) (\zeta^n + h) \tag{4.83}$$

Therefore, Eq. (4.81) can be represented in the following form:

$$\begin{aligned} &\frac{\zeta^{n+1} - \zeta^n}{\Delta t} - \frac{\Delta t}{2} g \left(1 - \frac{\Delta\rho_{sur}^{n+1}}{\rho_0} \right) (\zeta^n + h) \left(\frac{\partial^2 \zeta^{n+1}}{\partial x^2} + \frac{\partial^2 \zeta^{n+1}}{\partial y^2} \right) \\ &- \frac{1}{2} \sum_{ktbm}^{ksur} \left(\frac{\Delta u^{**}}{\Delta x} + \frac{\Delta v^{**}}{\Delta y} + \frac{\Delta u^n}{\Delta x} + \frac{\Delta v^n}{\Delta y} \right) \Delta z - \Delta p_{\text{inf}}^{n+1} \end{aligned} \tag{4.84}$$

If the next expressions are used,

$$A = \frac{\Delta t}{2} g \left(1 - \frac{\Delta\rho_{sur}^{n+1}}{\rho_0} \right) (\zeta^n + h) \tag{4.85}$$

$$FF^{n+1} = -\frac{1}{2} \sum_{kbtm}^{ksur} \left(\frac{\Delta u^{**}}{\Delta x} + \frac{\Delta v^{**}}{\Delta y} + \frac{\Delta u^n}{\Delta x} + \frac{\Delta v^n}{\Delta y} \right) \Delta z - \Delta p_{\text{inf}}^{n+1} \tag{4.86}$$

Equation (4.84) can be rewritten as

$$\frac{\zeta^{n+1} - \zeta^n}{\Delta t} - A \left(\frac{\delta^2 \zeta^{n+1}}{\partial x^2} + \frac{\partial^2 \zeta^{n+1}}{\partial y^2} \right) = FF^{n+1} \tag{4.87}$$

Equation (4.87) is a Poisson equation and can be solved by various methods. In this section, the resulting Poisson ratio of ζ is split into two equations and solved by the double-sweep method:

$$\frac{\tilde{\zeta} - \zeta^n}{\Delta t} - A \frac{\partial^2 \zeta^{n+1}}{\partial x^2} = 0 \tag{4.88}$$

$$\frac{\zeta^{n+1} - \tilde{\zeta}}{\Delta t} - A \frac{\partial^2 \zeta^{n+1}}{\partial y^2} = FF^{n+1} \tag{4.89}$$

Equations (4.88) and (4.89) are discretized into the following difference equations:

$$\frac{\tilde{\zeta}_{i+1/2,j+1/2}-\zeta^{n}_{i+1/2,j+1/2}}{\Delta t}-A_{i+1/2,j+1/2}\left\{\frac{\tilde{\zeta}_{i+3/2,j+1/2}-\tilde{\zeta}_{i+1/2,j+1/2}}{(\Delta x)^2}-\frac{\tilde{\zeta}_{i+1/2,j+1/2}-\tilde{\zeta}_{i+1/2,j+1/2}}{(\Delta x)^2}\right\}=0 \tag{4.90}$$

$$\frac{\zeta_{i+1/2,j+1/2^{n+1}}-\tilde{\zeta}_{i+1/2,j+1/2}}{\Delta t}-A_{i+1/2,j+1/2}\left\{\frac{\zeta^{n+1}_{i+1/2,j+3/2}-\zeta^{n+1}_{i+1/2,j+1/2}}{(\Delta y)^2}-\frac{\zeta^{n+1}_{i+1/2,j+1/2}-\zeta^{n-1}_{i+1/2,j-1/2}}{(\Delta y)^2}\right\}=FF^{n+1} \tag{4.91}$$

$\zeta^{n+1}_{i+1/2,j+1/2}$ is calculated from the discussed Poisson equations by the same double-sweep method used in the second fractional step. By using $\zeta^{n+1}_{i+1/2,j+1/2}$, the pressure p is calculated from Eq. (4.74).

Moreover, the following equations can be obtained by discretizing Eqs. (4.71) and (4.72).

$$u^{k+1/2,n+1}_{i,j+1/2}=u^{k+1/2,**}_{i,j+1/2}-\frac{\Delta t}{2\rho_0}\left\{\left(p^{k,n+1}_{i+1/2,j+1/2}+p^{k+1,n+1}_{i+1/2,j+1/2}\right)\Delta y\Delta z^{k+1/2,n}_{i+1/2,j+1/2}-\left(p^{k,n+1}_{i-1/2,j+1/2}-p^{k+1,n+1}_{i-1/2,j+1/2}\right)\Delta y\Delta z^{k+1/2,n}_{i-1/2,j+1/2}\right\}\Delta V^{k+1/2,n}_{i,j+1/2} \tag{4.92}$$

$$v^{k+1/2,n+1}_{i,j+1/2}=v^{k+1/2,**}_{i,j+1/2}-\frac{\Delta t}{2\rho_0}\left\{\left(p^{k,n+1}_{i+1/2,j+1/2}+p^{k+1,n+1}_{i+1/2,j+1/2}\right)\Delta x\Delta z^{k+1/2,n}_{i+1/2,j+1/2}-\left(p^{k,n+1}_{i+1/2,j-1/2}-p^{k+1,n+1}_{i+1/2,j-1/2}\right)\Delta x\Delta z^{k+1/2,n}_{i+1/2,j-1/2}\right\}\Delta V^{k+1/2,n}_{i+1/2} \tag{4.93}$$

$u^{k+1/2,n+1}_{i,j+1/2}$ and $v^{k+1/2,n+1}_{i+1/2,j}$ are calculated from Eqs. (4.92) and (4.93).

Finally, the continuity equation is discretized into the following difference equation:

$$w^{k+1,n+1}_{i+1/2,j+1/2}=w^{k,n+1}_{i+1/2,j+1/2}-\frac{u^{k+1/2,n+1}_{i+1,j+1/2}\Delta z^{k+1/2,n+1}_{i+1,j+1/2}-u^{k+1/2,n+1}_{i,j+1/2}\Delta z^{k+1/2,n+1}_{i,j+1/2}}{\Delta x}-\frac{v^{k+1/2,n+1}_{i+1/2,j+1}\Delta z^{k+1/2,n+1}_{i+1/2,j+1}-v^{k+1/2,n+1}_{i+1/2,j}\Delta z^{k+1/2,n+1}_{i+1/2,j}}{\Delta y} \tag{4.94}$$

$w^{k+1,n+1}_{i+1/2,j+1/2}$ is calculated from the bottom to the water surface by the use of the boundary condition $w|_{\text{bottom}}=0$ at the bottom.

4.4 CONTROL VOLUME 3D MODEL (IDOR)

In this section, the following 3D numerical methods are discussed. In the momentum equations and temperature equation, the Adams-Bashforth scheme and leap-frog scheme for the temporal difference, and the Donor-cell scheme and QUICK (Quadratic Upstream Interpolation for Convection Kinematics) scheme for the advective terms are applied respectively. The pressure term is split into the barotropic and baroclinic modes by using the hydrostatic approximation, in which the baroclinic mode is calculated explicitly and the barotropic mode is computed by both explicit and implicit formulas, respectively. For the convenience of discussion, the control volume ($\Delta x \times \Delta y \times$ water depth) is called *column*, and the control volume ($\Delta x \times \Delta y \times \Delta z$) is called *cell*. Only the difference equations of the x-wise momentum equation, the temperature equation, and the continuity equation are illustrated in detail, and the y-wise momentum equation is omitted because of the symmetry (Kinoshita, 1989).

The x-wise momentum equation and temperature equation can be represented in the following way, after integrating them over the control volume (cell):

$$\int_{\Delta V} \frac{\partial u}{\partial t} dV = u_{\mathrm{adv}} + u_p + u_{\mathrm{diff}} + u_{\mathrm{co}} \tag{4.95}$$

$$\int_{\Delta \rho} \frac{\partial \Delta T}{\partial t} dV = T_q + T_{\mathrm{adv}} + T_{\mathrm{diff}} \tag{4.96}$$

where ΔV is the volume of the cell. When the x-wise momentum equation is discretized, the control volume (cell) should be shifted by $\Delta x/2$ in the x-direction, while the density deficit equation is discretized on the control volume without shift (see Fig. 4.4).

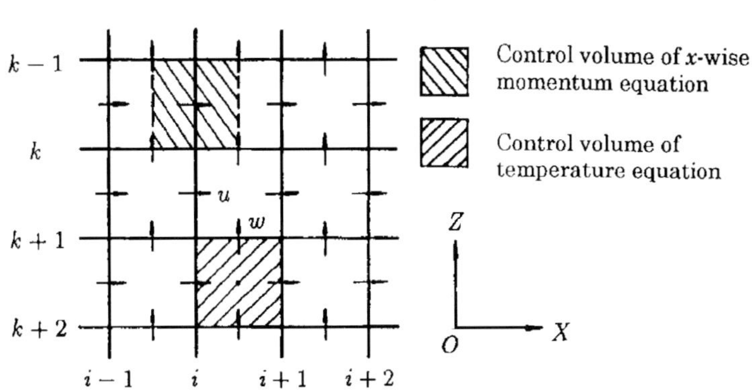

Fig. 4.4. Control volumes for momentum equations and temperature equation (Shen, 1991).

4.4.1 Momentum Equation

Temporal difference

The leap-frog scheme is a widely used scheme, with conditional stability and second-order accuracy concerning time. Because it produces numerical oscillation with a period of $2\Delta t$, it is necessary to improve this scheme in order to avoid the numerical oscillation (e.g., introduce a so-called Matsuno's scheme (Matsuno, 1966)). In this study, Matsuno's scheme is introduced at every 6 steps for the leap-frog scheme. The details on Matsuno's scheme are referred to in Iwasa and Inoue (1985) and Iwasa et al. (1983).

$$\frac{u^{n+2} - u^n}{2\Delta t}\Delta V^{n+1} = F^{(n+1,n)} \tag{4.97}$$

$$\frac{T^{n+2} - T^n}{2\Delta t}\Delta V^{n+1} = G^{(n+1,n)} \tag{4.98}$$

where F and G include the terms on the right–hand side of Eqs. (4.95) and (4.96) respectively.

On the other hand, the Adams–Bashforth scheme also has second-order accuracy in time, and has been proven to have a little weak dispersion by the Neumann's stability analysis (Roache, 1972), in case only advective terms exist. If diffusive terms are involved, this scheme will be stable.

$$\frac{u^{n+1} - u^n}{\Delta t}\Delta V^n = 1.5F^{n+1} - 0.5F^n \tag{4.99}$$

$$\frac{T^{n+1} - T^n}{\Delta t}\Delta V^n = 1.5G^{n+1} - 0.5G^n \tag{4.100}$$

Spatial difference

The spatial difference schemes in the leap-frog scheme and the Adams–Bashforth scheme are different. Only the spatial difference schemes of the Adams–Bashforth method are illustrated in the following, and those of the leap-frog method are referred to in Iwasa et al. (1983).

Advective terms. Advective terms are non-linear terms which often give rise to many numerical stability problems. According to Neumann's stability analysis, the non-linear stability criterion that follows must be satisfied if advective terms are treated explicitly:

$$V_{\max}\Delta t\sqrt{\frac{1}{\Delta x^2} + \frac{1}{\Delta y^2} + \frac{1}{\Delta z^2}} \leq 1 \tag{4.101}$$

where $V_{\max}$ is the maximal velocity of lake flow. Donor-cell and QUICK schemes are employed for advective terms.

(i) Donor-cell scheme

The Donor-cell scheme is a second-order upstream scheme that deals with space. The formula which expresses upwind difference is defined as

$$(u, v_m, \theta)_{m=i}^{m=i+1} = uv_{i+1/2} + \frac{\theta}{2}|u|(v_i - v_{i+1}) \tag{4.102}$$

where θ is the parameter of upstream difference. When $\theta = 1$, the formula is ordinary upwind difference; if $\theta = 0$, it becomes center difference. The advective terms are discretized into the following formula by the use of a Donor-cell scheme:

$$\begin{aligned}
u_{\mathrm{adv}}(i, j+1/2, k+1/2) = &-\left(u_{i+1/2,j+1/2}^{k+1/2,n}, u_{m,j+1/2}^{k+1/2,n}, \theta_h\right)_{m=i}^{m=i+1} \Delta y \Delta z_{i+1/2,j+1/2}^{k+1/2,n} \\
&+ \left(u_{i+1/2,j+1/2}^{k+1/2,n}, u_{m,j+1/2}^{k+1/2,n}, \theta_h\right)_{m=i-1}^{m=i} \Delta y \Delta z_{i+1/2,j+1/2}^{k+1/2,n} \\
&- \left(v_{i-1/2,j+1}^{k+1/2,n}, u_{i-1/2,m}^{k+1/2,n}, \theta_h\right)_{m=j+1/2}^{m=j+3/2} \Delta x \Delta z_{i-1/2,j+1}^{k+1/2,n}/2 \\
&- \left(v_{i+1/2,j+1}^{k+1/2,n}, u_{i+1/2,m}^{k+1/2,n}, \theta_h\right)_{m=j+1/2}^{m=j+3/2} \Delta x \Delta z_{i+1/2,j+1}^{k+1/2,n}/2 \\
&+ \left(v_{i-1/2,j}^{k+1/2,n}, u_{i-1/2,m}^{k+1/2,n}, \theta_h\right)_{m=j-1/2}^{m=j+1/2} \Delta x \Delta z_{i-1/2,j}^{k+1/2,n}/2 \\
&+ \left(v_{i+1/2,j}^{k+1/2,n}, u_{i+1/2,m}^{k+1/2,n}, h\right)_{m=j-1/2}^{m=j+1/2} \Delta x \Delta z_{i+1/2,j}^{k+1/2,n}/2 \\
&- \left(w_{i+1/2,j+1/2}^{k,n}, u_{i+1/2,j+1/2}^{m,n}, \theta_v\right)_{m=k+1/2}^{m=k-1/2} \Delta x \Delta y/2 \\
&- \left(w_{i-1/2,j+1/2}^{k,n}, u_{i-1/2,j+1/2}^{m,n}, \theta_v\right)_{m=k+1/2}^{m=k-1/2} \Delta x \Delta y/2 \\
&+ \left(w_{i+1/2,j+1/2}^{k+1,n}, u_{i+1/2,j+1/2}^{m,n}, \theta_v\right)_{m=k+3/2}^{m=k+1/2} \Delta x \Delta y/2 \\
&+ \left(w_{i=1/2,j+1/2}^{k+1,n}, u_{i-1/2,j+1/2}^{m,n}, \theta_v\right)_{m=k+3/2}^{m=k+1/2} \Delta x \Delta y/2
\end{aligned} \tag{4.103}$$

(ii) QUICK scheme

The QUICK scheme is a third-order upwind scheme about space which evaluates the advective terms by interpolation between three points. Compared to the Donor-cell scheme, it not only has higher accuracy but also is more available to flows with high Reynolds numbers. The advective terms are discretized by means of the QUICK scheme as

$$\begin{aligned}
u_{\mathrm{adv}}(i, j+1/2, j+1/2) = &-\left(u_{i+1/2,j+1/2}^{k+1/2,n}, u_{i+1/2,j+1/2}^{k+1/2,n}\right) \Delta y \Delta z_{i+1/2,j+1/2}^{k+1/2,n} \\
&+ \left(u_{i-1/2,j+1/2}^{k+1/2,n}, u_{i-1/2,j+1/2}^{k+1/2,n}\right) \Delta y \Delta z_{i-1/2,j+1/2}^{k+1/2,n}
\end{aligned}$$

$$
\begin{aligned}
&-\left(v_{i-1/2,j+1/2}^{k+1/2,n}, u_{i-1/2,j+1}^{k+1/2,n}\right)\Delta x \Delta z_{i-1/2,j+1}^{k+1/2,n}/2 \\
&-\left(v_{i+1/2,j+1}^{k+1/2,n}, u_{i+1/2,j+1}^{k+1/2,n}\right)\Delta x \Delta z_{i+1/2,j+1}^{k+1/2,n}/2 \\
&+\left(v_{i-1/2,j}^{k+1/2,n}, u_{i-1/2,j}^{k+1/2,n}\right)\Delta x \Delta z_{i-1/2,j}^{k+1/2,n}/2 \\
&-\left(v_{i+1/2,j}^{k+1/2,n}, u_{i+1/2,j}^{k+1/2,n}\right)\Delta x \Delta z_{i+1/2,j}^{k+1/2,n}/2 \\
&-\left(w_{i-1/2,j+1/2}^{k,n}, u_{i+1/2,j+1/2}^{k,n}\right)\Delta x \Delta y/2 \\
&-\left(w_{i+1/2,j+1/2}^{k,n}, u_{i-1/2,j+1/2}^{k,n}\right)\Delta x \Delta y/2 \\
&+\left(w_{i+1/2,j+1/2}^{k+1,n}, u_{i+1/2,j+1/2}^{k+1,n}\right)\Delta x \Delta y/2 \\
&+\left(w_{i-1/2,j+1/2}^{k+1,n}, u_{i-1/2,j+1/2}^{k+1,n}\right)\Delta x \Delta y/2
\end{aligned}
\tag{4.104}
$$

As an example, only the $\partial(uu)/\partial x$ term is illustrated in the following:

$$
\begin{aligned}
&\left(u_{i+1/2,j+1/2}^{k+1/2,n}, u_{i+1/2,j+1/2}^{k+1/2,n}\right) = \frac{u_{i+1,j+1/2}^{k+1/2,n} + u_{i,j+1/2}^{k+1/2,n}}{2} \\
&\left\{\frac{1}{2}\left(u_{i+1,j+1/2}^{k+1/2,n} + u_{i,j+1/2}^{k+1/2,n}\right) - \frac{1}{8}\left(u_{i+1,j+1/2}^{k+1/2,n} - 2u_{i,j+1/2}^{k+1/2,n} + u_{i-1,j+1/2}^{k+1/2,n}\right)\right. \\
&+\frac{1}{24}\left(u_{i,j+3/2}^{k+1/2,n} - 2u_{i,j+1/2}^{k+1/2,n} + u_{i,j-1/2}^{k+1/2,n}\right) \\
&\left.+\frac{1}{24}\left(u_{i,j+1/2}^{k-1/2,n} - 2u_{i,j+1/2}^{k+1/2,n} + u_{i,j+1/2}^{k+3/2,n}\right)\right\}
\end{aligned}
\tag{4.105}
$$

$$
\begin{aligned}
&\left(u_{i+1/2,j+1/2}^{k+1/2,n}, u_{i+1/2,j+1/2}^{k+1/2,n}\right) = \frac{u_{i+1,j+1/2}^{k+1/2,n} + u_{i,j+1/2}^{k+1/2,n}}{2} \\
&\left\{\frac{1}{2}\left(u_{i+1,j+1/2}^{k+1/2,n} + u_{i,j+1/2}^{k+1/2,n}\right) - \frac{1}{8}\left(u_{i+2,j+1/2}^{k+1/2,n} - 2u_{i+1,j+1/2}^{k+1/2,n} + u_{i,j+1/2}^{k+1/2,n}\right)\right. \\
&+\frac{1}{24}\left(u_{i+1,j+3/2}^{k+1/2,n} - 2u_{i+1,j+1/2}^{k+1/2,n} + u_{i+1,j-1/2}^{k+1/2,n}\right) \\
&\left.+\frac{1}{24}\left(u_{i,j+1/2}^{k-1/2,n} - 2u_{i,j+1/2}^{k+1/2,n} + u_{i,j+1/2}^{k+3/2,n}\right)\right\}
\end{aligned}
\tag{4.106}
$$

Pressure term. Because pressure terms are the most important in momentum equations for lake flows, they must be carefully treated. Pressure p can be represented as follows:

$$
p = g\int_z^{\zeta} \rho \; dz = \rho_0 g(\zeta - z) - g\int_z^{\zeta} \Delta\rho \; dz \tag{4.107}
$$

where ζ is the water stage. The first part of Eq. (4.107) is the component due to the variation in water stage that is, the barotropic component, and the second component is due to the density variation (the baroclinic component). The term $\rho_0 gz$ in Eq. (4.107) is independent of both x and y and does not contribute to $\partial p/\partial x$ and $\partial p/\partial x$ at all. Therefore, it can be removed from the preceding equation so that the pressure can be simplified to the following equation:

$$p = \rho_0 g\zeta - g\int_z^\zeta \Delta\rho \ dz \tag{4.108}$$

If ζ is treated explicitly, the calculating formula is very simple, but the Courant stability criterion must be satisfied:

$$\sqrt{gH_{\max}}\Delta t\sqrt{\frac{1}{\Delta x^2}+\frac{1}{\Delta y^2}+\frac{1}{\Delta z^2}} \le 1 \tag{4.109}$$

where $H_{\max}$ is the maximal water depth. It means a very small time step is required and the computational time will become longer. If an implicit scheme about ζ is used, it need not satisfy Eq. (4.109) so that a larger Δt can be chosen. But it does become necessary to carry out iterative calculations in order to solve the Poisson equation concerning ζ, and the variations of flows with short periods cannot be reproduced. In this section, both explicit and implicit schemes are applied.

(i) Implicit scheme

The water stage ζ is represented as follows by use of weight parameter θ:

$$\zeta = (1-\theta)\zeta^n + \theta\zeta^{n+1} \tag{4.110}$$

and Eq. (4.108) can be discretized by substituting ζ into it.

$$\tilde{p}^k_{i+1/2,j+1/2} = \rho_0 g\tilde{\zeta} - \sum_{m-1}^{k-1} g\Delta\rho^{m+1/2,n}_{i+1/2,j+1/2}\Delta z^{m+1/2,n}_{i+1/2,j+1} \tag{4.111}$$

The pressure term is expressed as:

$$u_p(i,j+/1,k+1/2) = -\frac{1}{\rho_0}\left\{\frac{1}{2}\left(P^k_{i+1/2,j+1/2}+p^{k+1}_{i+1/2,j+1/2}\right)\right.$$
$$\left.\Delta y\Delta yz^{k+1/n,}_{i+1/2,j+1/2} - \frac{1}{2}\left(p^k_{i-1/2,j+1/2}+p^{k+1}_{i-1/2,j+1/2}\right)\Delta y\Delta z^{k+1/2,n}_{i=1/2,j+1/2}\right\} \tag{4.112}$$

Equation (4.111) can be divided into explicit and implicit parts:

$$\tilde{p}^k_{i+1/2,j+1/2} = p'^{k+1}_{i+1/2,j+1/2} + \rho_0 g\theta\zeta^{n+1}_{i+1/2,j+1/2} \tag{4.113}$$

where the explicit part is

$$\begin{aligned} p'^{k}_{i+1/2,j+1/2} &= \rho_0 g(1-\theta)\xi^{n}_{i+1/2,j+1/2} \\ &\quad - \frac{1}{\rho_0}\left\{\left(p'^{k}_{i+1/2,j+1/2} + p'^{i+1}_{k+1/2,j+1/2}\right)\Delta y \Delta z^{k+1/2,n}_{i+1/2,j+1/2}\right. \\ &\quad \left. - \frac{1}{2}\left(p'^{k}_{i-1/2,j+1/2} + p'^{k+1}_{i-1/2,j+1/2}\right)\Delta y \Delta z^{k+1/2,n}_{i-1/2,j+1/2}\right\} \end{aligned} \tag{4.114}$$

The continuity equation integrated from the bottom to the water surface can be expressed by the following equation:

$$\frac{\partial}{\partial t}\int_A \zeta dA_s = -\tilde{u}A_x|^{x_{i+1}}_{x_i} - \tilde{v}A_y|^{y_{i+1}}_{y_i} \tag{4.115}$$

where A_x and A_y are the areas of cells in the yz plane and xz plane. $\tilde{u}$ and $\tilde{v}$ are defined by the next equations.

$$\tilde{u} = (1-\theta)u^n + \theta u^{n+1} \tag{4.116}$$

$$\tilde{v} = (1-\theta)v^n + \theta v^{n+1} \tag{4.117}$$

By using Leibnitz's rule, Eq. (4.115) can be rewritten in the following form.

$$\begin{aligned} \frac{\zeta^{n-1}_{i+1/2,j+1/2} - \zeta^{n}_{i+1/2,j+1/2}}{\Delta t} &= -\sum \tilde{u}^{k+1/2}_{i+1/2,j+1/2}\Delta z^{k+1/2,n}_{i+1/2,j+1/2}/\Delta x \\ &\quad + \sum \tilde{u}^{k+1/2}_{i,j+1/2}\Delta z^{k+1/2,n}_{i,j+1/2}/\Delta x \\ &\quad - \sum \tilde{v}^{k+1/2}_{i+1/2,j+1}\Delta z^{k+1/2,n}_{i+1/2,j+1}/\Delta y \\ &\quad + \sum \tilde{v}^{k+1/2}_{i+1/2,j}\Delta z^{k+1/2,n}_{i+1/2,j}/\Delta y \end{aligned} \tag{4.118}$$

In the preceding equation, the unknown components of velocity at $(n+1)\Delta t$ time: (u^{n+1} and v^{n+1}) are included, and they can be expressed by rewriting Eq. (4.99) as

$$\begin{aligned} u^{k+1/2,n+1}_{i,j+1/2} &= u^{*k+1/2}_{i,j+1/2} - g\theta\frac{\Delta t}{\Delta x}\left(\Delta z^{k+1/2,n}_{i+1/2,j+1/2}\zeta^{n+1}_{i+1/2,j+1/2}\right. \\ &\quad \left. - \Delta z^{k+1/2,n}_{i-1/2,j+1/2}\zeta^{n+1}_{i-1/2,j+1/2}\right)\Delta z^{k+1/2,n}_{i,j+1/2} \end{aligned} \tag{4.119}$$

where

$$u^{*k+1/2}_{i,j+1/2} = u^{k+1/2,n}_{i,j+1/2} + (1.5G^m - 0.5G^{m-1})\Delta t/\Delta v^{k+1/2,n}_{i,j+1/2} \tag{4.120}$$

and

$$\begin{aligned} G^n = &+ u_{\text{co}}(i,j+1/2,k+1/2) + u_{\text{adv}}(i,j+1/2,k+1/2) \\ &+ u_{\text{diff}}(i,j+1/2,k+1/2) + u'_p(i,j+1/2,k+1/2) \end{aligned} \tag{4.121}$$

Substituting Eqs. (4.119) and (4.120) into Eq. (4.118), the following Poisson equation concerning the water stage ζ results:

$$\begin{aligned}
\zeta^{n+1}_{i+1/2,j+1/2} &= \zeta^{*}_{i+1/2,j+1/2} + g\left(\frac{\theta\Delta t}{\Delta x}\right)^2 \Big\{(\eta_{i+1,u+1/2} + \zeta^{n}_{i+3/2,j+1/2})\zeta^{n+1}_{i+3/2,j+1/2} \\
&+ (\eta_{i,j+1/2} + \zeta^{n}_{i-1/2,j+1/2})\zeta^{n+1}_{i+3/2,j+1/2} \\
&-(\eta_{i+1,j+1/2} + \eta_{i,j+1/2} + 2\zeta^{n}_{i+1/2,j+1/2})\zeta^{n+1}_{i+1/2,j+1/2}\Big\} \\
&+ g\left(\frac{\theta\Delta t}{\Delta y}\right)^2 \Big\{(\eta_{i+1/2,j+1} + \zeta^{n}_{i+1/2,j+3/2})\zeta^{n+1}_{i+1/2,j+3/2} \\
&+ (\eta_{i+1/2,j} + \zeta^{n}_{i+1/2,j-1/2})\zeta^{n+1}_{i+1/2,j-1/2} \\
&-(\eta_{i+1/2,j+1} + \eta_{i+1/2,j} + 2\zeta^{n}_{i+1/2,j+1/2})\zeta^{n+1}_{i+1/2,j+1/2}\Big\}
\end{aligned} \tag{4.122}$$

where

$$\begin{aligned}
\zeta^{*}_{i+1/2,j+1/2} &= \zeta^{n}_{i+1/2,j+1/2} - \frac{\Delta t}{\Delta x \Delta y}\Big[\Delta y \sum \Delta z^{k+1/2,n}_{i+1,j+1/2} \\
&\times\Big\{(1-\theta)u^{k+1/2,n}_{i+1,j+1/2} + \theta u^{*k+1/2}_{i+1,j+1/2}\Big\} \\
&- \Delta y \sum \Delta z^{k+1/2,n}_{i,j+1/2}\Big\{(1-\theta) + \Delta x \sum \Delta z^{k+1/2,n}_{i+1/2,j+1} \\
&\times\Big\{(1-\theta)v^{k+1/2,n}_{i+1/2,j+1} + \theta v^{*k+1/2}_{i+1/2,j+1}\Big\}\Big\} \\
&- \Delta x \sum \Delta z^{k+1/2,n}_{i+1/2,j}\Big\{(1-\theta)v^{k+1/2,n}_{i+1/2,j} + \theta v^{*k+1/2}_{i+1/2,j}\Big\}\Big]
\end{aligned} \tag{4.123}$$

and η is the term resulting from the difference of $\sum \Delta z$ between columns. When the lake bed is uneven, in order to avoid calculating the unreal flow between the cell in the shallower column, which has reduced the lake bed, and the cell in the deeper column which has not, η is introduced and defined as

$$\eta_{i,j+1} = \min(\mathrm{H}_{i-1/2,j+1/2}, \mathrm{H}_{i+1/2,j+1/2}) \tag{4.124}$$

where $H_{i-1/2,j+1/2}$ is the water depth from the reference plane, whose centerline is located at $\{(i-1/2)\Delta x, (j+1/2)\Delta y\}$.

(ii) Explicit scheme

If the weight parameter θ is given to be zero, $\zeta = \zeta^n$ is obtained from Eq. (4.110). Therefore, the explicit scheme can be considered as one case of the implicit scheme. Equation (4.111) can be represented, as follows, by substituting $\zeta = \xi^n$ into it:

$$p^{k}_{i+1/2,j+1/2} = \rho_0 g \zeta^{n}_{i+1/2,j+1/2} - \sum_{m=0}^{k-1} g \Delta\rho^{m+1/2,n}_{i+1/2,j+1/2} \Delta z^{m+1/2,n}_{i+1/2,j+1/2} \tag{4.125}$$

and the velocity component due to the pressure term is

$$\begin{aligned} u_p(i,j+1/2,k+1/2) &= -\frac{1}{\rho_0}\left\{\frac{1}{2}(p^k_{i+1/2,j+1/2}+p^{k+1}_{i+1/2,j+1/2})\Delta y\Delta z^{k+1/2,n}_{i+1/2,j+1/2}\right. \\ &\left. -\frac{1}{2}(p^k_{i-1/2,j+1/2}+p^{k+1}_{i-1/2,j+1/2}\Delta y\Delta z^{k+1/2,n}_{i-1/2,j+1/2}\right\} \end{aligned} \tag{4.126}$$

In contrast to Eq. (4.112), all quantities in the preceding equation are already known, so that u_p can be calculated directly.

Coriolis and Diffusive Terms
The Coriolis term and diffusive terms are

$$\begin{aligned} u_{\mathrm{co}}(i,j+1/2,k+1/2) = \frac{1}{4}\Big(& v^{k+1/2,n}_{i+1/2,j}+v^{k+1/2,n}_{i+1/2,j+1} \\ &+v^{k+1/2,n}_{i-1/2,j}+v^{k+1/2,n}_{i-1/2,j+1}\Big)\Delta V^{k+1/2,n}_{i,j+1/2} \end{aligned} \tag{4.127}$$

$$\begin{aligned} u_{\mathrm{diff}}(i,j+1/2,k+1/2) = k_h\Bigg[&\frac{1}{\Delta x}\Big\{\Big(u^{k+1/2,n}_{i+1,j+1/2}-u^{k+1/2,n}_{i,j+1/2}\Big)\Delta y\Delta z^{k+1/2,n}_{i+1/2,j+1/2} \\ &-\Big(u^{k+1/2,n}_{i,j+1/2}-u^{k+1/2,n}_{i-1,j+1/2}\Big)\Delta y\Delta z^{k+1/2,n}_{i-1/2,j+1/2}\Big\} \\ &+\frac{1}{\Delta y}\Big\{\Big(u^{k+1/2,n}_{i+1/2,j+3/2}-u^{k+1/2,n}_{i+1/2,j+1/2}\Big)\Delta x\Delta z^{k+1/2,n}_{i+1/2,j+1/2}/2 \\ &+\Big(u^{k+1/2,n}_{i-1/2,j+3/2}-u^{k+1/2,n}_{i-1/2,j+1/2}\Big)\Delta x\Delta z^{k+1/2,n}_{i-1/2,j+1}/2 \\ &-\Big(u^{k+1/2,n}_{i+1/2,j+1/2}-u^{k+1/2,n}_{i+1/2,j-1/2}\Big)\Delta x\Delta z^{k+1/2,n}_{i+1/2,j}/2 \\ &-\Big(u^{k+1/2,n}_{i-1/2,j+1/2}-u^{k+1/2,n}_{i-1/2,j-1/2}\Big)\Delta x\Delta z^{k+1/2,n}_{i-1/2,j}/2\Big\}\Bigg] \\ &+k_v\Big\{\frac{1}{\Delta z_{i+1/2,j+1/2}}\Big(u^{k+1/2,n}_{i+1/2,j+1/2} \end{aligned} \tag{4.128a}$$

$$\begin{aligned} -u^{k+1/2,n}_{i+1/2,j+1/2}\Big)\Delta x\Delta y/2 &+ \frac{1}{\Delta z_{i-1/2,j+1/2}}\Big(u^{k-1/2,n}_{i-1/2,j+1/2}-u^{k+1/2,n}_{i-1/2,j+1/2}\Big)\Delta x\Delta y/2 \\ &-\frac{1}{\Delta z^{k+1,n}_{i+1/2,j+1/2}}\Big(u^{k+1/2,n}_{i+1/2,j+1/2}-u^{k+3/2,n}_{i+1/2,j+1/2}\Big)\Delta x\Delta y/2 \\ &-\frac{1}{\Delta z^{k+1,n}_{i-1/2,j+1/2}}\Big(u^{k+1/2,n}_{i-1/2,j+1/2}-u^{k+3,n}_{i-1/2,j+1/2}\Big)\Delta x\Delta y/2\Big\} \end{aligned} \tag{4.128b}$$

4.4.2 Equation of Temperature

As there is no pressure term in the temperature equation, the advective terms play greater roles than those in the momentum equations and need to be carefully treated.

Advective terms

$$
\begin{aligned}
& T_{\mathrm{adv}}(i+1/2, j+1/2, k+1/2) \\
& = -\left(u_{i+1,j+1/2}^{k+1/2,n}, T_{m,j+1/2}^{k+1/2,n}, \theta_h\right)_{m=i+1/2}^{m-i+3/2} \Delta y \Delta z_{i+1,j+1/2}^{k+1/2,n} \\
& + \left(u_{i,j+1/2}^{k+1/2,n}, T_{m,j+1/2}^{k+1/2,n}, \theta_h\right)_{m=i-1/2}^{m=i+1/2} \Delta y \Delta z_{i,j+1/2}^{k+1/2,n} \\
& - \left(v_{i+1/2,j+1}^{k+1/2,n}, T_{i+d1/2,m}^{k+1/2,n}, \theta_h\right)_{m=j+1/2}^{m=j+3/2} \Delta x \Delta z_{i+1/2,j+1}^{k+1/2,n} \\
& + \left(v_{i+1/2,j}^{k+1/2,n}, T_{i+1/2,m}^{k+1/2,n}, \theta_h\right)_{m=j-1/2}^{m=j+1/2} \Delta x \Delta z_{i+1/2,j}^{k+1/2,n} \\
& - \left(w_{i+1/2,j+1/2}^{k,n} T_{i+1/2,j+1/2}^{m,n}, \theta_v\right)_{m=k+1/2}^{m=k-1/2} \Delta x \Delta y \\
& + \left(w_{i+1/2,j+1/2}^{k+1,n}, T_{i+1/2,j+1/2}^{m,n}, \theta_v\right)_{m=k+3/2}^{m=k+1/2} \Delta x \Delta y
\end{aligned}
\tag{4.129}
$$

Diffusive terms

$$
\begin{aligned}
& T_{\mathrm{diff}}(i+1/2, j+1/2, k+1/2) \\
& = N_h \Bigg[\frac{1}{\Delta x} \Big\{ \left(T_{i+3/2,j+1/2}^{k+1/2,n} - T_{i+3/2,j+1/2}^{k+1/2,n}\right) \Delta y \Delta z_{i+1,j+1/2}^{k+1/2,n} \\
& - \left(T_{i+1/2,j+1/2}^{k+1/2,n} - T_{i-1/2,j+1/2}^{k+1/2,n}\right) \Delta y \Delta z_{i,j+1/2}^{k+1/2,n} \Big\} \\
& + \frac{1}{\Delta y} \Big\{ \left(T_{i+1/2,j+3/2}^{k+1/2,n} - T_{i+1/2,j+1/2}^{k+1/2,n}\right) \Delta x \Delta z_{i+1/2,j+1}^{k+1/2,n} \\
& - \left(T_{i+1/2,j+1/2}^{k+1/2,n} - T_{i+1/2,j-1/2}^{k+1/2,n}\right) \Delta x \Delta z_{i+1/2,j}^{k+1/2,n} \Big\} \Bigg] \\
& + N_v \left\{ \frac{1}{\Delta z_{i+1/2,j+1/2}^{k,n}} \left(T_{i+1/2,j+1/2}^{k-1/2,n} - T_{i+1/2,j+1/2}^{k+1/2,n}\right) \Delta x \Delta y \right. \\
& \left. - \frac{1}{\Delta z_{i+1/2,j+1/2}^{k+1,n}} \left(T_{i+1/2,j+1/2}^{k+1/2,n} - T_{i+1/2,j+1/2}^{k+3/2,n}\right) \Delta x \Delta y \right\}
\end{aligned}
\tag{4.130}
$$

Production terms

$$
T_q(i+1/2, j+1/2, k+1/2) = q_{i+1/2,j+1/2}^{k+1/2,n} \Delta V_{i+1/2,j+1/2}^{k+1/2,n}
\tag{4.131}
$$

4.4.3 Continuity Equation

The continuity equation is discretized in the control volume without shift.

$$w_{i+1/2,j+1/2}^{k+1,n+1} = w_{i+1/2,j+1/2}^{k,n+1} - \frac{u_{i+1,j+1/2}^{k+1/2,n+1}\Delta z_{i+1,j+1/2}^{k+1/2,n+1} - u_{i,j+1/2}^{k+1/2,n+1}\Delta z_{i,j+1/2}^{k+1/2,n}}{\Delta z} - \frac{v_{i+1/2,j+1}^{k+1/2,n+1}\Delta z_{i+1/2,j+1}^{k+1/2,n+1} - v_{i+1/2,j}^{k+1/2,n+1}\Delta z_{i+1/2,j}^{k+1/2,n+1}}{\Delta y} \tag{4.132}$$

4.4.4 Calculation Procedures

According to the treatment of the barotropic component of pressure terms, the calculation procedures for the implicit and explicit schemes are illustrated in the following.

Implicit scheme

1. The advective terms, diffusive terms, Coriolis terms, and the components of pressure terms, which are not related to the water stage ζ^{n+1}, are calculated from the hydraulic quantities of time $n\Delta t$.
2. The water stage ζ^{n+1} is obtained by Successive Over Relaxation (SOR) method from Eq. (4.122).
3. The velocity u^{n+1} is calculated in Eq. (4.119) and v^{n+1} is calculated in the same way.
4. The vertical velocity w^{n+1} is calculated in Eq. (4.132) by substituting the obtained velocities u^{n+1} and v^{n+1} into this equation. One of the boundary conditions w^{n+1} at the bottom is given to be zero.
5. The temperature T^{n+1} is calculated in Eq. (4.100).

Explicit scheme

1. Because the water stage is evaluated only by the hydraulic quantities of time $n\Delta t$, the pressure terms are known. Therefore, the velocity u^{n+1} can be calculated from Eq. (4.119) directly and v^{n+1} is calculated in the same way.
2. The vertical velocity w^{n+1} is calculated in Eq. (4.132) by using u^{n+1} and v^{n+1}.
3. The water stage ζ^{n+1} is obtained from the vertically integrated continuity equation.
4. The temperature T^{n+1} is obtained from Eq. (4.100).

4.4.5 Numerical Experiments with the IDOR 3D Model

Because the flows induced by wind on a lake depend strongly on the bathymetry of the basin and stratification, several enclosed water basins with different shapes are adopted and the flows in non-stratified and stratified cases are analyzed separately.

Non-stratified cases

In non-stratified lakes, the horizontal circulation and vertical circulation induced by wind are considered as the main large-scaled flows. In order to investigate these two flows, the numerical simulations are made in the following simplified water basins. To reproduce the horizontal circulation, the first one has the stair distribution of water depth in both the x-direction (longitudinal direction) and the y-direction (traverse direction), as shown in Figure 4.5, and is called basin A. The second one, basin B, has the stair distribution of water depth in the x-direction and uniform depth in the y-direction (shown in Fig. 4.6) for investigating the vertical circulation (Shen, 1991).

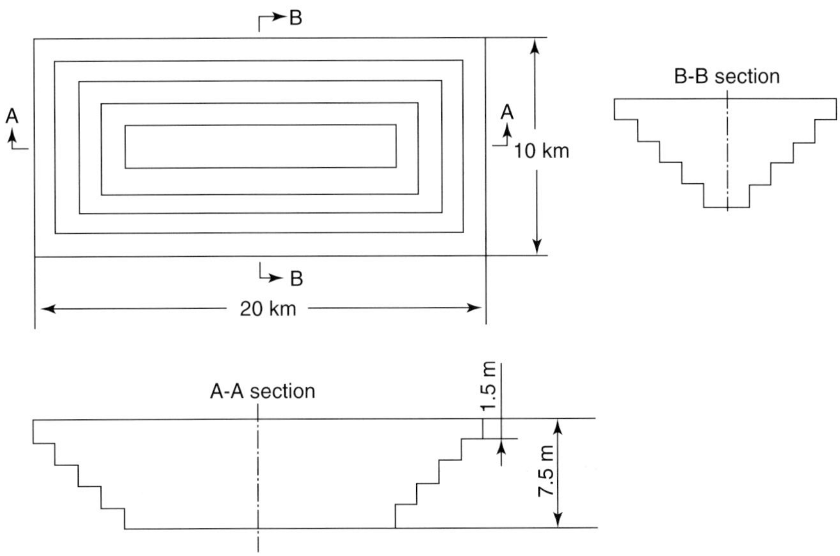

Fig. 4.5. Sketch of simplified water basin A (Shen, 1991).

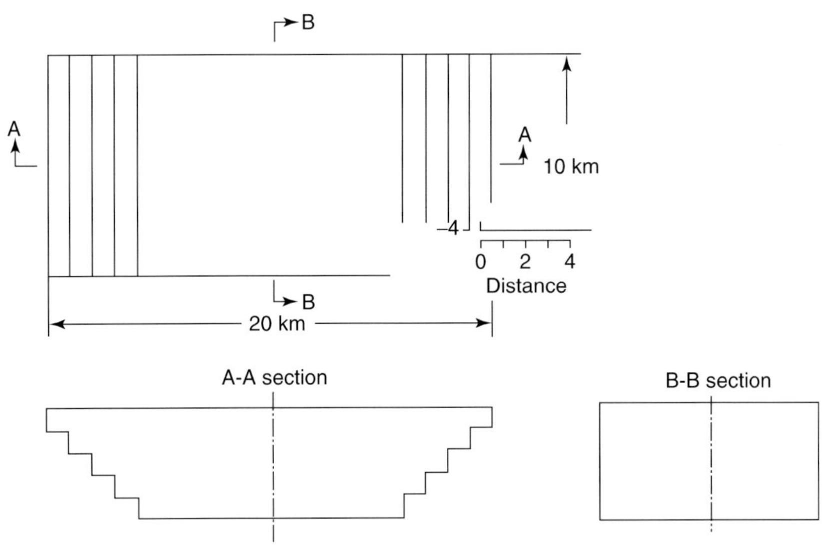

Fig. 4.6. Sketch of simplified water basin B (Shen, 1991).

Computational conditions. Water basins A and B are both divided by the grid net whose mesh intervals are 1 km in both the x- and y-directions and 1.5 m in the vertical direction. The parameters used in the computations are given as follows: the Coriolis parameter $f = 8.34 \times 10^{-5}$; the air density $\rho_a = 1.2\,\mathrm{kg/m^3}$; the water density at 4°C $\rho_0 = 1000\,\mathrm{kg/m^3}$; the drag coefficient at the water surface $C_D = 1.3 \times 10^{-3}$; the drag coefficient at the bottom $C_B = 2.6 \times 10^{-3}$; the parameters of the upstream schemes $\theta_h = \theta_v = 0.7$; and the eddy viscosity and diffusivity coefficients in the neutral state $k_h = 10\,\mathrm{m^2/s}$, $k_{v0} = 2 \times 10^{-3}\,\mathrm{m^2/s}$, $N_h = 30\,\mathrm{m^2/s}$, and $N_{v0} = 1 \times 10^{-4}\mathrm{m^2}$. A uniform wind with a velocity of 5 m/s is assumed to blow along the x-direction for 5 h from the still state. The time increment is chosen as 120 s and the computational period is 24 h. The calculations of the two cases are made by the operator-splitting method described earlier.

Simulation results and discussions. Figure 4.7 illustrates the velocity distributions in the surface layer of water basin A. At the onset of blowing wind, the current along the wind direction occurs in the shore areas, whereas no current of predominant direction could be observed in the central area (see Figure 4.7a). When the wind continues to blow, the current along the wind direction in the shore areas gradually develops and the current in the opposite direction appears in the central area, that is, horizontal circulation occurs (see Fig. 4.7b). The mechanism of this phenomena can be explained as follows. At the onset of the wind, the water surface is set up by the wind. In the shore areas, the wind shear stress is larger than the pressure gradient due to the set-up of the water surface and therefore the current along the wind direction is observed but as the water depth in the central area is large and the pressure gradient is very close to the wind shear stress, the current with predominant direction cannot be observed. If the wind continues to blow, the wind shear stress is still larger than the pressure gradient in the shore areas and therefore, the leeward current becomes gradually stronger, whereas the pressure gradient in the central area is larger than the wind stress, and thus the windward current occurs.

It is known from Figures 4.7c and 4.7d that the attenuation of velocity is faster in the shore areas after the wind calms. The reason for this is the wind shear stresses balance with the sum of the pressure gradient and bottom friction after the velocity becomes large (steady state). If the pressure gradient along the shore areas is small, the influence of bottom friction in the shore areas is large.

Figure 4.8 shows the velocity distributions in the central longitudinal section of water basin A. In this figure, it can be seen that the velocity in the opposite direction to the wind is small near the water surface while the wind is blowing, and the velocities of all layers become gradually uniform after the wind stops. However, the vertical circulation in the longitudinal section cannot be observed. It is because the wind shear stresses act on the water surface while the wind blows, the current in the opposite direction to the wind makes the surface layer small.

Figures 4.9 and 4.10 illustrate the velocity distributions in the surface layer and longitudinal section of the water basin B, respectively. It is known from these figures that horizontal circulation does not occur in water basin B and only vertical circulation appears. This is because the cross-section is a rectangle and the pressure

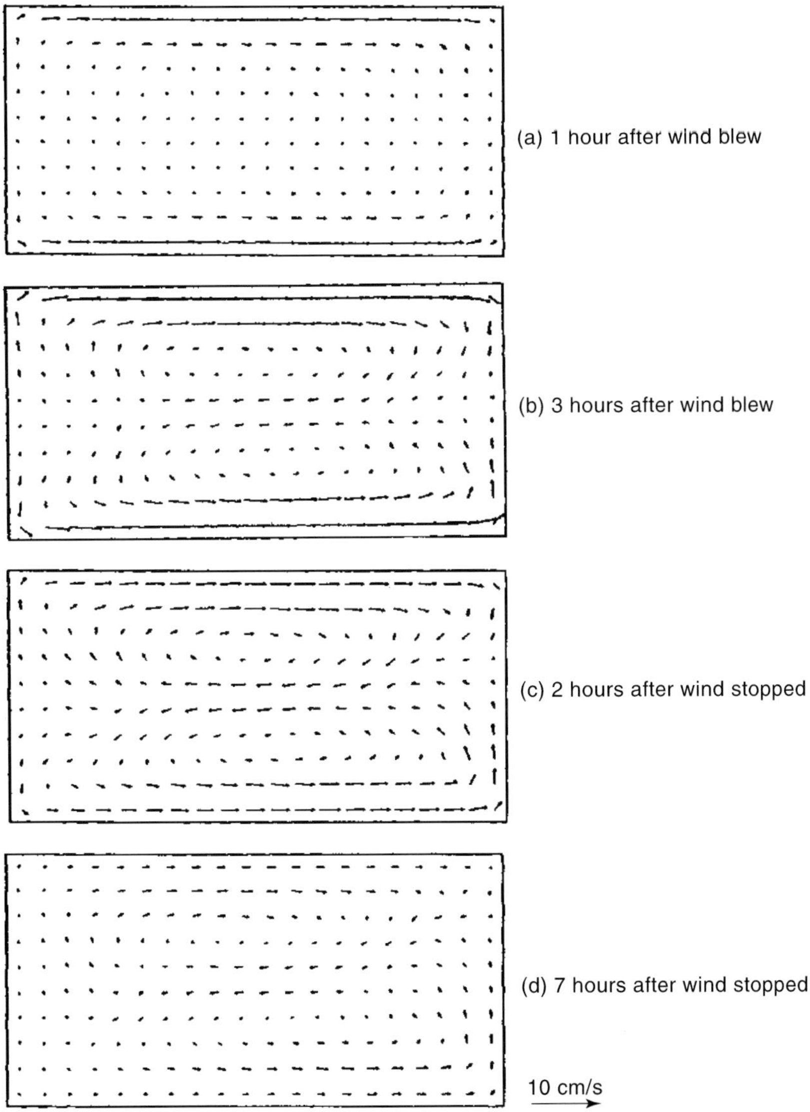

Fig. 4.7. Velocity distributions in the surface layer of water basin A (a) 1 h after wind blew; (b) 3 h after wind blew; (c) 2 h after wind stopped; (d) 7 h after wind stopped (Shen, 1991).

gradient is uniform in the same cross section; therefore, the amplitude and direction of the velocity in the horizontal plane are almost the same. Moreover, because the wind stress is larger than the pressure gradient in the upper layers, current along the wind direction occurs. Whereas the wind stress is small in the lower layers, the pressure gradient is larger than the wind stress, thus, the backward current appears. Therefore, on the basis of this discussion it can be concluded that large-scaled flows in lakes without stratification are strongly dependent on the shapes of the lakes.

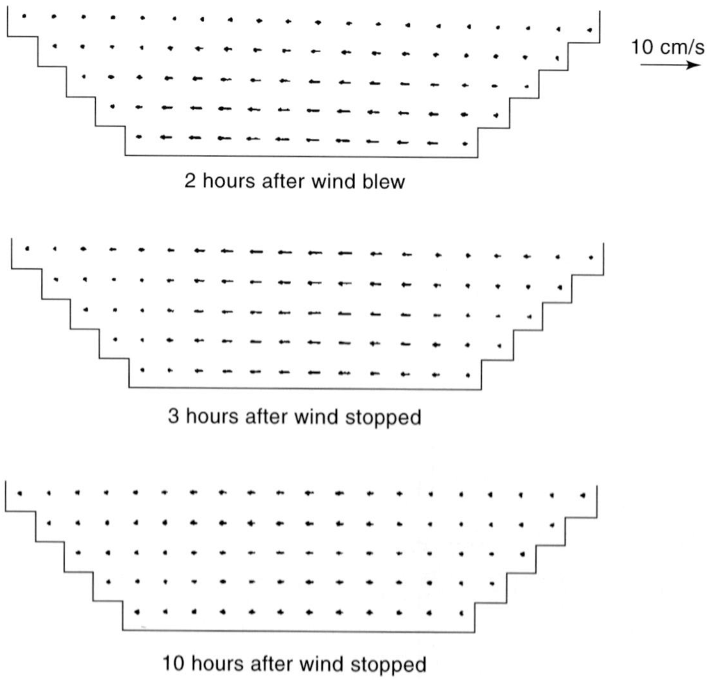

Fig. 4.8. Velocity distributions in the central longitudinal section of water basin A (Shen, 1991).

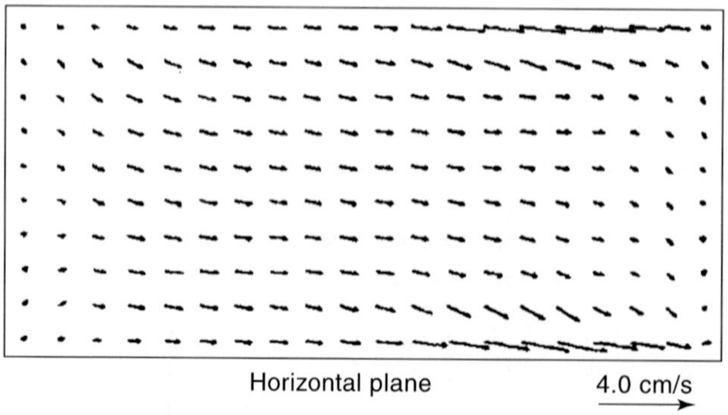

Fig. 4.9. Velocity distributions in the surface layer of water basin B (Shen, 1991).

Stratified cases

In stratified lakes, the periodic variations of temperature near the metalimnion, which are called internal waves, are caused by the variations of wind or air pressure. In internal waves, various modes with periods from several minutes to several days are included. The periods of internal waves are determined by the shape, scale, and metalimnion's position in the lake, and their amplitudes depend mainly on the wind shear stress and blowing time. In order to investigate internal waves, which are

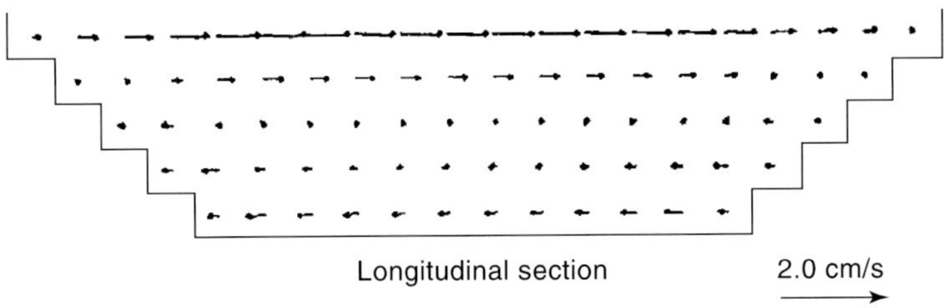

Fig. 4.10. Velocity distributions in the longitudinal section of water basin B (Shen, 1991).

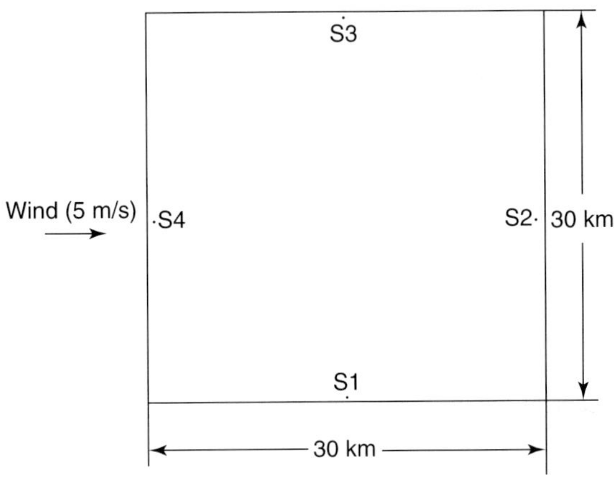

Fig. 4.11. Sketch of simplified water basin C (Shen, 1991).

presented by the temperature variations of the metalimnion, considerable water depth is required. Moreover, since the periods of internal waves possibly coincide with the period of the topographic wave and are difficult to separate, a water basin with a uniform distribution of water depth is to be adopted to remove the topographic wave. This water basin is assumed to be square and 40 m in water depth, as shown in Figure 4.11, and is later called water basin C.

Computational conditions. Water basin C is divided by a grid system whose mesh intervals are 1.5 km in both the x- and y-directions and 2 m in the vertical direction. It is assumed that uniform west wind with a velocity of 5 m/s blows for 12 h from the still state and a continuously varying temperature distribution (whose metalimnion is located at the water depth 15 m at the initial state) is given.

The time increment given is 60 s and the computational period is 120 h. The other parameters are the same as in water basins A and B and the operator-splitting method is adopted (Shen, 1991).

Simulation results and discussions. Figure 4.12 illustrates the longitudinal (x-direction) distribution of the water stage and the isotherm near the metalimnion. It is observed from this figure that the water surface is set up by the wind, and the water stage rises at the leeward side, whereas the metalimnion rises at the windward

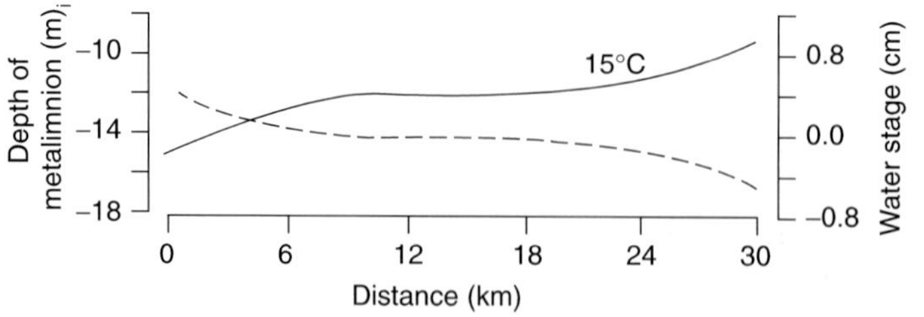

Fig. 4.12. Longitudinal distributions of the water stage and the isotherms near the metalimnion (dotted line: water stage, solid line: isothermal line) (Shen, 1991).

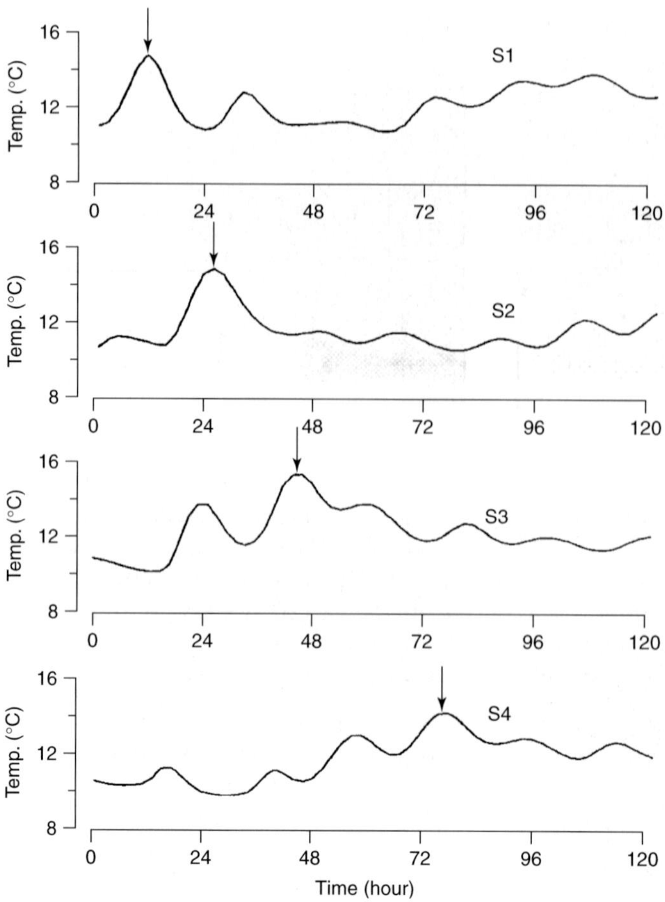

Fig. 4.13. Temporal variations of the temperature near the metalimnion (Shen, 1991).

side. Also in this figure, the amplitude of the metalimnion variation amounts to 3 m, whereas the oscillation amplitude of the water surface is less than 1 cm. Figure 4.13 shows the temporal variations of temperature near the metalimnion of points S1, S2, S3, and S4 which are located in the shore area (as shown in Fig. 4.11). In this figure, it is observed that a large rise of temperature happens due to wind stoppage at the right side of the wind blowing direction, and this water front of high temperature propagates counterclockwise with time around the water basin. Again, from the horizontal plane distributions of temperature near the metalimnion, shown in Figure 4.14, it is verified that the portion with high temperature (the light tone in the figure) rotates counterclockwise along the shore areas. The propagation period of the large internal wave in the shore areas due to the stopping of wind is estimated to be about 80 h from these figures.

According to the internal Kelvin wave theory, which uses the two-layer model by Csanady (1967, 1975), the internal Rossby radius R_2 of the water basin and the

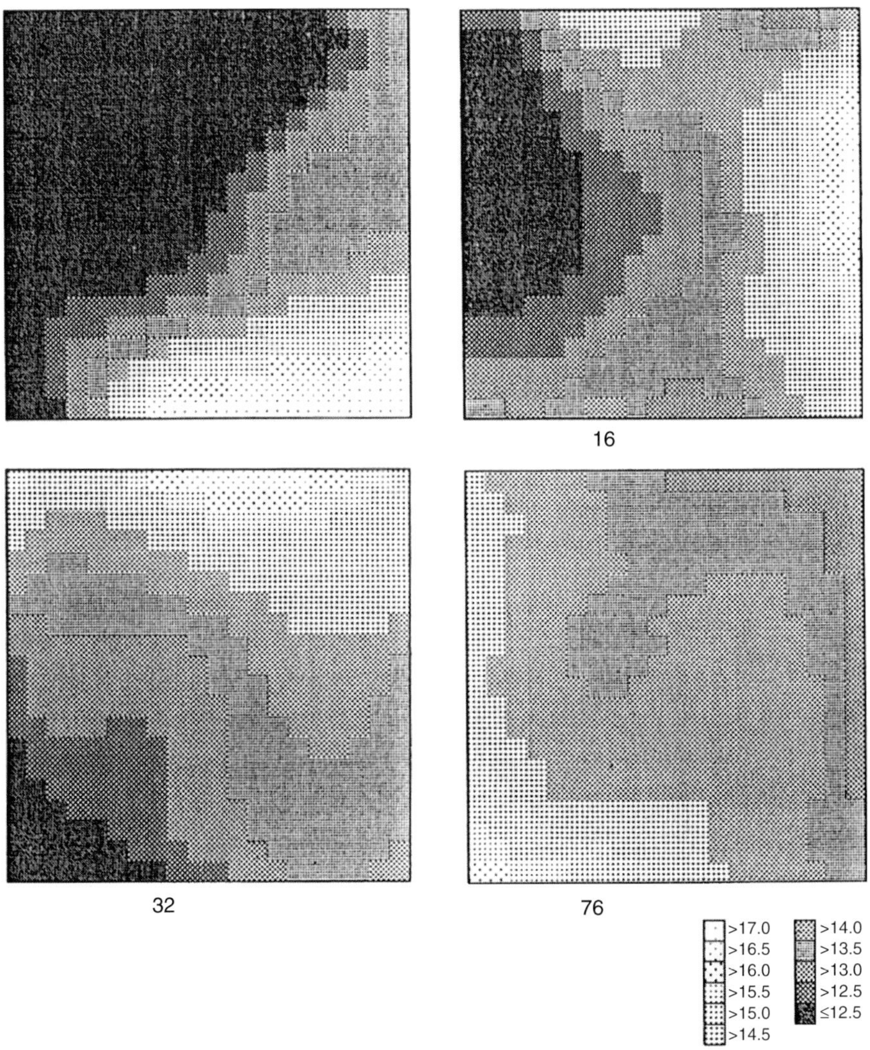

Fig. 4.14. Horizontal plane distributions of temperature near the metalimnion (Shen, 1991).

speed of internal Kelvin wave c are given by the following formulas:

$$R_2 = \frac{1}{f}\sqrt{\frac{\rho_2 - \rho_1}{\rho_2}\frac{gh_1h_2}{h_1 + h_2}} \tag{4.133}$$

$$c = \sqrt{\frac{\rho_2 - \rho_1}{\rho_2}\frac{gh_1h_2}{h_1 + h_2}} \tag{4.134}$$

where ρ_1 and ρ_2 are the densities of the upper layer and lower layer, respectively, and h_1 and h_2 are the water depths of the upper layer and lower layer, respectively. From the distribution of water temperature obtained by this calculation, if the values $\rho_1 = 998.234\,\text{kg/m}^3$ (corresponding to 20°C), $\rho_2 = 999.848\,\text{kg/m}^3$ (corresponding to 8°C), $h_1 = 1.5\,\text{m}$, and $h_2 = 25\,\text{m}$ are given, $R_2 = 4.61\,\text{km}$ and $c = 0.385\,\text{m/s}$ are calculated. Furthermore, the equivalent radius is

$$r_e = \sqrt{\frac{30 \times 30}{\pi}} = 16.93\ (\text{km}) \tag{4.135}$$

Because $R_2 << r_e$, the internal Kelvin wave may appear obvious, and its period is calculated as:

$$T = \frac{2\pi r_e}{c} = \frac{2\pi \times 16.93 \times 10^3}{0.385 \times 3600} = 76.8\ (\text{hours}) \tag{4.136}$$

Since the period of the internal Kelvin wave is almost the same as the period of the internal wave obtained in the aforementioned calculation, the internal wave along the shore areas observed in Figure 4.14 can be considered as the internal Kelvin wave.

Figure 4.15 shows the temporal variations of the velocities at the surface layer and metalimnion near the shore area. From this figure, it is known that the velocity of the surface layer in the shore area due to the internal Kelvin wave is large, while

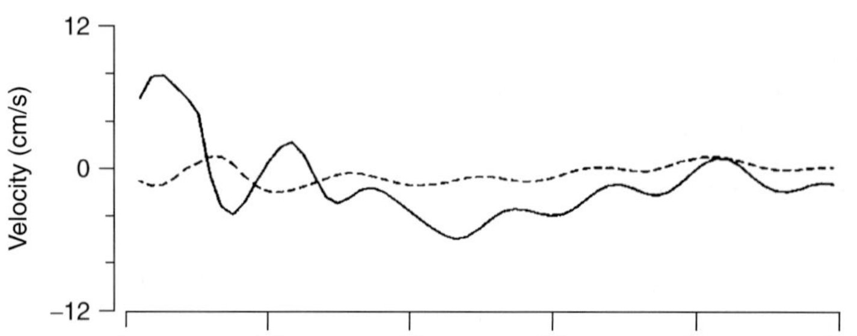

Fig. 4.15. Temporal variations of the velocity component u at the surface layer and metalimnion at point S1 (solid line: velocity at the surface layer, dotted line: velocity at the metalimnion (Shen, 1991)).

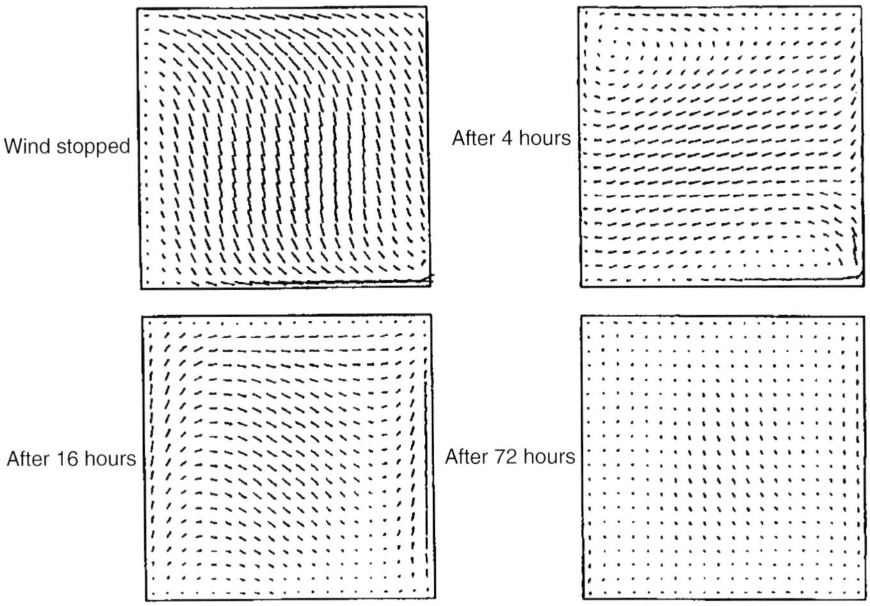

Fig. 4.16. Velocity distributions in the surface layer (Shen, 1991).

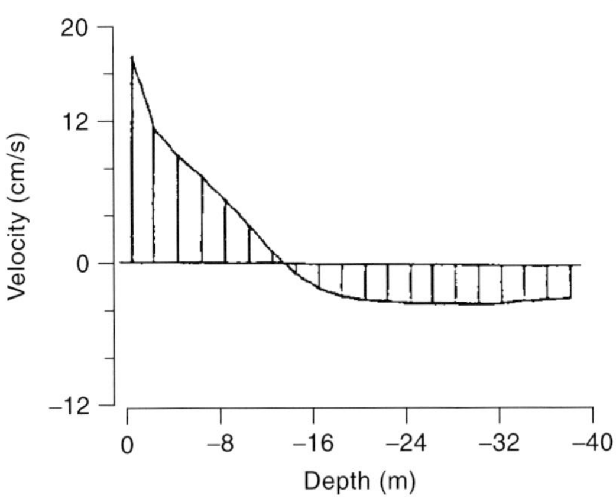

Fig. 4.17. Vertical distribution of the velocity along the shore at S1 (Shen, 1991).

the velocity below the metalimnion is small. This large velocity at the surface layer appears only in shore areas and is widely known as *coastal jet flow*. It can be observed in Figure 4.16 that the coastal jet flow along the shore rotates counterclockwise with the internal Kelvin wave.

Figure 4.17 is the vertical distribution of the velocity along the shore at the point S1. It can be seen from this figure that the velocity of the lower layers is in the opposite direction to that of the upper layers.

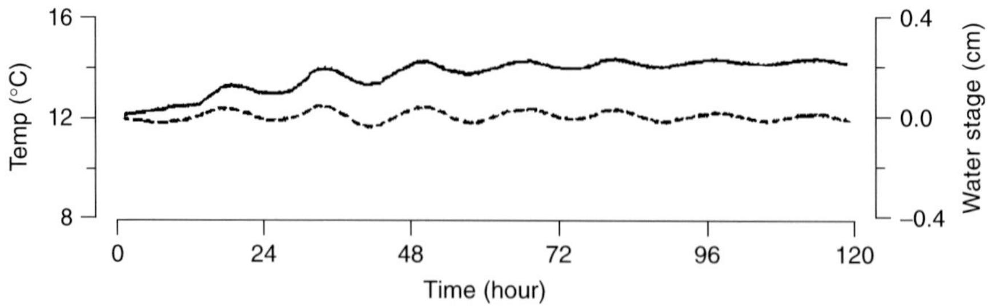

Fig. 4.18. Temporal variations of the water temperature near the metalimnion in the central area [dotted line: water stage, solid line: temperature (Shen, 1991)].

Figure 4.18 illustrates the temporal variation of the water temperature near the metalimnion in the central area of the water basin. It is known from this figure that the variation with a period of 17 h is included in the water temperature variation, which is a little shorter than the inertia period $T_p = 2\rho/f = 20.88$ h. According to the Poincaré wave theory using the two-layer model (Csanady, 1967), the frequency of the internal Poincaré wave, which is predominant in the central area, is given by

$$\sigma_n^2 = f^2 + \frac{n^2\pi^2c^2}{b^2} \quad (n = 1, 3, \ldots.) \tag{4.137}$$

where b is the width of the water basin and c is the same as the speed of the internal Kelvin wave. The basic mode is presented in the following formula by setting $n = 1$ in Eq. (4.137):

$$\sigma_1^2 = f^2 + \frac{\pi^2c^2}{b^2} \tag{4.138}$$

If $b = 30$ km and $c = 0.383$ m/s are substituted into the preceding formula,

$$\sigma_1 = 9.28 \times 10^{-1} \ \text{(l/sec)}$$

is calculated and the period of internal Poincaré wave's basic mode is,

$$T = \frac{2\pi}{\sigma_1} = 18.8 \ \text{(hours)}$$

Because this value is approximately in agreement with the period of water temperature variation observed in Figure 4.18, the variation of temperature in the figure can be considered to result from the internal Poincaré wave. Furthermore, Figure 4.16 shows that the flow direction in the central area is rotating clockwise. Therefore, it is concluded that this rotating flow, like the variation of water temperature, results from the internal Poincaré wave. Figure 4.19, which shows the

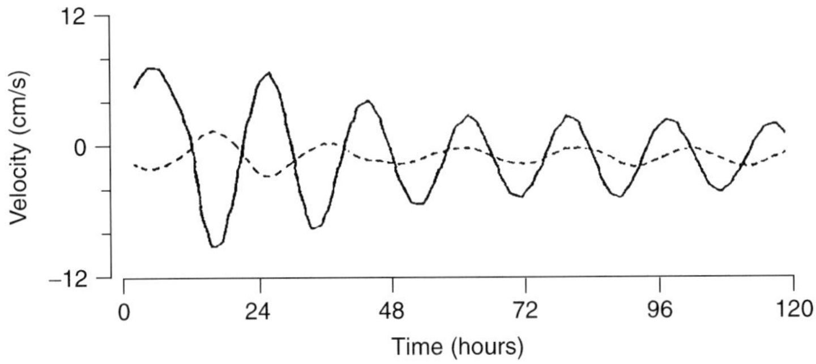

Fig. 4.19. Velocity component, u, at the surface layer and metalimnion in the control area (dotted line: velocity near the metalimnion; solid line: velocity at the surface layer (Shen, 1991)).

velocity distributions at the surface layer and metalimnion in the central area, also verifies this conclusion.

Figure 4.20 is the velocity distribution of the surface layer while the wind is blowing. Reviewing this figure, it is known that the flow direction at the surface layer in the central area is gradually rotating clockwise as the wind blows. If the wind continues to blow, the flow tends to arrive at a steady state and the direction of the flow is almost at right angles to the wind direction. The Coriolis force acting on the flow parallel to the wind and the pressure gradient perpendicular to the wind arrives at the geostrophic balance, and the pressure gradient perpendicular to the wind in the central area is nearly zero. Therefore, the current along the wind direction barely occurs. The Coriolis force acting on the flow perpendicular to the wind is balanced by the sum of the pressure gradient parallel to the wind and the wind stress, and since the pressure gradient in the central area is small, a large flow perpendicular to the wind occurs. As a result, it is concluded that the wind stresses are in balance with the Ekman transport in the central area. Moreover, this flow is attenuating gradually while rotating clockwise after the wind has stopped (see Fig. 4.16).

Figure 4.21 is the velocity distribution of the surface layer, in case there is no stratification and the other conditions are the same as the previous case. In this case, when the wind continues to blow, the flow in the central area will rotate approximately 45° right to the leeward direction as explained by the Ekman spiral and eventually arrive at a steady state. However, it never rotates over 45° as in the stratification case. Therefore, it is interesting that the characteristics of the Ekman spiral are obviously different depending on the existence of stratification. Furthermore, the velocity direction begins to rotate clockwise after the wind has stopped, just like in the case of stratification. Its rotating period is much longer, however, than the stratification case. It can be verified by Eq. (4.137) that $c = \sqrt{gH}$ is the speed of the barotropic wave. If $H = 40\,\text{m}$ is given, obviously,

$$\frac{n^2\pi^2c^2}{b^2} \gg f^2$$

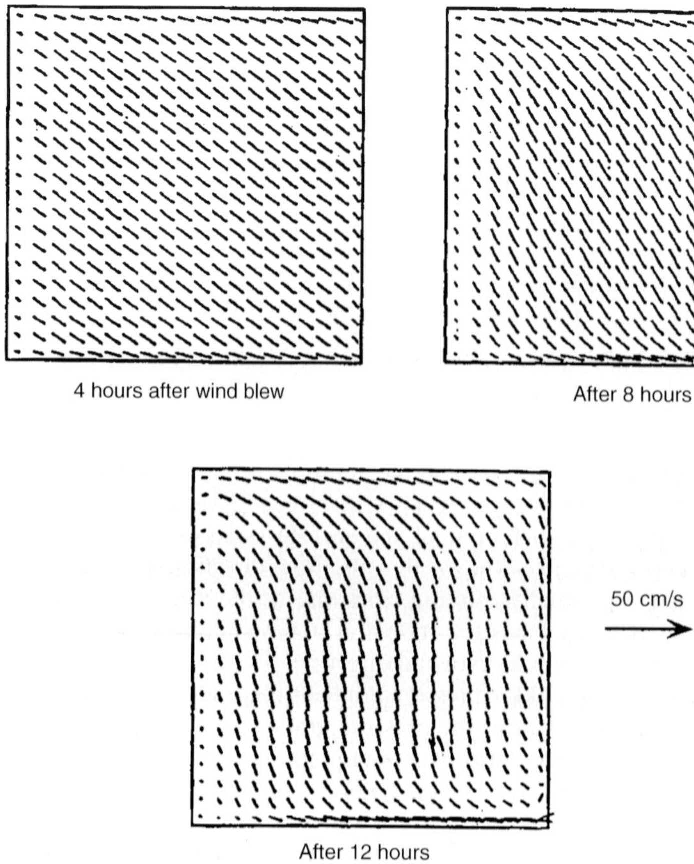

Fig. 4.20. Velocity distributions of the surface layer when the wind is blowing (Shen, 1991).

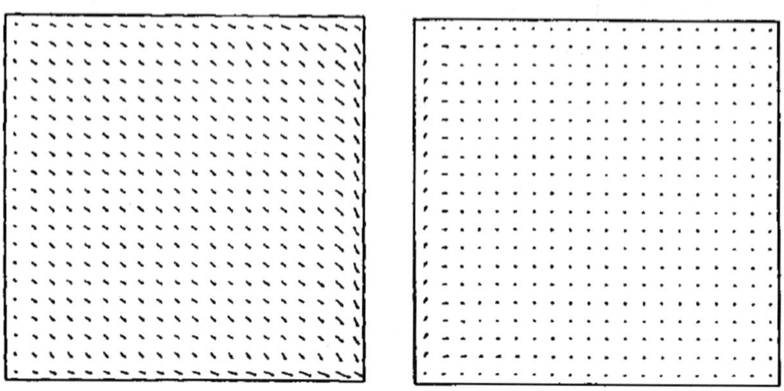

Fig. 4.21. Velocity distributions of the surface layer without stratification (Shen, 1991).

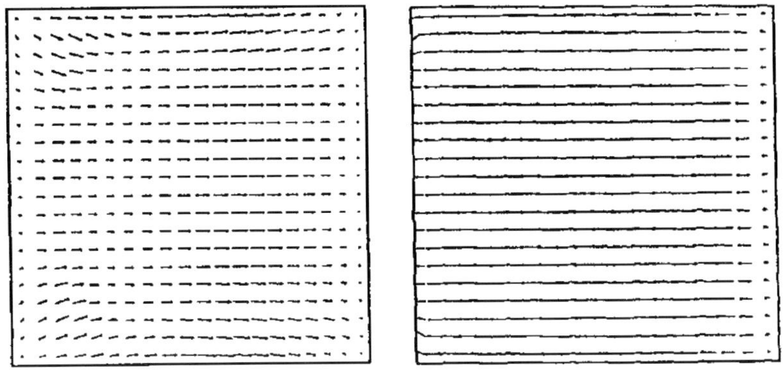

Fig. 4.22. Velocity distributions of the surface layer without the Coriolis force (Shen, 1991).

Therefore, the Coriolis force gives little influence to the period of seiche in the case of non-stratification. Moreover, by comparing Figures 4.16 and 4.21, it is known that the velocity of the surface layer in the case of stratification is larger than that in the case of non-stratification, and the velocity of attenuation in the former is later than that in the latter after the wind has stopped. This verifies that the internal wave is predominant in the case of stratification.

When the Coriolis factor, f, is assumed to be zero, in other words, the influence of the earth's rotation is not taken into account, the velocity distributions are illustrated in Figure 4.22. Obviously, the rotation of flow cannot be observed and the velocity direction turns over regularly due to seiche. The turning period is longer than that of the case considering the Coriolis force as shown in Eq. (4.137). Therefore, it is concluded from the earlier discussions that the internal Kelvin wave and Poincaré wave results from Coriolis force.

4.5 FULLY THREE-DIMENSIONAL MODEL

For the large-scale flows in most lakes and reservoirs, the vertical accelerations are usually much smaller than the horizontal ones because of the extreme discrepancy in scale between the vertical and horizontal directions. Therefore, the approximation of hydrostatic pressure is available for the numerical analysis of lake or reservoir flows. It is possible, however, that vertical acceleration has a value of the same order as horizontal ones locally. For instance, destratification of lake water in the winter, upwelling and downwelling of metalimnion in lakes, and aeration or the intake of water from deep layers in reservoirs are examples when such a situation arises. In these cases, the approximation of hydrostatic pressure loses its applicability and it becomes necessary to solve the momentum equation of vertical direction. The purpose of this section is to develop a fully 3D model, which does not use the approximation of hydrostatic pressure, and test a numerical simulation in a simplified water basin.

4.5.1 Approximation of Hydrostatic Pressure

In the previous models, the approximation of hydrostatic pressure was used. As mentioned earlier, this approximation is conditionally appropriate and its availability is investigated by the analysis of order magnitude as follows.

If the pressure p is divided into two parts, one of which is the hydrostatic pressure p_s and another is the dynamic pressure p_d ($p = p_s + p_d$), the momentum equation of vertical direction can be represented as:

$$\frac{\partial w}{\partial t} + \frac{\partial(uw)}{\partial y} + \frac{\partial(vw)}{\partial y} + \frac{\partial(ww)}{\partial z} = -\frac{1}{\rho}\frac{\partial p_d}{\partial z} + \frac{\partial}{\partial x}\left(K_x\frac{\partial w}{\partial x}\right) + \frac{\partial}{\partial y}\left(K_y\frac{\partial w}{\partial y}\right) + \frac{\partial}{\partial z}\left(K_z\frac{\partial w}{\partial z}\right) \tag{4.139}$$

Moreover, it is assumed that the representative scale of the flows in the horizontal plane is L, the representative scale in the depth-wise direction is H and the representative velocity in the horizontal plane is U. The next formula results from the continuity equation.

that is,

$$\Delta p_d \sim O\left(\frac{H^2}{L}\frac{dU}{dt}\right) \tag{4.140}$$

If $L \gg H$, $\Delta p_d \to 0$, which implies that the approximation of hydrostatic pressure is appropriate. Since the flows of different scales have different values of L and H, and the scales of the large-scaled flows in a lake are of the same order as the scale of the lake basin ($L \gg H$), the approximation of hydrostatic pressure is applicable to the large-scaled flows in lakes. On the other hand, for the smaller scale flows or local flows, the values of L are not necessarily much larger than H, therefore the availability of the hydrostatic approximation is questionable and it becomes necessary to solve the vertical momentum equation in these cases.

4.5.2 Procedure of Numerical Calculation

By using the Boussinesq approximation and assuming that density is a function of temperature only, the governing equations of the fully 3D model, which include the continuity equation, momentum equations of horizontal plane, temperature equation, and the momentum equation of vertical direction are obtained.

Substituting the formula $p = p_s + p_d$ into the momentum equations, the following equations yield:

$$\frac{u^{n+1} - u^*}{\Delta t} = -\frac{1}{\rho_0}\frac{\partial p_d}{\partial x} \tag{4.141}$$

$$\frac{v^{n+1} - v^*}{\Delta t} = -\frac{1}{\rho_0}\frac{\partial p_d}{\partial y} \tag{4.142}$$

$$\frac{w^{n+1} - w^*}{\Delta t} = -\frac{1}{\rho}\frac{\partial p_d}{\partial z} \tag{4.143}$$

where

$$\begin{aligned} u^* = u^n + \Bigg\{ & fu - \frac{1}{\rho_0}\frac{\partial p_s}{\partial x} - \frac{\partial(uu)}{\partial x} - \frac{\partial(vu)}{\partial y} - \frac{\partial(wu)}{\partial z} \\ & + \frac{\partial}{\partial x}\left(k_x \frac{\partial u}{\partial x}\right) + \frac{\partial}{\partial y}\left(k_y \frac{\partial u}{\partial y}\right) + \frac{\partial}{\partial z}\left(k_z \frac{\partial u}{\partial z}\right)\Bigg\}\Delta t \end{aligned} \tag{4.144}$$

$$\begin{aligned} v^* = v^n + \Bigg\{ & -fv - \frac{1}{\rho_0}\frac{\partial p_s}{\partial y} - \frac{\partial(uv)}{\partial x} - \frac{\partial(vv)}{\partial y} - \frac{\partial(wv)}{\partial z} \\ & + \frac{\partial}{\partial x}\left(K_x \frac{\partial v}{\partial x}\right) + \frac{\partial}{\partial y}\left(K_y \frac{\partial v}{\partial y}\right) + \frac{\partial}{\partial z}\left(K_z \frac{\partial v}{\partial z}\right)\Bigg\}\Delta t \end{aligned} \tag{4.145}$$

$$\begin{aligned} w^* = w^n + \Bigg\{ & -\frac{\partial(uw)}{\partial x} - \frac{\partial(vw)}{\partial y} - \frac{\partial(ww)}{\partial z} \\ & + \frac{\partial}{\partial x}\left(K_x \frac{\partial w}{\partial x}\right) + \frac{\partial}{\partial w}\left(K_y \frac{\partial w}{\partial y}\right) + \frac{\partial}{\partial z}\left(K_z \frac{\partial w}{\partial z}\right)\Bigg\}\Delta t \end{aligned} \tag{4.146}$$

By substituting $\tilde{p} = p_d/\rho_0$ into Eqs. (4.144)–(4.146) and discretizing them, the following difference equations are obtained:

$$\begin{aligned} & \frac{u_{i+1,j+1/2}^{k+1/2,n+1} - u_{i,j+1/2}^{k+1/2,n+1}}{\Delta x} + \frac{v_{i+1/2,j+1}^{k+1/2,n+1} - v_{i+1/2,j}^{k+1/2,n+1}}{\Delta y} \\ & + \frac{w_{i+1/2,j+1/2}^{k,n+1} - w_{i+1/2,j+1/2}^{k+1,n+1}}{\Delta z_{i+1/2,j+1/2}^{k+1/2,n}} = 0 \end{aligned} \tag{4.147}$$

$$u_{i+1,j+1/2}^{k+1/2,n+1} = u_{i+1,j+1/2}^{k+1/2,*} - \frac{\tilde{p}_{i+3/2,j+1/2}^{k+1/2,n+1} - \tilde{p}_{i+1/2,j+1/2}^{k+1/2,n+1}}{\Delta x}\Delta t \tag{4.148}$$

$$v_{i+1/2,j+1}^{k+1/2,n+1} = v_{i+1/2,j+1}^{k+1/2,*} - \frac{\tilde{p}_{i+1/2,j+3/2}^{k+1/2,n+1} - \tilde{p}_{i+1/2,j+1/2}^{k+1/2,n+1}}{\Delta y}\Delta t \tag{4.149}$$

$$w_{i+1/2,j+1/2}^{k,n+1} = w_{i+1/2,j+1/2}^{k,*} - \frac{\rho_0}{\rho}\frac{\tilde{p}_{i+1/2,j+1/2}^{k-1/2,n+1} - \tilde{p}_{i+1/2,j+1/2}^{k+1/2,n+1}}{\Delta z_{i+1/2,j+1/2}^{k,n}}\Delta t \tag{4.150}$$

Although substituting Eqs. (4.148), (4.149) and (4.150) into Eq. (4.147), Eq. (4.147) may not be satisfactory, if the residual is defined as:

$$R = \frac{u_{i+1,j+1/2}^{k+1/2,n+1} - u_{i,j+1/2}^{k+1/2,n+1}}{\Delta x} + \frac{v_{i+1/2,j+1}^{k+1/2,n+1} - v_{i+1/2,j}^{k+1/2,n+1}}{\Delta y} + \frac{w_{i+1/2,j+1/2}^{k,n+1} - w_{i+1/2,j+1/2}^{k+1,n+1}}{\Delta z_{i+1/2,j+1/2}^{k+1/2,n}} \tag{4.151}$$

R can be made to approach zero by adjusting the dynamic pressure $\tilde{p}$ for all cells. The problem is that if the pressure of one cell is adjusted, the pressures of others will change (Hirt and Cook, 1972). Therefore, it is necessary to carry out iterative calculations over all the cells in order to make all of them satisfy Eq. (4.147). The following Poisson equation can be obtained by substituting Eqs. (4.148)–(4.150) into Eq. (4.147):

$$\begin{aligned}
&\frac{u_{i+1,j+1/2}^{k+1/2,*} - u_{i,j+1/2}^{k+1/2,*}}{\Delta x} + \frac{v_{i+1/2,j+1}^{k+1/2*} - v_{i+1/2,j}^{k+1/2,*}}{\Delta y} + \frac{w_{i+1/2,j+1/2}^{k,*} - w_{i+1/2,j+1/2}^{k+1,*}}{\Delta z_{i+1/2,j+1/2}^{k+1/2,n}} \\
&= \frac{\Delta t}{(\Delta x)^2}\left(\tilde{p}_{i+3/2,j+1/2}^{k+1/2,n+1} - 2\tilde{p}_{i+1/2,j+1/2}^{k+1/2,n+1} + \tilde{p}_{i+1/2,j+1/2}^{k+1/2,n+1}\right) \\
&+ \frac{\Delta t}{(\Delta y)^2}\left(\tilde{p}_{i+3/2,j+1/2}^{k+1/2,n+1} - 2\tilde{p}_{i+1/2,j+1/2}^{k+1/2,n+1} + \tilde{p}_{i+1/2,j-1/2}^{k+1/2,n+1}\right) \\
&+ \frac{\rho_0}{\rho}\frac{\Delta t}{\Delta z_{i+1/2,j+1/2}^{k+1/2,n}}\left(\frac{\tilde{p}_{i+1/2,j+1/2}^{k-1/2,n+1} - \tilde{p}_{i+1/2,j+1/2}^{k+1/2,n+1}}{\Delta z_{i+1/2,j+1/2}^{k,n}}\right) \\
&- \frac{\rho_0}{\rho}\frac{\Delta t}{\Delta z_{i+1/2,j+1/2}^{k+1/2,n}}\left(\frac{\tilde{p}_{i+1/2,j+1/2}^{k+1/2,n+1} - \tilde{p}_{i+1/2,j+1/2}^{k+3/2,n+1}}{\Delta z_{i+1/2,j+1/2}^{k+1,n}}\right)
\end{aligned} \tag{4.152}$$

The preceding equation can be represented as:

$$\begin{aligned}
&A_{k+1/2}\tilde{p}_{i+3/2,j+1/2}^{k+1/2,n+1} + A_{k+1/2}\tilde{p}_{i-1/2,j+1/2}^{k+1/2,n+1} + B_{k+1/2}\tilde{p}_{i+1/2,j+3/2}^{k+1/2,n+1} \\
&+ B_{k+1/2}\tilde{p}_{i+1/2,j-1/2}^{k+1/2,n+1} + C_{k+1/2}\tilde{p}_{i+1/2,j+1/2}^{k-1/2,n+1} \\
&+ D_{k+1/2}\tilde{p}_{i+1/2,j+1/2}^{k+3/2,n+1} - \tilde{p}_{i+1/2,j+1/2}^{k+1/2,n+1} - E_{k+1/2} = 0
\end{aligned} \tag{4.153}$$

where

$$\begin{aligned}
A_{k+1/2} &= \frac{\Delta t}{(\Delta x)^2}/F_{k+1/2} \\
B_{k+1/2} &= \frac{\Delta t}{(\Delta y)^2}/F_{k+1/2} \\
C_{k+1/2} &= \frac{\rho_0}{\rho}\frac{\Delta t}{\Delta z_{i+1/2,j+1/2}^{k+1/2,n}\Delta z_{i+1/2,j+1/2}^{k+1,n}}/F_{k+1/2}
\end{aligned} \tag{4.154a}$$

$$D_{k+1/2} = \frac{\rho_0}{\rho} \frac{\Delta t}{\Delta z^{k+1/2,n}_{i+1/2,j+1/2} \Delta z^{k+1,n}_{i+1/2,j+1/2}} / F_{k+1/2}$$

$$E_{k+1/2} = \left\{ \frac{u^{k+1/2,*}_{i+1,j+1/2} - u^{k+1/2,*}_{i,j+1/2}}{\Delta z^{k+1/2,n}_{i+1/2,j+1/2}} + \frac{v^{k+1/2,*}_{i+1/2,j+1} - v^{k+1/2,*}_{i+1/2,j}}{\Delta y} \right.$$
$$\left. + \frac{w^{k,*}_{i+1/2,j+1/2} - w^{k+1,*}_{i+1/2,j+1/2}}{\Delta z^{k+1/2,n}_{i+1/2,j+1/2}} \right\} / F_{k+1/2}$$

$$F_{k+1/2} = \left(1 + \frac{2\Delta t}{(\Delta x)^2} + \frac{2\Delta t}{(\Delta y)^2} \right.$$
$$\left. + \frac{\rho_0}{\rho} \frac{2\Delta t}{\Delta z^{k,n}_{i+1/2,j+1/2} \Delta z^{k+1,n}_{i+1/2,j+1/2}} \right) \tag{4.154b}$$

This Poisson equation is solved by the SOR method. In solving the equation, the gradient of dynamic pressure p_d is given as zero at the bottom, shore and water surface of the water basin. After $\tilde{p}^{k+1/2,n+1}_{i+1/2,j+1/2}$ is calculated, velocities u^{n+1}, v^{n+1} and w^{n+1} can be obtained from Eqs. (4.148), (4.149), and (4.150). If an explicit scheme about water stage ζ is used, ζ is calculated from the vertically integrated continuity equation. In order to carry out this calculation, it is necessary to solve u^*, v^*, and w^* in advance. As in the plane multi-layer model, the Adams–Bashforth scheme is adopted for the temporal difference, the Donor-cell scheme for the advective terms, the central difference scheme for the diffusive terms, and the explicit and implicit schemes for the hydrostatic pressure (concerning water state ζ). The discretizations of the continuity equation, momentum equations of horizontal plane, and the temperature equation, will not be described here, and are referred to in the plane multi-layer model. Only the discretization of the vertical momentum equation is illustrated here.

When discretizing the vertical momentum equation, the control volume is shifted by $\Delta z/2$ along the z-direction, and w^* is calculated in the following equation:

$$w^{k,*}_{i+1/2,j+1} = w^{k,n}_{i+1/2,j+1} + (-w_{adv} + w_{diff}) \Delta t / \Delta V^{k,n}_{i+1/2,j+1/2} \tag{4.155}$$

Define the following formula:

$$(u, w_m, \theta)^{m=i+1}_{m=i} = uw_{i+1/2} + \frac{\theta}{2} |u| (w_i - w_{i+1}) \tag{4.156}$$

The advective terms are then discretized into the following equation by the use of the Donor-cell scheme:

$$w_{\text{adv}}(i + 1/2, j + 1/2, k)$$
$$= -\left(u^{k+1/2,n}_{i+1,j+1/2}, w^{k+1/2,n}_{m,j+1/2}, \theta_h \right)^{m=i+3/2}_{m=i+1/2} \Delta y \Delta z^{k+1/2,n}_{i+1,j+1/2} / 2$$

$$
\begin{aligned}
&- \left(u^{k-1/2,n}_{i+1,j+1/2}, w^{k-1/2,n}_{m,j+1/2}, \theta_h\right)^{m=i+3/2}_{m=i+1/2} \Delta y \Delta z^{k-1/2,n}_{i+1,j+1/2}/2 \\
&+ \left(u^{k+1/2,n}_{i,j+1/2}, w^{k+1/2,n}_{m,j+1/2}, \theta_h\right)^{m=i+3/2}_{m=i+3/2} \Delta y \Delta z^{k+1/2,n}_{i,j+1/2}/2 \\
&+ \left(u^{k-1/2,n}_{i,j+1/2}, w^{k-1/2,n}_{m,j+1/2}, \theta_h\right)^{m=i+1/2}_{m=i-1/2} \Delta y \Delta z^{k-1/2,n}_{i,j+1/2}/2 \\
&- \left(v^{k+1/2,n}_{i+1/2,j+1}, w^{k+1/2,n}_{i+1/2,m}, \theta_h\right)^{m=j+3/2}_{m=j+1/2} \Delta x \Delta z^{k+1/2,n}_{i+1/2,j+1}/2 \\
&- \left(v^{k-1/2,n}_{i+1/2,j+1}, w^{k+1/2,n}_{i+1/2,m}, \theta_h\right)^{m=j+3/2}_{m=j+1/2} \Delta x \Delta z^{k-1/2,n}_{i+1/2,j+1}/2 \\
&+ \left(v^{k+1/2,n}_{i+1/2,j}, w^{k+1/2,n}_{i_1/2,m}, \theta_h\right)^{m=j+1/2}_{m=j-1/2} \Delta x \Delta z^{k+1/2,n}_{i+1/2,j}/2 \\
&+ \left(v^{k+1/2,n}_{i+1/2,j}, w^{k-1/2,n}_{i+1/2,m}, \theta_h\right)^{m=j+1/2}_{m=j-1/2} \Delta x \Delta z^{k-1/2,n}_{i+1/2,j}/2 \\
&- \left(w^{k-1/2,n}_{i+1/2,j+1/2}, w^{m,n}_{i+1/2,j+1/2}, \theta_v\right)^{m=k-3/2}_{m=k+1/2} \Delta x \Delta y \\
&+ \left(w^{k+1/2,n}_{i+1/2,j+1/2}, w^{m,n}_{i+1/2,j+1/2}, \theta_v\right)^{m=k+1/2}_{m=k+3/2} \Delta x \Delta y
\end{aligned} \tag{4.157}
$$

The diffusive terms are discretized in the following difference equation:

$$
\begin{aligned}
&w_{\text{diff}}(i+1/2, j+1/2, k) \\
&= K_h\left[\frac{1}{\Delta x}\left\{\left(w^{k-1/2,n}_{i+3/2,j+1/2} - w^{k-1/2,n}_{i+1/2,j+1/2}\right)\Delta y \Delta z^{k-1/2,n}_{i+1,j+1/2}/2\right.\right. \\
&+ \left(w^{k+1/2,n}_{i+3/2,j+1/2} - w^{k+1/2,n}_{i+1/2,j+1/2}\right)\Delta y \Delta z^{k+1/2,n}_{i+1,j+1/2}/2 \\
&- \left(w^{k-1/2,n}_{i+1/2,j+1/2} - w^{k-1/2,n}_{i+1/2,j+1/2}\right)\Delta y \Delta z^{k-1/2,n}_{i,j_1/2}/2 \\
&\left. - \left(w^{k+1/2,n}_{i+1/2,j+1/2} - w^{k+1/2,n}_{i-1/2,j+1/2}\right)\Delta y \Delta z^{k+1/2,n}_{i,j+1/2}/2\right\}
\end{aligned} \tag{4.158a}
$$

$$
\begin{aligned}
&+ \frac{1}{\Delta y}\left\{\left(w^{k-1/2,n}_{i+1/2,j+3/2} - w^{k-1/2,n}_{i+1/2,j+1/2}\right)\Delta x \Delta z^{k-1/2,n}_{i+1/2,j+1}/2\right. \\
&+ \left(w^{k+1/2,n}_{i+1/2,j+3/2} - w^{k+1,n}_{i+1/2,j+1/2}\right)\Delta x \Delta z^{k+1/2,n}_{i+1/2,j+1}/2 \\
&- \left(w^{k-1/2,n}_{i+1/2,j+1/2} - w^{k-1/2,n}_{i+1/2,j-1/2}\right)\Delta x \Delta z^{k-1/2,n}_{i+1/2j}/2 \\
&\left.\left. - \left(w^{i+1/2,j-1/2}_{i+1/2,j+1/2} - w^{k+1/2,n}_{i_1/2,j-1/2}\right)\Delta x \Delta z^{k+1/2,n}_{i+1/2,j}/2\right\}\right] \\
&+ \frac{K_v}{\Delta z^{k-1/2,n}_{i+1/2,j+1/2}}\left(w^{k-1,n}_{i+1/2,j+1/2} - w^{k,n}_{i+1/2,j+1/2}\right)\Delta x \Delta y \\
&- \frac{K_v}{\Delta z^{k+1/2,n}_{i+1/2,j+1/2} w^{k+1,n}_{i+1/2,j+1/2}} \Delta x \Delta y
\end{aligned} \tag{4.158b}
$$

Obviously, w in the surface cells cannot be obtained from the earlier formulas. They are calculated in the continuity equation,

$$
\begin{aligned}
w_{i+1/2,j+1/2}^{s+1,n+1} = w_{i+1/2,j+1/2}^{s,n+1} &- \frac{u_{i+1,j+1/2}^{s+1/2,n+1}\Delta z_{i+1,j+1/2}^{s+1/2,n} - u_{i,j+1/2}^{s+1/2,n+1}\Delta z_{i,j+1/2}^{s+1/2,n}}{\Delta x} \\
&- \frac{v_{i+1/2,j+1}^{s+1/2,n+1}\Delta z_{i+1/2,j+1}^{s+1/2,n+1} - v_{i+1/2,j}^{s+1/2,n+1}\Delta z_{i+1/2,j}^{s+1/2,n}}{\Delta y}
\end{aligned}
\tag{4.159}
$$

where s denotes variables of the surface cells.

Therefore, the procedure for fully 3D calculation can be summarized as follows (Shen et al., 1990b):

1. u^* and v^* are calculated by the same procedure as in the plane multi-layer model.
2. The vertical velocity w^* is obtained from Eq. (4.155).
3. The Poisson equation concerning the dynamic pressure, Eq. (4.152), is solved by the SOR method. The velocities u^{n+1}, v^{n+1}, and w^{n+1} are obtained by substituting the calculated value of $\tilde{p}$ into Eqs. (4.148)–(4.150) and w^{n+1} of surface cells is calculated in Eq. (4.159).
4. In the explicit scheme case concerning the barotropic component, the water stage ζ is calculated from the vertically integrated continuity equation. The procedure is slightly modified for the implicit scheme case.
5. The temperature is calculated in the equation of temperature.

4.5.3 Numerical Example

As described in the last subsection, the fully 3D model needs to not only solve the momentum equation of vertical direction, but also carry out iterative calculations about the cell, in order to make all cells satisfy the continuity equation. Therefore, longer computational time and larger computer memory capacity are required in comparison to the plane multi-layer model. For this reason, it is neither economical nor necessary to apply the fully 3D model to some lakes. Usually, the model is only used in the subdomain of a lake, where the water movements in the vertical direction are predominant and the hydrostatic approximation is not available. Naturally, there is a problem specifying the open boundary conditions if only a portion of a lake is treated. Since the objective of this section is not to discuss the treatment of open boundary conditions, but to develop a fully 3D model, a simple, small water basin is assumed and the whole water basin is adopted as the computational domain.

For this example, an enclosed water basin with a vertical non-uniform temperature distribution is chosen. By drawing the water from deep layers through a pipe, by pumps, the warm water from the surface layer is pulled down to the lower layers and a vertical circulation is created in order to improve the water quality of the basin (see Fig. 4.23). A square water basin with uniform water depth is assumed, and its side length and water depth are given as 20 and 10 m, respectively. The pipe is assumed to be a square one. A square pipe may be rare, in which case, a cylindrical coordinate system is recommended. Another assumption is that the

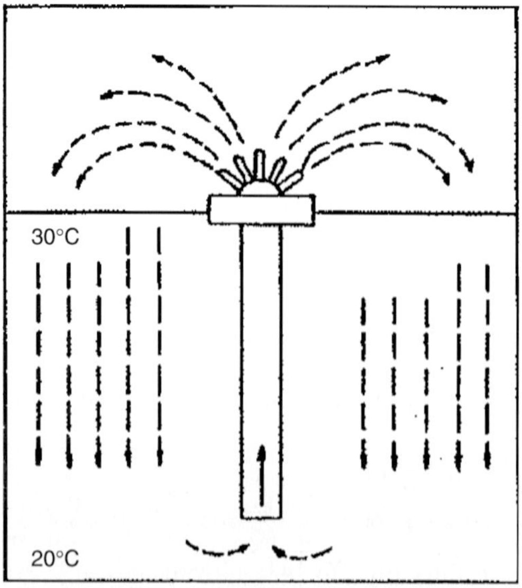

Fig. 4.23. Mechanism sketch (Shen, 1991).

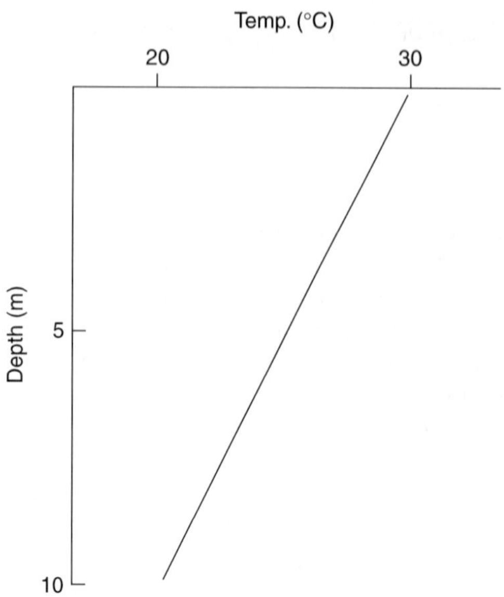

Fig. 4.24. Initial distribution of the temperature (Shen, 1991).

water drawn up by the pipe is not brought back into the basin. This is to take into account the withdrawal of water which remains a difficult problem.

The computational domain is divided by the grid system, with the mesh intervals $\Delta x = \Delta y = \Delta z = 1$ m so that 400 columns and 4000 cells are generated. The initial condition, shown in Figure 4.24 is a still state (the initial pressure distribution is the

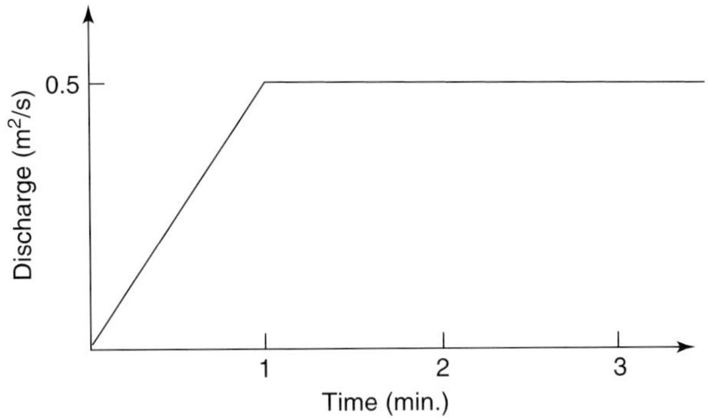

Fig. 4.25. Drawing discharge through the pipe (Shen, 1991).

hydrostatic pressure distribution) with a linear stratification. The drawing discharge of the pipe is shown in Figure 4.25. The time increment is chosen as 0.05 s for the explicit scheme and 0.5 s for the implicit scheme, and the computational period is 3 min. The turbulence closure conditions are specified in the same way as in Chapter 3 and the eddy viscosity coefficients and eddy diffusivity coefficients of three directions are given as $K_x = K_y = 0.05\,m^2/s$, $K_{z0} = 0.01\,m^2/s$, $N_x = Ny = 0.03\,m^2/s$, $N_{z0} = 0.001\,m/s$, in consideration of the mesh sizes. Because the computational domain is small, the Coriolis terms can be ignored. The other parameters used are mentioned in the last section.

Figure 4.26 illustrates the velocity vector distributions of the A–A vertical section. This figure shows that the velocity of water movement gradually becomes larger with time and finally arrives at a steady state. Moreover, the water of the upper layers moves to the lower layers, due to withdrawal, and a vertical circulation occurs. Because the drawing discharge is very small compared to the water volume of the whole water basin, only the velocity near the mouth of the drawing pipe is large, and a large water circulation covering the whole water basin does not appear. If the discharge of drawing increases, a larger vertical circulation is expected to occur. The problem is that because the water withdrawn through the pipe is not moved back into the water basin, a large drawing discharge will result in rapid drop of the water surface and, therefore, a computation for long periods cannot be carried out. Figure 4.27 shows the temperature distribution of the A–A cross section. In this figure, the temperature near the drawing mouth is lower than that of the surroundings, which means that the cooler water in the deep layer is withdrawn by the pipe.

Figure 4.28 expresses the temporal variations of u and w at the point near the pipe mouth. In this figure, u and w increase gradually with the increase in drawing discharge (Fig. 4.25) and reach steady state at some time t_1. However, the time t_1 to reach steady state is a little later than that of the drawing discharge. Although the water stage is gradually falling down because the water is being withdrawn, the water volume of the basin is large, and therefore, the falling of the stage is very small during the computational period. That is to say, the falling of the water stage will greatly influence the flows in the water basin, and thus the flows can reach a steady state if the computational period is not too long.

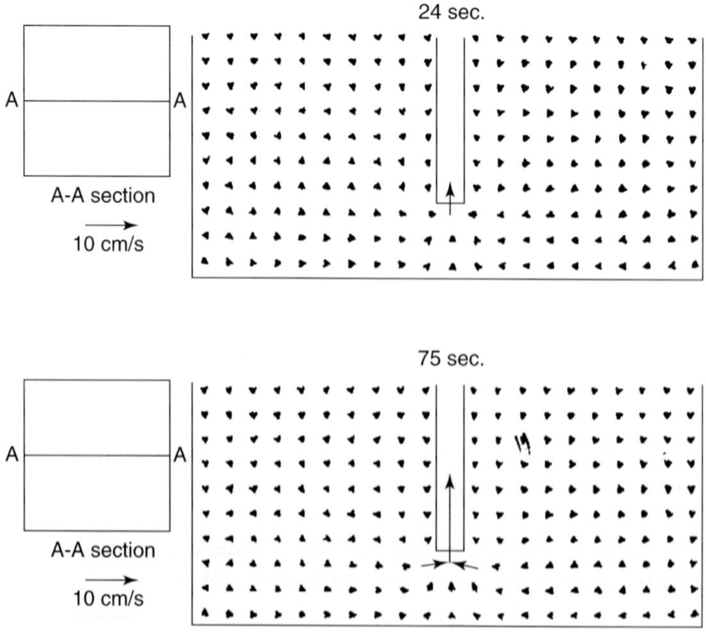

Fig. 4.26. Velocity distributions in A–A section (Shen, 1991).

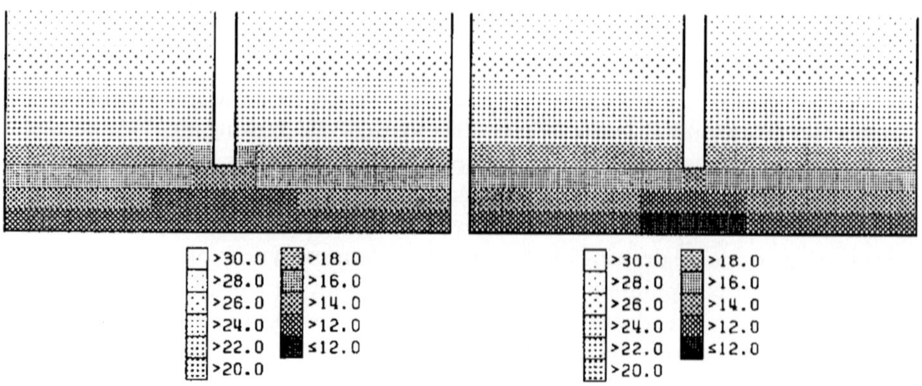

Fig. 4.27. Temperature distribution in A–A section (Shen, 1991).

Figure 4.29 shows the vertical distributions of the dynamic pressure p_d near the drawing pipe at different times. This figure shows that the dynamic pressure near the drawing mouth gradually decreases with time. Because the pressures near the drawing mouth are lower than those of the surroundings, the surrounding water is absorbed into the pipe. Figure 4.30 shows the contour lines of dynamic pressure p_d in the A–A section, in which the decrease in pressure near the drawing mouth can be verified. In the approximation of hydrostatic pressure, the pressure falling near the

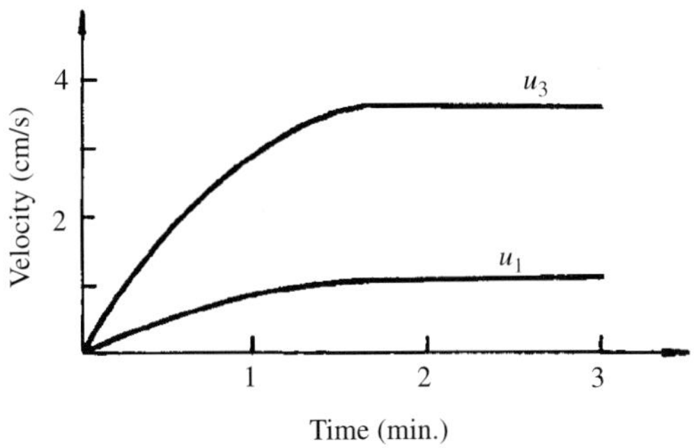

Fig. 4.28. Temporal variations of the velocities near the pipe mouth (Shen, 1991).

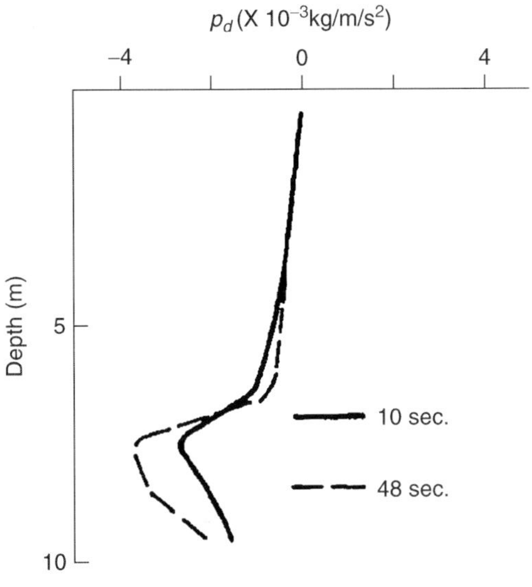

Fig. 4.29. Vertical distributions of the p_d near the pipe (Shen, 1991).

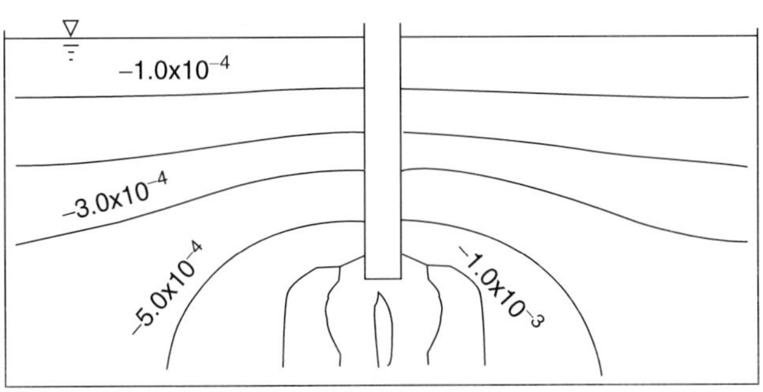

Fig. 4.30. Contour lines of the p_d in A–A section (Shen, 1991).

drawing mouth cannot be calculated, and hence the water being concentrated near the drawing mouth cannot be reproduced. On the other hand, in view of the numerical calculation, the vertical velocities are calculated from the continuity equation by assuming the vertical velocity at the bottom to be zero. Therefore, the velocity at the drawing mouth becomes an excess boundary condition and the continuity equation in the cell close to the drawing mouth is not satisfied by the plane multi-layer model. Therefore, it is almost impossible to treat this kind of flow by using the plane multi-layer model which includes the approximation of hydrostatic pressure.

In the fully 3D model, the treatment of the barotropic component is the most important factor in the stability of calculation, as in the plane multi-layer model. Even if an implicit scheme is adopted for the barotropic component, the time increment Δt is limited by the stability condition $w\Delta t/\Delta z \leq 1$ due to the vertical advective term. Furthermore, because the vertical momentum equation has to be solved in the fully 3D model and the iterative calculation must be made on all the cells instead of the columns, the CPU time of the fully 3D model is much larger than that of the plane multi-layer model.

4.6 LARGE EDDY SIMULATION MODEL

4.6.1 An Application of the LES Model

In this section, Lake Biwa is employed as the computational domain, with the aim of evaluating the applicability of the Large Eddy Simulation Model (LES) model to an actual lake (Iwasa and Inoue, 1989). The whole lake is divided into mesh intervals $\Delta x = \Delta y = 1$ km in the horizontal plane and $\Delta z = 2.5$ m at depths shallower than 50 m and $\Delta z = 10$ m at depths deeper than 50 m in the vertical direction. For simplicity, no inflow or outflow rivers are taken into account and the heat exchanges at the water surface are also ignored. At the initial condition, the lake is assumed to be in a still state, and a metalimnion with a temperature difference of 7°C is located at a depth of 15 m. A constant southwestern wind with the velocity of 5 m/s is assumed to blow for 12 h. The results obtained by the numerical simulation are those under a calm condition after the wind has stopped. Reference point B and section A–A in Lake Biwa are illustrated in Figure 4.31.

4.6.2 Comparison of LES and Non-LES Analysis

Figure 4.32 illustrates the velocity distributions in the surface layer at different times calculated from LES and non-LES analysis, respectively. From these figures, it can be seen that there is almost no difference between the velocities obtained from the two analyses. The LES analysis gives slightly faster flows and the difference in velocity gradually becomes large with passing time.

This small difference can be explained as follows: When the wind is blowing, the wind shear stresses act on the water surface and the mean flow component induced by the wind stress is much larger than the turbulence component, thus the eddy

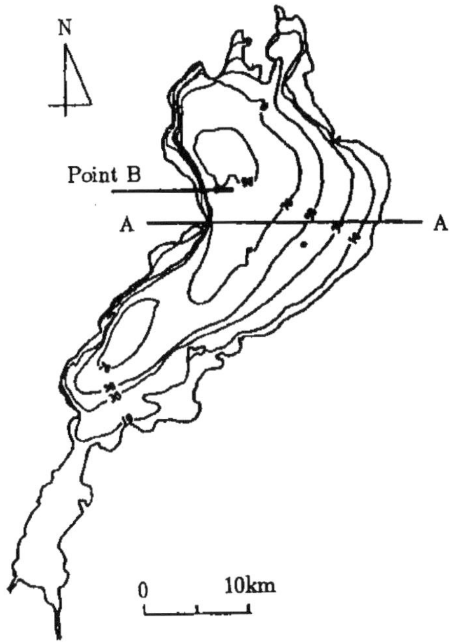

Fig. 4.31. Point B and section A–A in Lake Biwa (Shen, 1991).

viscosity contributes very little. Hence, there is no significant difference between the results of the two analyses, which are different in the treatment of eddy viscosity. On the other hand, after the wind has calmed, the mean flow component gradually attenuates and the weight of the turbulence component increases relatively. Therefore, the difference between the results of these two analyses diminishes.

The values of eddy viscosity coefficients used in the non-LES analysis are 30 m^2/s in the horizontal plane and 0.01 m^2 in the depth-wise direction. These values are commonly adopted in the numerical simulations of Lake Biwa, and are considerably larger than those used in the LES analysis. Therefore, non-LES analysis shows faster momentum transfer than LES analysis. However, very little difference between these two analyses gives less significance in viscous actions to the numerical simulation. Furthermore, as the wind blows, the current along the wind direction gradually becomes stronger, and, finally, arrives at a steady state in the results of both analyses. After the wind has stopped, the current in the central area of the north basin rotates clockwise with time. The rotation shown in Figure 4.32 results from the inertia oscillations.

Figure 4.33 shows the velocity distributions at 20-m depths below the metalimnion. General hydraulic characteristics obtained in both analyses are similar to those at the lake surface, with not much difference between the two analyses due to the small velocities. The directions of current in this layer are approximately opposite to those in the surface layer. Moreover, after the wind calms, an area of large velocity appears in the north edge, and it rotates counterclockwise along the west shore with time, possibly as a result of the internal Kelvin wave.

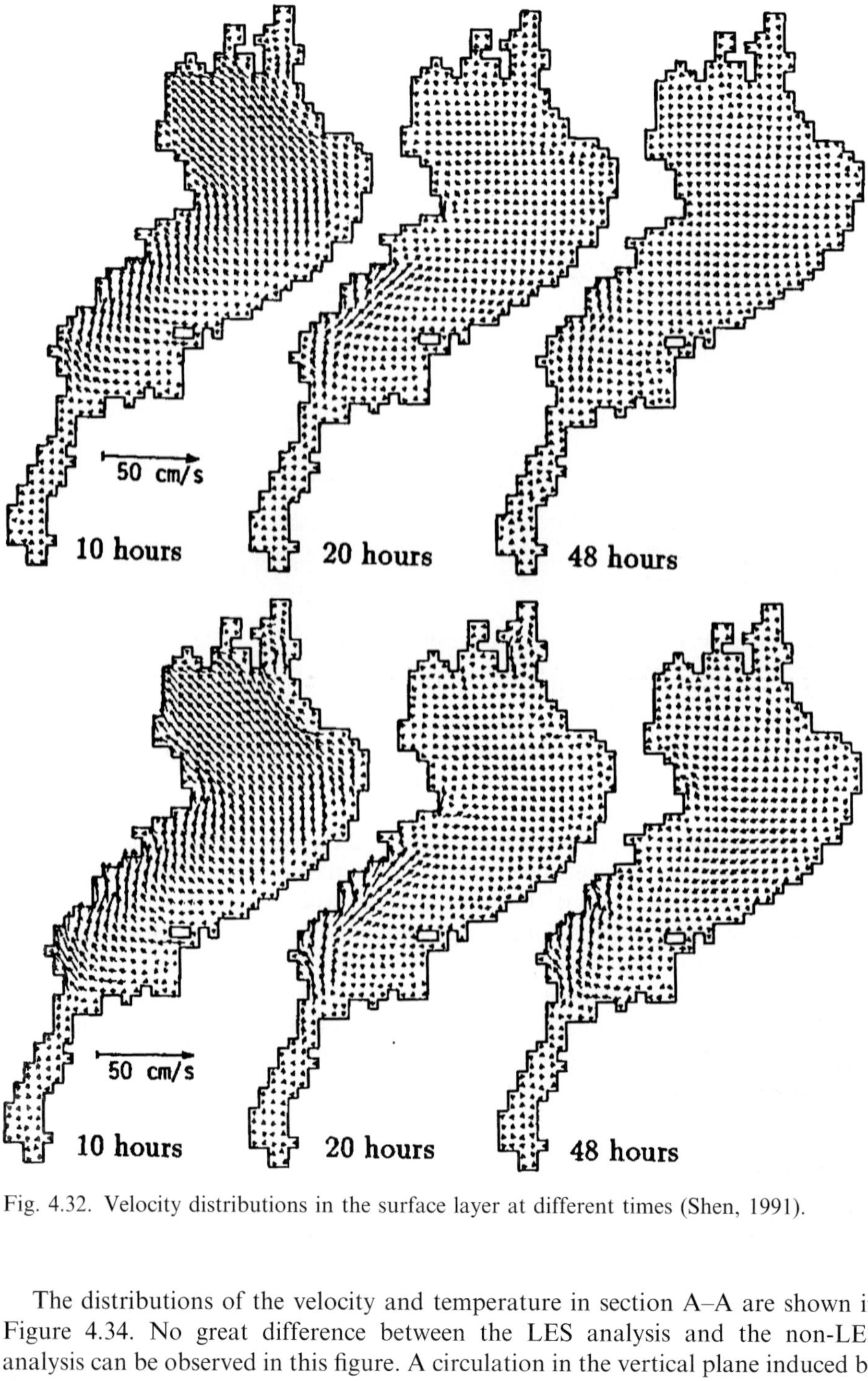

Fig. 4.32. Velocity distributions in the surface layer at different times (Shen, 1991).

The distributions of the velocity and temperature in section A–A are shown in Figure 4.34. No great difference between the LES analysis and the non-LES analysis can be observed in this figure. A circulation in the vertical plane induced by the wind can be seen, which gradually decreases with time, after calming. The

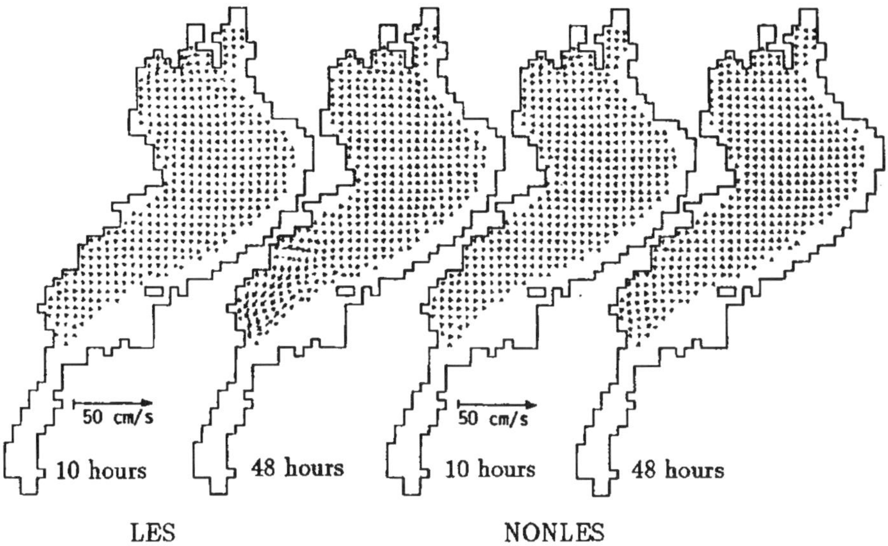

Fig. 4.33. Velocity distributions.

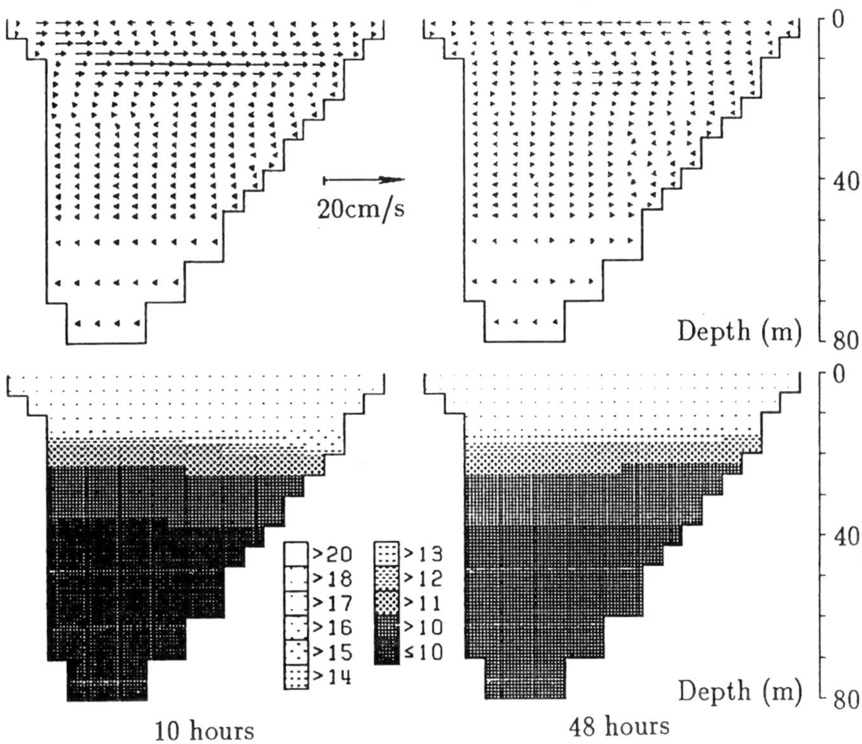

Fig. 4.34. Distributions of the velocity and temperature in A–A section (Non-LES) (Shen, 1991).

inclined metalimnion at 10 h is reversed in phase compared with that at 48 h, which does not result from the x-wise internal seiche in the A–A section but from the rotational Kelvin wave.

Figure 4.35 illustrates the temperature distributions of the eighth layer. In this figure, no significant difference can be observed, and the light tone expresses the

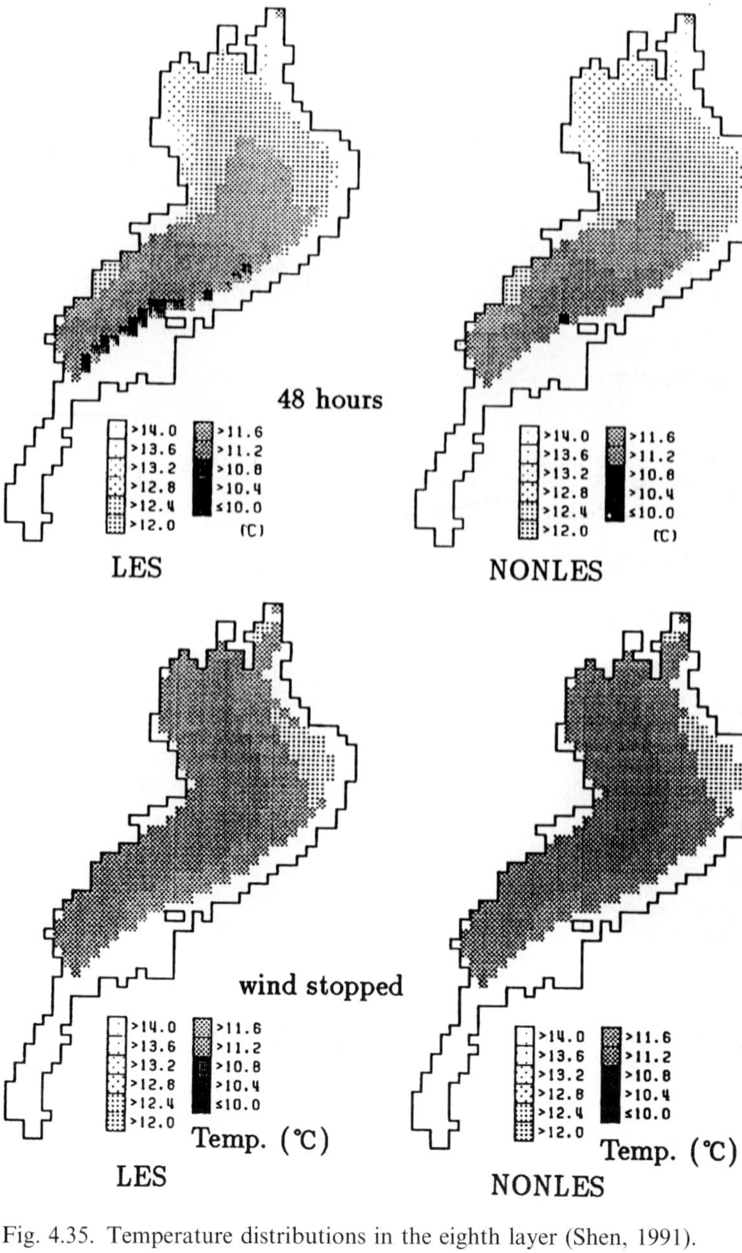

Fig. 4.35. Temperature distributions in the eighth layer (Shen, 1991).

area of high water temperature where the metalimnion is falling down. It can be seen that the metalimnion oscillates while rotating counterclockwise after the wind calms down.

Figure 4.36 is the vertical eddy viscosity v_t^v in the third layer (5.0–7.5 m in depth) at point B. It is obvious that the vertical eddy viscosity v_t^v increases with the initiation of wind blowing and conversely decreases with the disappearing of the lake flows after calming. Being in the order of 10^{-4} m^2/s, the value of v_t^v is relatively small compared to the vertical eddy viscosity used in the non-LES analysis. This is the reason why the velocity calculated in the LES analysis is slightly smaller than that of the non-LES analysis.

The depth-wise distribution of the water temperature, v_t^v and v_t^v (R_i) (modified v_t^v by the Richardson number) at the same point B obtained from the LES analysis, is illustrated in Figure 4.37. At the initial onset of the wind blowing, the values of v_t^v and $v_t^v(R_i)$ at the water surface are large. As the wind continues to blow, the v_t^v and

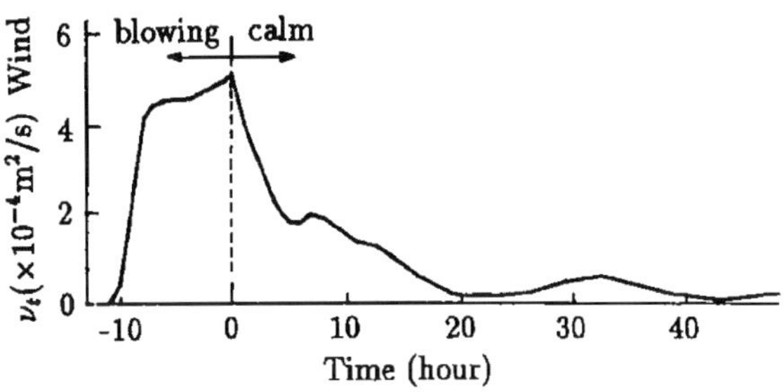

Fig. 4.36. Temporal variations of v_t^v in the third layer at point B (Shen, 1991).

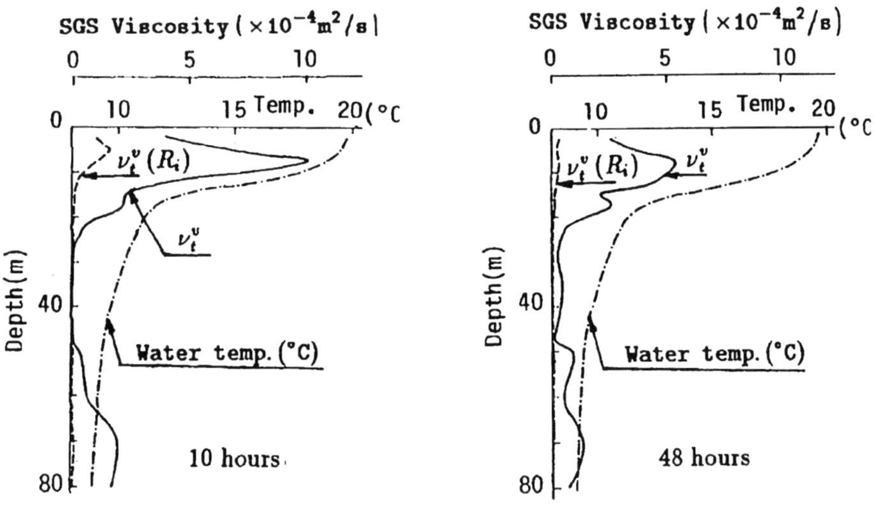

Fig. 4.37. Depth-wise distributions of the water temperature, v_t^v and v_t^v (R_i) at point B (Shen, 1991).

$\nu_t^v(R_i)$ in deep areas become gradually large. This is because a large velocity gradient in the vertical direction occurs near the water surface and gradually transmits to deeper areas. The value of ν_t^v near the depth of 15 m in the metalimnion is large because of the large depth-wise gradient of velocity. The value of $\nu_t^v(R_i)$ is very small. However, because the depth-wise gradient of water temperature is large and the velocity gradient near the metalimnion, large Richardson numbers are calculated. The values of ν_t^v and $\nu_t^v(R_i)$ at the water depth below 50 m are relatively large. The reason is that the thickness of the mesh at the water depth below 50 m is 10 m, which is four times the mesh thickness at depths shallower than 50 m; therefore, large ν_t^v are calculated. From this figure, it is also observed that the vertical eddy viscosity ν_t^v is in the order of $10^{-4}\,m^2/s$ and decreases with time. Finally, it is also observed from the vertical distribution of water temperature in Figure 4.37 that the stratification is gradually smoothed because of the internal heat exchanges.

Although the large-scaled flows in lakes could be reproduced by the use of a LES model, there are several problems with the application of a LES model to lakes.

Firstly, it is not obvious whether the approximation of hydrostatic pressure is valid or not for the large eddy components of flows dealt within LES analysis. Although Bedford (1981) discussed this by making the governing equations dimensionless, the influences of depth-wise accelerations are not disclosed. Another problem with the approximation of hydrostatic pressure is that it is not clear as to how much influence it gives to the modelling of the sub-grid scale component. The only solution to the problem is to clarify the significance of making Eq. (2.53) dimensionless and approximating Eq. (2.53) by Eq. (2.54).

Secondly, in modelling the sub-grid scale component, the filter width is expressed by the mesh sizes as in Eqs. (2.55) and (2.58). When using LES, the filter width must correspond to the spatial scales of turbulence possibly expressed by Eqs. (2.50) and (2.51). In other words, the mesh sizes must be equal to the scales of turbulence and the LES analysis is expected to provide the physical evidence for the selection of the mesh size. A similar conclusion can be obtained for the temporal difference increment. However, this calculation is not referred to in the discussion. A review of the results obtained by LES analysis must be made in view of turbulent theory. In this study, the mesh sizes are taken as 1 km in the horizontal plane, which is too large for a LES model; in other words, if a too large mesh size is chosen, the LES model will lose its applicability in a strict sense. But small mesh sizes will result in enormous computations, which is one of the difficulties with the application of LES models to lakes. Moreover, it is also necessary to examine in terms of LES theory, the validity of the extreme difference of the mesh sizes between the horizontal plane and the depth-wise direction.

Finally, the sub-grid scale component is modelled by using eddy viscosity in LES analysis. This coefficient, defined as a variable dependent on the flow field, is calculated as a function of time and space in the computation process. Therefore, the difference between LES analysis and non-LES analysis is only connected to the difference in values of eddy viscosity between the two models (if only reviewing the computational results). Further discussion about the basic characteristics of LES analysis are required.

For the large-scaled components of flows in lakes, no significant difference between the results of a LES model and a non-LES model was observed; whereas, there were some differences for the small scale components. Moreover, several prominent flows in Lake Biwa, which have been verified by non-LES models, were also reproduced by the LES model. Therefore, it is possible to simulate the flows in lakes by a LES model as well as a non-LES model. The CPU time consumed in the LES analysis was only 1.1–1.2 times than that in the non-LES analysis (constant eddy viscosity/diffusivity model) under the same computational conditions. In this sense, the LES model is more favourable than other turbulence models whose CPU times become much longer due to the increase in the number of equations.

Chapter 5

QUASI-THREE-DIMENSIONAL CIRCULATION MODELS

The depth-averaged 2D model, which is obtained by integrating the 3D model equations over the depth, has been extensively used for simulating storm surges, long-term circulation, and pollutant transport in lakes and estuaries. It is relatively easy to use and may provide reasonable results when the detailed flow in the third dimension is relatively unimportant.

Problems were noticed by many researchers when conventional 2D models were used for simulating wind-induced circulation (Wang and O'Connor, 1975; Ramming and Kowalik, 1980; Koutitas and Gousidou-Koutita, 1986; Simons and Schertzer, 1989b; Schwab et al., 1989). Simons and Schertzer (1989a, b) pointed out that the conventional 2D circulation models underestimated the observed current speeds and wind-induced water set-up. From a practical viewpoint, the simulation results may be adjusted by increasing the wind-drag coefficient or decreasing the bottom friction coefficient. From a physical viewpoint, however, the underestimation is highly significant. The researchers have acknowledged the physically unrealistic formulation of bottom friction terms in the 2D model. However, no solution to this problem has been proposed in their work. Efforts have been undertaken to improve the formulation of bottom friction stress and depth-averaged advective terms by introducing momentum dispersion into advective terms, and relating the bottom friction stress to the wind shear stress (Wang and O'Connor, 1975; Ramming and Kowalik, 1980). The results are not entirely satisfactory, because of the lack of a sound physical mechanism. Koutitas and Gousidou-Koutita (1986) pointed out that velocity profiles develop in an entirely different way for the two cases of tidal and wind-induced circulation. They developed the "quasi-depth-varying" (Q3D) model to account for the variation in horizontal velocity in the vertical. Then, the bottom shear stress was properly evaluated, using the given velocity profile. However, the parabolic profile used in the model is only consistent with a laminar type flow, and is not adequate for turbulent shear-induced flow (Tsanis, 1986).

Fully 3D models (Sheng, 1983; Swanson, 1986; Lynch and Werner, 1991; Shen, 1991; Wu, 1993) can be employed to solve these problems. However, the conventional depth-averaged circulation model is still used either as an external model in external–internal mode-splitting 3D models (Simons, 1980; Sheng, 1983, 1990; Swanson, 1986), or as the first module in vertical/horizontal splitting (VHS) 3D models (Lardner and Cekirge, 1988; Al-Rabeh and Gunay, 1992). As a result, a new 2D model or an efficient 3D model needs to be developed.

An efficient 3D wind-induced model (VHI3D) was developed by Wu and Tsanis (1995b). The model was based on previous theoretical analysis, and validated with field and laboratory measurements of turbulent shear-induced flows. In contrast to the VHS approach, which was developed in recent years (Lardner and Cekirge, 1988), the vertical/horizontal integrating (VHI) approach integrates the depth-averaged velocity and vertical-velocity structure. To evaluate the VHI3D model, the analytical solutions and laboratory data (Tsuruya et al., 1985) of wind-induced countercurrent flow was used. The model was compared with the depth-averaged 2D, the quasi-three-dimensional model (Q3D) developed by Koutitas and Gousidou-Koutita (1986), and the multi-layered 3D model by Wu (1993).

5.1 Q3D MODEL[4]

In order to incorporate the effects of non-uniform velocity in the vertical plane, Koutitas and Gousidou-Koutita (1986) developed a Q3D wind-driven circulation model. This model is presented here with some modifications. A parabolic velocity distribution in the z-direction is assumed:

$$u(z) = \alpha z^2 + \beta z + \gamma \tag{5.1}$$

where z is measured upward from the lake bottom for the convenience of model derivation. The wind shear stress at the free water surface, no-slip condition at the bottom, and the definition of depth-averaged velocities are all used to determine the coefficients α, β, γ, Then the velocity profiles are obtained as a function of depth-mean velocity and wind shear stress:

$$u = \left(-\frac{3}{2}U + \frac{3}{4}\frac{\tau_{sx}}{\rho}\frac{h}{K_v}\right)\left(\frac{z}{h}\right)^2 + \frac{\tau_{sx}}{\rho}\frac{h}{K_v}\frac{z}{h} + \left(\frac{3}{2}U + \frac{\tau_{sx}}{\rho}\frac{h}{4K_v}\right) \tag{5.2}$$

$$v = \left(-\frac{3}{2}V + \frac{3}{4}\frac{\tau_{sy}}{\rho}\frac{h}{K_v}\right)\left(\frac{z}{h}\right)^2 + \frac{\tau_{sy}}{\rho}\frac{h}{K_v}\frac{z}{h} + \left(\frac{3}{2}V + \frac{\tau_{sy}}{\rho}\frac{h}{4K_v}\right) \tag{5.3}$$

In order to be consistent with the parabolic velocity distribution, a constant viscosity in the vertical direction is assumed. The depth-averaged advective terms are evaluated through Eqs. (5.2) and (5.3). For example,

$$\frac{1}{h}\int_0^h \frac{\partial u}{\partial x}\,dz = U\frac{\partial U}{\partial x} + \left(0.2U + \frac{\tau_{sxh}}{40\rho K_v}\right)\frac{\partial U}{\partial x} \tag{5.4}$$

The bottom friction shear stresses are evaluated as

$$\tau_{bx} = 3\bar{\lambda}\;\frac{V}{h}\sqrt{\tau_{sx}/\rho} - 0.5\frac{\tau_{sx}}{\rho h} \tag{5.5}$$

[4]Sections in this Chapter are reproduced from Wu and Tsanis (1995b), copyright with permission from Elsevier Science Ltd., The Boulevard, Langford Lane, Kidlington OX5 1GB, UK.

$$\tau_{by} = 3\bar{\lambda}\ \frac{V}{h}\sqrt{\tau_{sy}/\rho} - 0.5\frac{\tau_{sy}}{\rho h} \tag{5.6}$$

Finally, the 2D model, improved with respect to horizontal momentum dispersion and bottom friction for wind-driven circulation, becomes

$$\frac{\partial U}{\partial t} + \left(U + 0.2U + \frac{\tau_{sxh}}{40\rho K_v}\right)\frac{\partial U}{\partial x} + \left(V + 0.2V + \frac{\tau_{sy}h}{40\rho K_v}\right)\frac{\partial U}{\partial y}$$
$$= fV - g\frac{\partial \zeta}{\partial x} + \frac{\tau_{sx} - \tau_{bx}}{\rho H} \tag{5.7}$$

$$\frac{\partial V}{\partial t} + \left(U + 0.2U + \frac{\tau_{sxh}}{40\rho K_v}\right)\frac{\partial V}{\partial x} + \left(V + 0.2V + \frac{\tau_{sy}h}{40\rho K_v}\right)\frac{\partial V}{\partial y}$$
$$= -fU - g\frac{\partial \zeta}{\partial y} + \frac{\tau_{sy} - \tau_{by}}{\rho H} \tag{5.8}$$

$$\frac{\partial \zeta}{\partial t} + \frac{\partial HU}{\partial x} + \frac{\partial HV}{\partial y} = Q \tag{5.9}$$

The model also permits the computation of current patterns at different depths. The surface velocity is calculated through Eqs. (5.2) and (5.3) by setting $z = 0$:

$$u_s = 1.5U + \frac{\tau_{sx}h}{4\rho K_v}; \quad v_s = 1.5V + \frac{\tau_{sy}h}{4\rho K_v} \tag{5.10}$$

5.2 VERTICAL EDDY VISCOSITY DISTRIBUTION

For a typical lake with a horizontal length scale of 10^5 m and a velocity length scale of 10^{-1} m/s, the Reynolds number is in the order of 10^9. As a result, the flow in lakes and coastal waters is predominantly turbulent. Compared to various models such as the mixing length hypothesis model (Reid, 1957), the K–ε model (Svensson, 1978), the K–L model (Koutitas and O'Connor, 1980), and the eddy viscosity model (Pearce and Cooper, 1981; Tsanis, 1986), a simple parabolic distribution of eddy viscosity can produce a velocity profile close to that predicted by the sophisticated closure models discussed. The use of a parabolic distribution of eddy viscosity results in a double-logarithmic velocity distribution in the vertical plane (Spillane and Hess, 1978; Tsuruya et al., 1985; Tsanis, 1989). Including the viscous sub-layer thickness, the eddy viscosity distribution can be expressed as (Tsanis, 1989):

$$K_v = \frac{\lambda u_s^*}{h}(z + z_b)(z_s + h - z) \tag{5.11}$$

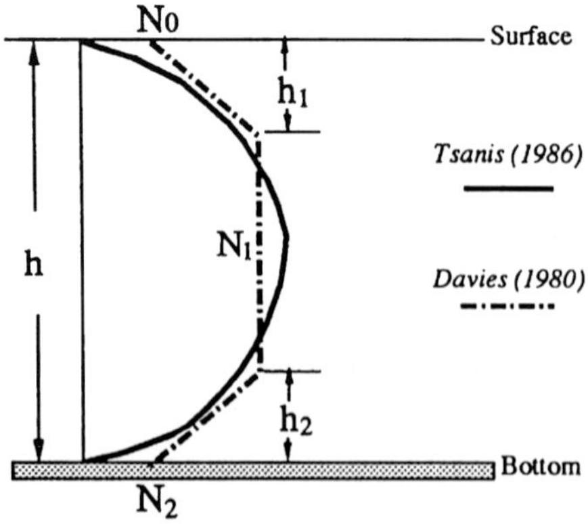

Fig. 5.1. Distributions of vertical eddy viscosity used in the VHI3D model and one in the coastal 3D circulation models (Reprinted from Wu and Tsanis (1995b), copyright with kind permission from Elsevier Science Ltd., The Boulevard, Langford Lane, Kidlington OX5 1GB,UK).

In the preceding equation, λ is a constant to characterize the intensity of the turbulence and is a function of the Reynolds number based on the surface velocity. By matching the experimental profiles with the theoretical predictions, and taking into account the variation of the surface velocity ratio, it was found that, for $R_s = 10^3$ to 10^5, λ varies from 0.2 to 0.5 and the average value is about 0.35 (Tsanis, 1989). z_b and z_s are characteristic lengths determined at the bottom $z = 0$ and at the surface $z = h$ (h is the water depth), respectively. They are a relative measure of the thickness of the viscous sublayers. It was found by Van Dorn (1953) that the bottom characteristic length $z_{bh} = z_b/h$ is of the order of 10^{-3}–10^{-4}, while Reid (1957) determined that $z_{sh} = z_s/h$ has a value around 1/3000. From experimental data, Tsanis (1989) found that the characteristic lengths z_{sh} and z_{bh} have values of 2.2×10^{-4} and 0.6×10^{-4}. u_s^* is the surface shear velocity which can be evaluated from the wind stress. A systematic study on the sensitivities of the parameters λ, z_{sh}, and z_{bh} on velocity profile can be found in Tsanis (1989).

As shown in Figure 5.1, the parabolic distribution of the vertical eddy viscosity is very similar to that widely used in coastal 3D circulation models (Davies, 1980a, b; Heaps, 1987; Al-Rabeh and Gunay, 1992). The latter has a top layer of depth h_1, in which viscosity N increases from N_0 to N_1, and a bottom layer of depth h_2, in which N decreases from N_1 to N_2. In between the two layers, N is uniform over the depth. Detailed values of h_1, h_2, N_0, and N_2 can be found in Davies (1980a, b).

5.3 VERTICAL–HORIZONTAL INTEGRATING (VHI) 3D MODEL

5.3.1 Model Development

Considering only the u-component equation, it is recognized from a number of experimental results that the velocity profile for a wind-driven countercurrent flow is well approximated by a double-logarithmic structure: one for the surface region and one for the bottom part of the return flow (Spillane and Hess, 1978; Tsuruya et al., 1985; Tsanis, 1989). In general, the profile can be described as (Wu and Tsanis, 1995b):

$$u = a \ \ln\left(1 + \frac{z}{z_s}\right) + b \ \ln\left(1 - \frac{z}{z_b + h}\right) + c \tag{5.12}$$

For the convenience of model derivation, the vertical coordinates are taken in such a way that the bottom is zero. The following conditions are used to determine the above coefficients *a*, *b*, *c*.

Condition 1
At the free surface ($z = h$), the shear stress is evaluated by the wind velocity:

$$K_v \frac{\partial u}{\partial z}\bigg|_{z=h} = \tau_s; \qquad \tau_s = \rho_a C_D W|W| \tag{5.13}$$

Condition 2
At the bottom ($z = 0$)

$$u|_{z=0} = 0 \tag{5.14}$$

Condition 3
The depth-averaged velocity definition

$$\frac{1}{H}\int_0^H u \ dz = U \tag{5.15}$$

These three conditions form an algebraic system involving the three coefficients in Eq. (5.12). The solution is

$$c = 0 \tag{5.16}$$

$$a = a_1 U + a_2 u_s^* \tag{5.17}$$

$$b = b_1 U + b_2 u_s^* \tag{5.18}$$

In the preceding solution, the following notation is introduced in order to simplify the expressions:

$$\begin{aligned} a_1 &= -p_2/(p_1 q_2 - q_1 p_2), \quad & a_2 &= q_2/(p_1 q_2 - q_1 p_2), \\ b_1 &= p_1/(p_1 q_2 - q_1 p_2), \quad & b_2 &= -q_1/(p_1 q_2 - q_1 p_2) \end{aligned} \tag{5.19}$$

where,

$$p_1 = \lambda z_{sh} \tag{5.20}$$

$$p_2 = \lambda z_{sh}/z_{bh} \tag{5.21}$$

$$q_1 = (1 + z_{sh})[\ln(1 + 1/z_{sh}) - 1] + z_{sh} \tag{5.22}$$

$$q_2 = -z_{bh}[\ln(z_{bh}) - \ln(1 + z_{bh}) - 1] - (1 + z_{bh}) \tag{5.23}$$

Therefore, the velocities at any depth can be expressed in terms of the depth-averaged velocity and the wind stress acting on the surface.

$$u = (a_1 U + a_2 u_s^*)\ln\left(1 + \frac{z}{z_s}\right) + (b_1 U + b_2 u_s^*)\ln\left(1 - \frac{z}{z_b + h}\right) \tag{5.24}$$

Suppose that the wind shear stress at the surface can be divided into two components in the x and y directions, such that

$$\tau_{sx} = \rho_a C_D W_x \sqrt{W_x^2 + W_y^2} \quad \tau_{sy} = \rho_a C_D W_y \sqrt{W_x^2 + W_y^2} \quad \tau_s = \sqrt{\tau_{sx}^2 + \tau_{sy}^2} \tag{5.25}$$

Define

$$\tau_x = \tau_{sx}/\tau_s, \quad \tau_y = \tau_{sy}/\tau_s \tag{5.26}$$

Then, the general profiles for both velocity components u and v are written as

$$u = (a_1 U + a_2 \tau_x)\ln\left(1 + \frac{z}{z_s}\right) + (b_1 U + b_2 \tau_x)\ln\left(1 - \frac{z}{z_b + h}\right) \tag{5.27}$$

$$v = (a_1 V + a_2 \tau_y)\ln\left(1 + \frac{z}{z_s}\right) + (b_1 V + b_2 \tau_y)\ln\left(1 - \frac{z}{z_b + h}\right) \tag{5.28}$$

The bottom shear stress can then be evaluated using these velocity profiles:

$$\begin{aligned}\tau_{bx} &= K_v \frac{\partial u}{\partial z}\bigg|_{z=0} \\ &= \lambda\sqrt{\tau_s/\rho} z_{bh}(z_{sh} + 1)[(a_1 U + a_2 \tau_x)/z_{sh} - b_1 U + b_2 \tau_x)/(z_{bh} + 1)]\end{aligned} \tag{5.29}$$

$$\begin{aligned}\tau_{by} &= K_v \frac{\partial v}{\partial z}\bigg|_{z=0} \\ &= \lambda\sqrt{\tau_s/\rho}\, z_{bh}(z_{sh} + 1)[(a_1 V + a_2 \tau_y)/z_{sh} - (b_1 V + b_2 \tau_y)/(z_{bh} + 1)]\end{aligned} \tag{5.30}$$

Substituting the discussed velocity profiles into the advective terms and then depth averaging leads to

$$\frac{1}{H}\int_0^H u\frac{\partial u}{\partial x}\,dz = (X_1U + X_2\tau_x)\frac{\partial U}{\partial x} \tag{5.31}$$

$$\frac{1}{H}\int_0^H v\frac{\partial u}{\partial y}\,dz = (X_1V + X_2\tau_y)\frac{\partial U}{\partial y} \tag{5.32}$$

$$\frac{1}{H}\int_0^H u\frac{\partial v}{\partial x}\,dz = (X_1U + X_2\tau_x)\frac{\partial V}{\partial x} \tag{5.33}$$

$$\frac{1}{H}\int_0^H v\frac{\partial v}{\partial y}\,dz = (X_1V + X_2\tau_y)\frac{\partial V}{\partial y} \tag{5.34}$$

where

$$X_1 = a_1^2 I_1 + 2a_1b_1I_2 + b_1^2 I_3 \tag{5.35}$$

$$X_2 = a_1a_2I_1 + (a_2b_1 + a_1b_2)I_2 + b_1b_2I_3 \tag{5.36}$$

and

$$I_1 = (1 + z_{sh})\left[\ln^2\left(1 + \frac{1}{z_{sh}}\right) - 2\ln\left(1 + \frac{1}{z_{sh}}\right) + 2\right] - 2z_{sh} \tag{5.37}$$

$$I_2 = \frac{1}{2}\left(\ln z_{sh} + \frac{1}{2}\right) - \frac{1}{h}z_{sh}(\ln z_{sh} + 1) + \frac{3}{4}z_{sh}^2 \tag{5.38}$$

$$I_3 = -z_{sh}\left(\ln^2\frac{z_{bh}}{1 + z_{sh}} - 2\,\ln\frac{z_{bh}}{1 + z_{sh}} + 2\right) + 2 \tag{5.39}$$

Based on these expressions, the new 2D circulation model for wind-driven flows is described by the following (Wu and Tsanis, 1995b):

$$\begin{aligned}&\frac{\partial U}{\partial t} + (X_1U + X_2\tau_x)\frac{\partial U}{\partial x} + (X_1V + X_2\tau_y)\frac{\partial U}{\partial y} + A_h\left(\frac{\partial^2 U}{\partial x^2} + \frac{\partial^2 U}{\partial y^2}\right)\\ &= fV - g\frac{\partial \zeta}{\partial x} + \frac{\tau_{sx} - \tau_{bx}}{\rho H}\end{aligned} \tag{5.40}$$

$$\begin{aligned}&\frac{\partial V}{\partial t} + (X_1U + X_2\tau_x)\frac{\partial V}{\partial x} + (X_1V + X_2\tau_y)\frac{\partial V}{\partial y} + A_h\left(\frac{\partial^2 V}{\partial x^2} + \frac{\partial^2 V}{\partial y^2}\right)\\ &= -fU - g\frac{\partial \zeta}{\partial y} + \frac{\tau_{sy} - \tau_{by}}{\rho H}\end{aligned} \tag{5.41}$$

$$\frac{\partial \zeta}{\partial t} + \frac{\partial HU}{\partial x} + \frac{\partial HV}{\partial y} = Q \tag{5.42}$$

5.3.2 Characteristics of the VHI3D Model

When one first glances at the model equations, the VHI3D model seems to be lengthy and complicated. But, in fact, the model equations only involve a handful of constants and coefficients, and their forms do not differ significantly from the 2D model. The calculation procedure is as follows (Wu and Tsanis, 1995b):

(1) Specify the viscous sub-layer thicknesses z_{sh}, z_{bh}, and λ in the eddy viscosity distribution. Input the wind velocity components and depth data and all the model parameters, such as wind-drag coefficient, time step, grid size, Coriolis parameters, etc.
(2) Calculate p_1, p_2, q_1 and q_2, a_1, a_2, b_1, and b_2 to determine the profile and bottom shear stress terms, and I_1, I_2, I_3, X_1, and X_2 to evaluate the advective terms.
(3) Employ a finite difference scheme (Wu, 1993) to numerically solve the model equations in a staggered-grid system. The predicted variables are two depth-averaged velocity components U, V, and the water elevation ζ. The scheme is stable under the Courant-Friedrichs-Levy (CFL) condition.
(4) Calculate the horizontal velocities at any depth by simply employing the velocity profile, Eqs. (5.27) and (5.28), and by using the depth-averaged velocity, wind shear stress and a_1, a_2, b_1, and b_2. If needed, the vertical velocity component w can be calculated from the original continuity equation. It is significant that the 3D current structure can easily be calculated at almost no additional computer cost.

5.3.3 Model Verifications with Analytical and Laboratory Data

Comparisons with analytical solutions

The shear-induced countercurrent flow is a steady wind-driven flow in a closed channel, subjected to a uniform longitudinal wind stress. The wind acting on the water body surface causes a drift in the direction it is blowing, while there is a bottom return flow in the opposite direction, resulting in a zero mass flux. Because this simple problem includes the turbulent mixing mechanism for geophysical wind-driven flows, it has been widely studied in field studies, laboratory experiments, and numerical models.

In this example, the channel is 9100 m long and 10 m deep. The horizontal grid size is 1 m and the vertical grid is (of) 0.1 m. The time step of 1 s is used. In the fully 3D model, 9 layers are used in the vertical for laminar type flow, while 15 uneven layers (enhanced both near the surface and the bottom, in order to simulate the sharp velocity gradient close to the interface) are used for turbulent flow. The horizontal eddy viscosity K_h of 0.1 m^2/s is used in all models. For the vertical eddy viscosity, the constant $\bar{\lambda}$ for the Q3D model is taken as 0.04; and λ for the VHI3D model is 0.35. z_{sh} and z_{bh} are 0.00022 and 0.00014, respectively.

The simulated results from the Q3D and 3D models, using the same constant vertical viscosity coefficient, are in good agreement with the analytical solutions, as shown in Figure 5.2(a). Figure 5.2(b) shows the predictions by the 3D and VHI3D models in the case of turbulent flow, together with the analytical solutions. It is observed that the model predictions are in good agreement with the analytical

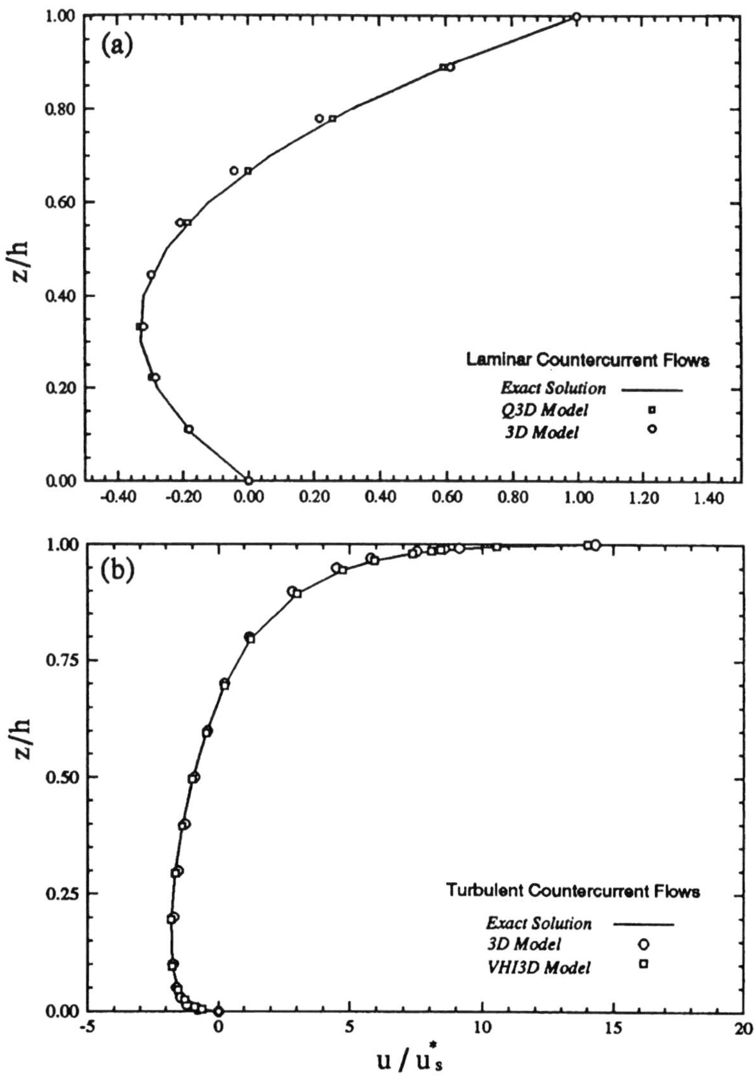

Fig. 5.2. Comparison of simulations with analytical solutions: (a) for laminar flow by the 3D and Q3D models; (b) for turbulent flow by the 3D and VHI3D models (Reprinted from Wu and Tsanis (1995b), copyright with kind permission from Elsevier Science Ltd., The Boulevard, Langford Lane, Kidlington OX5 1GB, UK).

solutions, and the VHI3D model is as capable as the 3D model in simulating the vertical current structure.

Verification with laboratory data

The experimental results of Tsuruya et al. (1985) are used for comparisons between the 3D and VHI3D models. As shown in Figure 5.3, the channel depth is 0.15 m and the length is 22 m long. The wind velocity is taken as 6.8 m/s, which is between the

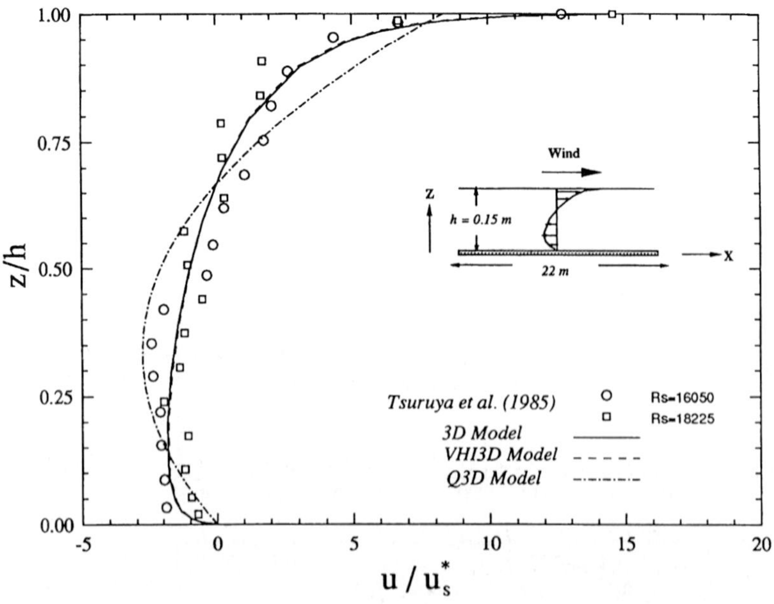

Fig. 5.3. Normalized velocity profile—calibration of the VHI3D, Q3D, and 3D models with the experimental data by Tsuruya et al. (1985) (Reprinted from Wu and Tsanis (1995b), copyright with kind permission from Elsevier Science Ltd., The Boulevard, Langford Lane, Kidlington OX5 1GB, UK).

Table 5.1. Experimental results (Tsuruya et al., 1985) and simulations by 3D and VHI3D models (Wu and Tsanis, 1995b).

Quantity		Test 1	Test 2	3D	VHI3D
Depth	H (m)	0.15	0.15	0.15	0.15
Wind velocity	W (m/s)	6.73	6.90	6.80	6.80
Surface shear velocity	U^*_s (cm/s)	0.815	0.835	0.823	0.823
Surface velocity	u_s (cm/s)	10.7	12.15	11.79	11.55
Reynolds number	$R_s = u_s h/\nu$	16050	18225	17685	17325
Normalized surface	u_s/u^*_s	12.74	14.55	14.32	14.32
Velocity	Z_{ow} (mm)	0.25	0.36	0.31	0.33
Absolute roughness					

two wind conditions (6.73 and 6.90 m/s) in the experiments. The surface velocity (u_s), normalized surface velocity, and absolute roughness, both from the experiments and model simulations by the VHI3D and 3D models, are presented in Table 5.1. The simulated velocity profiles by the 3D and VHI3D models, normalized by the surface shear velocity (u^*_s), are presented in Figure 5.3, together with the experimental results and the prediction of the Q3D model. It is observed that the model simulations are close to the experiment's range, indicating a good agreement between the model and experimental results. The logarithmic structure is found near both the surface and the bottom, reflecting the impact of boundary layers.

A better representation of the velocity profile uses the terms of the coordinates of the "smooth" inner law of velocity distribution, relative to either the surface or the bottom (Tsanis, 1986; Andreasson, 1991). In Figure 5.4(a), curve (A) represents the

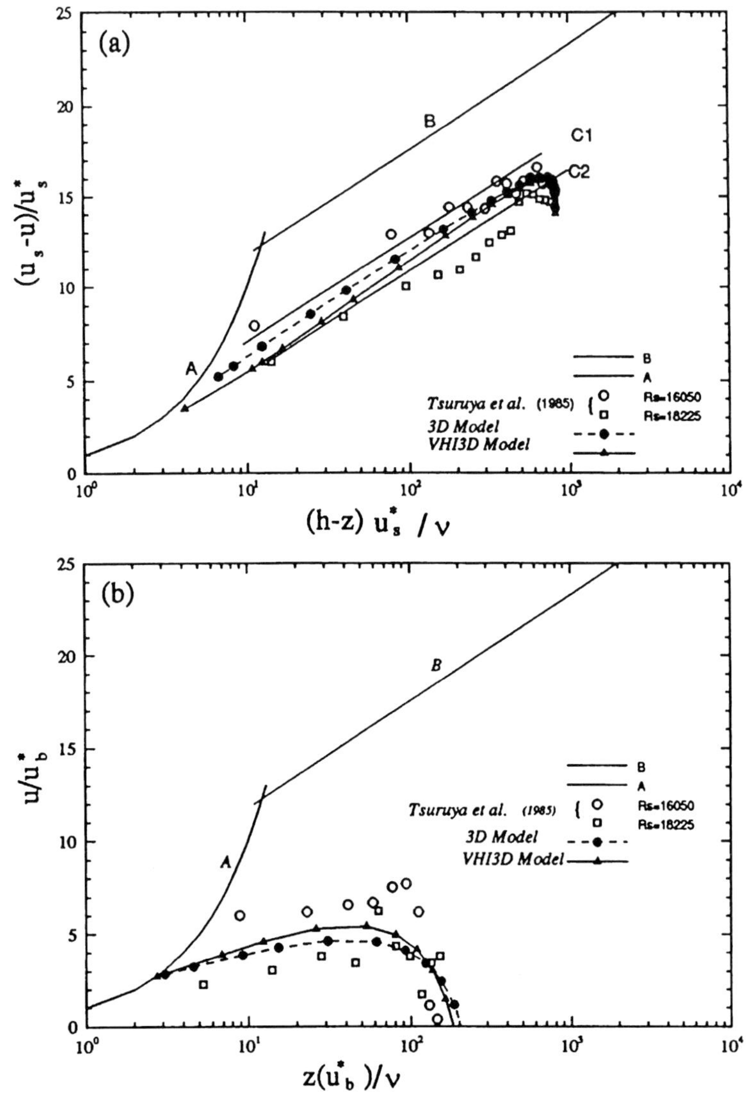

Fig. 5.4. "Smooth" inner law of velocity distribution of the experimental data by Tsuruya et al. (1985) and the model predictions by the VHI3D and 3D models: (a) relative to the surface interface; and (b) relative to the bottom (Reprinted from Wu and Tsanis (1995b), copyright with kind permission from Elsevier Science Ltd., The Boulevard, Langford Lane, Kidlington OX5 1GB, UK).

viscous sub-layer portion of the velocity profile,

$$\frac{u_s - u}{u_s^*} = \frac{(h - z)u_s^*}{v} \tag{5.43}$$

and curve (B), corresponding to the logarithmic portion of the velocity profile for hydrodynamically smooth conditions, is described by,

$$\frac{u_s - u}{u_s^*} = 5.75 \log \frac{(h - z)u_s^*}{v} + 6.0 \tag{5.44}$$

The curves labeled C_1 and C_2 represent the curves from the experimental results (Tsuruya et al., 1985) with Reynolds numbers of 16,050 and 18,225, respectively. These curves, as described by,

$$\frac{u_s - u}{u_s^*} = 5.75 \log \frac{(h - z)u_s^*}{v} + 8.5 - 5.75 \log R_{kw} \tag{5.45}$$

are parallel to curve (B), and lie below curve (B) by an amount equal to the roughness-induced velocity reduction. This value is proportional to the roughness Reynolds number, or von Karman number $R_{kw} = u^*_s Z_{ow}/\text{v}$, where v is the water viscosity and Z_{ow} is the absolute roughness of the wall (Schlichting, 1968). The roughness length Z_{ow} can then be calculated from R_{ow}. Both data and simulations are seen to be in very good agreement with the firmly established law for hydrodynamically "smooth" conditions in fully developed turbulence. The simulations are in good agreement with the experimental data in both wind conditions.

Similar presentations for near-bottom currents are plotted in inner law coordinates relative to the bottom, as shown in Figure 5.4(b). It is found that the linearity for model results (from the bottom $x = z(u^*_b)/v = 0.3$ to $x = 50$ in Figure 5.4(b)) is better than that for experimental data, indicating that well-developed return currents have been achieved in models.

Sensitivity study of parameters

The choice of parameters is not unique. They need to be calibrated in model applications. A study is performed to examine the sensitivities of three parameters in the vertical eddy viscosity distribution on the simulated velocity profile. A 6.8 m/s wind is applied to the channel used in Tsuruya et al. (1985). The tests are designed in such a way that in each test only one parameter is changed, with five different values, while the other two remain the same. For the purpose of comparison, the experimental data by Tsuruya et al. (1985) are also presented. It is found that the simulated surface velocity and velocity profile with parameters $\lambda = 0.35$, $z_{sh} = 0.00022$, and $z_{bh} = 0.00014$ are in good agreement with the experimental data.

Figure 5.5(a) presents the sensitivity of the results for λ of 0.20, 0.30, 0.35, 0.40, and 0.50 ($z_{sh} = 0.00022$ and $z_{bh} = 0.00014$). Because λ is the constant characterizing the intensity of the turbulence, it is not surprising that higher velocities (both drift

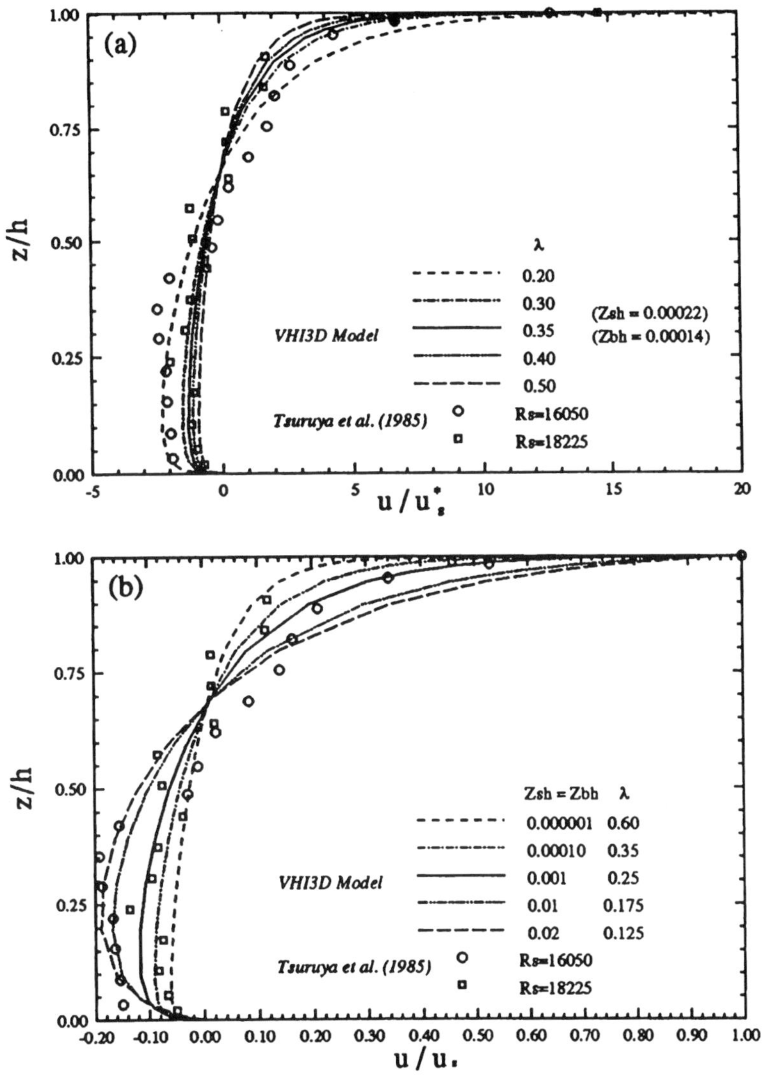

Fig. 5.5. Sensitivities of parameters in the vertical eddy viscosity distribution on the simulated velocity profile. (a) For various λ(set $z_{sh} = 0.00022$, $z_{bh} = 0.00014$). The velocities are normalized by the surface shear velocity u^*_s; (b) For various λ, z_{sh}, and z_{bh}. The velocities are normalized by the surface velocity u_s. Symbols refer to the experimental data by Tsuruya et al. (1985) (Reprinted from Wu and Tsanis (1995b), copyright with kind permission from Elsevier Science Ltd., The Boulevard, Langford Lane, Kidlington OX5 1GB, UK).

and return currents) are obtained for a smaller λ. When eddy viscosity is decreased, the internal shear stress and, consequently, the turbulence, is decreased. Therefore, the drift and return flows are increased. When λ becomes bigger, the velocities become smaller.

Physically, both the surface and bottom characteristic lengths z_{sh} and z_{bh} decrease, or velocity gradients increase with increasing Reynolds number or increasing λ. The surface viscous sub-layer is thinner than the bottom viscous sub-layer, that is, z_{sh} is larger than z_{bh}. However, there is little knowledge on the relationship between the two characteristic lengths. Tsanis (1989) tested different relationships and found that the change in the relationship influences mainly the return portion of the flow. z_{sh} and z_{bh} are about the same order of magnitude. Based on these considerations, another test is undertaken by changing λ and z_{sh} and z_{bh} (set $z_{sh} = z_{bh}$). Figure 5.5(b) shows the simulated velocity profiles for $\lambda = 0.60$, 0.35, 0.25, 0.175, 0.125 and $z_{sh} = z_{bh} = 0.000001$, 0.0001, 0.001, 0.01, and 0.02. The velocities are normalized with respect to the surface velocities to clearly show the velocity gradients near the surface and bottom. It is observed that a higher velocity gradient is presented both near the surface and bottom under a smaller z_{sh} and z_{bh} and a larger λ. This effect is more significant near the surface than near the bottom. The effect of these parameters on the current is less significant in the middle region (out of the surface and bottom boundary layers) than in the boundary layers.

5.3.4 Model Comparison in a Simplified Basin

Computational conditions

A 2D rectangular basin is employed to examine the model response in the presence of the wind forcing. As shown in Figure 5.6, the basin is square, with a side of 10 km. It is shallow near the north and south shores and gradually increases from 5 to 15 m in the central region. The basin is uniform in the west-east direction, which is suitable for investigating the effect of topography on the circulation pattern. A

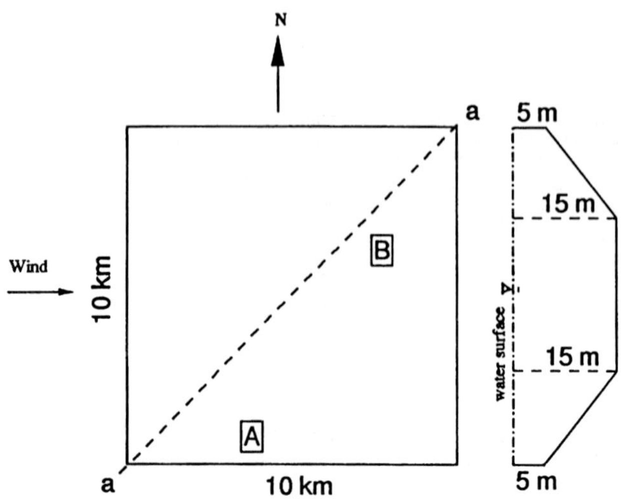

Fig. 5.6. A simplified test basin and its topography, used for model comparison. The basin is uniform along the *x*-direction and varied in the *y*-direction (Reprinted from Wu and Tsanis (1995b), copyright with kind permission from Elsevier Science Ltd., The Boulevard, Langford Lane, Kidlington OX5 1GB, UK).

uniform 10 m/s west wind is applied in the test. The grid size of 500 m and time step of 20 s is used in all models. Twelve layers, with enhanced resolution near both the surface and bottom are used in the multi-layered, 3D model in order to obtain a good velocity profile in the vertical plane. $\lambda = 0.35$, $z_{sh} = z_{bh} = 0.002$ are used in the VHI3D model in the simulations.

Simulated results and model comparison

The steady depth-averaged circulation patterns, as shown in Figure 5.7, predicted by the 2D, Q3D, VHI3D, and 3D models are similar. The strong current (about 10 cm/s) appears along the wind direction in the shallow shores, while, in the central deep basin, the current is in the opposite direction. It produces two symmetric gyres in the basin, one clockwise and another counterclockwise. This is due to the topography of the basin (Csanady, 1982). The surface drift is important in the study of surface oil spills and wave–current interactions.

Although it is very difficult to measure the surface drift in lakes or oceans, many laboratory experiments have been conducted to measure this velocity. The

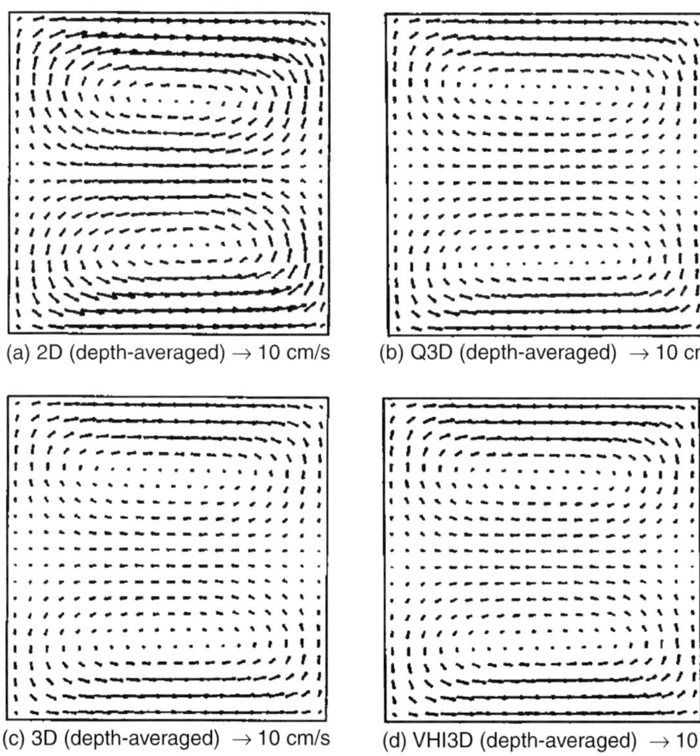

(a) 2D (depth-averaged) → 10 cm/s (b) Q3D (depth-averaged) → 10 cm/s (c) 3D (depth-averaged) → 10 cm/s (d) VHI3D (depth-averaged) → 10 cm/s

Fig. 5.7. Depth-averaged circulation patterns simulated by the 2D, Q3D, 3D, and VHI3D models. A 10 m/s west wind is applied and steady state of current is presented. (a) 2D (depth-averaged); (b) Q3D (depth-averaged); (c) 3D (depth-averaged); (d) VHI3D (depth-averaged) (Reprinted from Wu and Tsanis (1995b), copyright with kind permission from Elsevier Science Ltd., The Boulevard, Langford Lane, Kidlington 0X5 1GB, UK).

commonly accepted value of the ratio of water surface velocity to wind velocity is about 0.3 (Wu, 1975). A 10 m/s wind, according to Wu (1975), produces a surface drift of 30 cm/s. Figures 5.8(a), (b), and (c), show the surface velocity predicted by the models. Obviously, no solution can be obtained from the 2D model. The

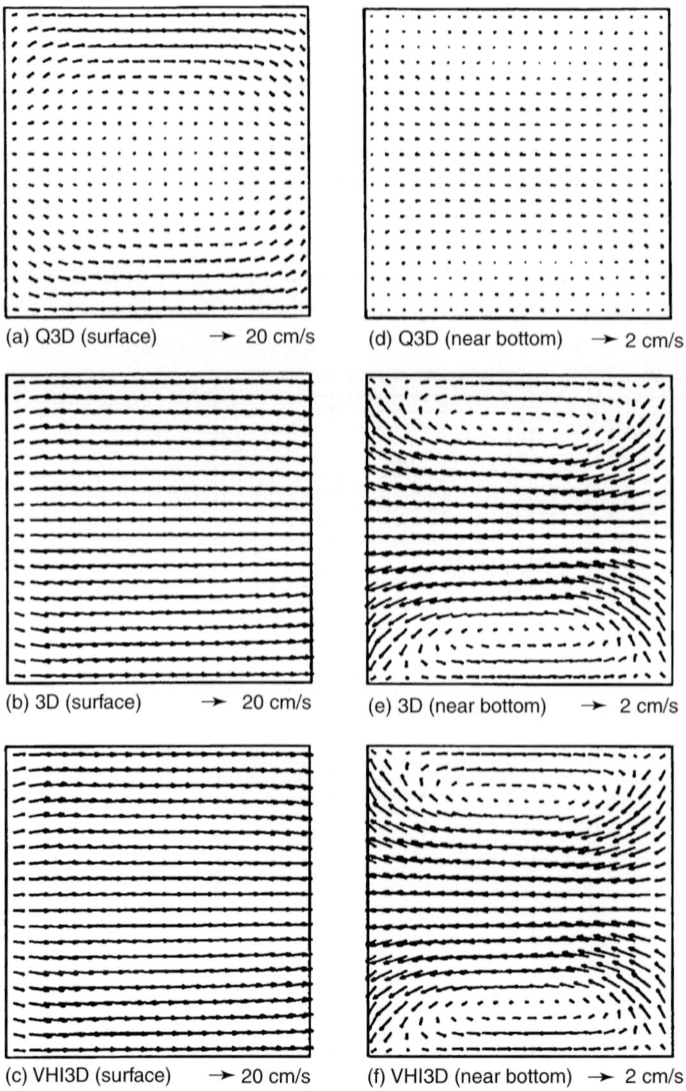

Fig. 5.8. The simulated surface and near-bottom circulation patterns simulated by the Q3D, 3D, and VHI3D models. A 10 m/s west wind is applied and steady state of current is presented. (a) Q3D (surace); (b) 3D (surface); (c) VHI3D (surface); (d) Q3D (near bottom); (e) 3D (near bottom); (f) VHI3D (near bottom) (Reprinted from Wu and Tsanis (1995b), copyright with kind permission from Elsevier Science Ltd., The Boulevard, Langford Lane, Kidlington OX5 1GB, UK).

currents predicted by the VHI3D and 3D models are almost exactly the same, about 27–30 cm/s in the same direction with the wind. The currents predicted by the Q3D model are about 15–17 cm/s in the shallow shore, while it is only 1–3 cm/s in the central deep basin. The near bottom velocity becomes very crucial in suspended solid transport models because the re-suspension rate is largely dependent on the near bottom shear stress, and, thus, the bottom velocity. The circulation patterns at the first layer near the bottom (e.g., $\sigma = 0.02$) are illustrated in Figures 5.8(e), (f), and (g). The circulation patterns are similar to the Q3D, 3D, and VHI3D models. However, the magnitude in the Q3D model is smaller than in the 3D and VHI3D models. The bottom current has the same direction as the wind in the shallow shores, except in the corners. This difference is due to the lateral boundaries. It has an opposite direction to the wind in the central deep basin.

The vertical current structure is illustrated in Figure 5.9. The velocity components u and v at two locations, **A** and **B** (see Fig. 5.6), are included. The v component is uniform because a west wind is applied. In the shallow shores,

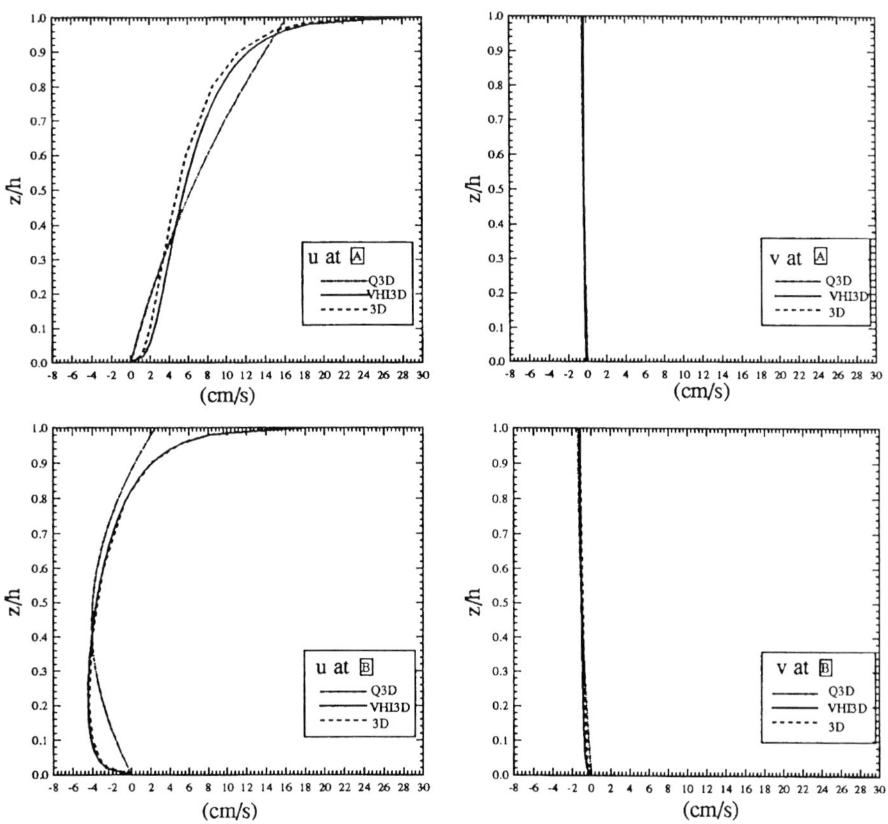

Fig. 5.9. Vertical profiles of horizontal velocity at points A and B simulated by the Q3D, 3D, and VHI3D models. A 10 m/s west wind is applied and steady state of current is presented (Reprinted from Wu and Tsanis (1995b), copyright with kind permission from Elsevier Science Ltd., The Boulevard, Langford Lane, Kidlington OX5 1GB, UK).

presented by point **A**, there is no return flow in the lower layers, and the current is along the wind direction in the whole depth. In the deep basin, presented by point **B**, there is a return flow opposite to the wind. The prediction by the Q3D model is a parabolic profile, nearly linear and decreasing from the surface to the bottom in the shallow shore. The prediction of the VHI3D is very close to that of the 3D model; both have a double-logarithmic profile. The velocity distributions near the surface and bottom are well simulated. Although both VHI3D and 3D models simulate the current structure well, a high-resolution grid for the vertical in the 3D model must be used (Wu, 1993), while there are no layers applied to the VHI3D model. The current, at any depth and at any location, can be directly calculated with little computational work. In contrast, the multi-layered 3D model current outputs are only made at those depths where the vertical layer is assigned.

Figure 5.10 shows the set-up in section a–a (see Fig. 5.6). The water goes down about 1 cm in the upwind west shore, and rises about 1 cm in the downwind east shore. The set-up predicted by the 2D model is smaller than that of the other three models. However, the set-up predicted by the Q3D model is larger than that of the VHI3D and 3D models. From a numerical viewpoint, these differences can be adjusted. For example, by decreasing the bottom friction coefficients, the set-up becomes higher. From a physical viewpoint, however, the differences are highly significant and depend largely on the different formulations of the bottom shear stress. Supposing that steady state is reached and the Coriolis force and advective terms are ignored, the set-up between the two locations ($\Delta\zeta$) can be calculated by

$$\Delta\zeta = \frac{L}{g\bar{H}}(\tau_s - \tau_b) \tag{5.46}$$

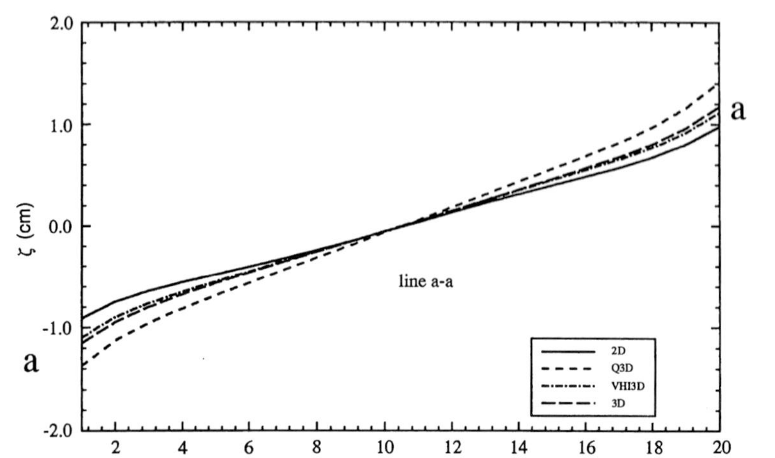

Fig. 5.10. The simulated wind set-up along a–a section (see Fig. 5.5) simulated by the 2D, Q3D, VHI3D, and 3D models. A 10 m/s west wind is applied and steady state of current is presented (Reprinted from Wu and Tsanis (1995b), copyright with kind permission from Elsevier Science Ltd., The Boulevard, Langford Lane, Kidlington OX5 1GB, UK).

where $\bar{H}$ is the effective depth in the set-up line (separated by a length L), determined by the topography. Therefore, for a given surface stress, which is determined by the wind velocity and wind drag coefficient, the set-up only depends on the formulation of bottom shear stress τ_b. Underestimation of bottom stress will result in overestimation of the wind set-up, and vice versa.

In the Q3D and VHI3D models, the bottom shear stress is evaluated using this formula based on the velocity profile, while, in the 3D model, it is calculated from the velocity at the first layer from the bottom. The conventional 2D model, however, computes the bottom stress based on the depth-averaged velocity, by using either the quadratic law (Leendertse, 1967) or the linear relationship (Simons, 1980, 1985; Schwab, 1983). The direction of the depth-averaged currents is not always the same as the current near the bottom. The depth-averaged velocity is usually larger than the bottom velocity. As a result, the bottom shear stresses are overestimated; then, the set-up is underestimated. In the Q3D model, the bottom shear stress is underestimated because the steep velocity gradient could not be well simulated; due to the parabolic velocity profile used. Therefore, the wind set-up predicted by this model is higher than the one for the VHI3D or 3D models.

Another important observation that should be emphasized is the computational efficiency of these models. Assuming the computational time (CPU) for the 2D model (2.1 min for 24 h simulation in SPARC station 10) as one unit, then the times for the Q3D, VHI3D, and 3D models are 1.5, 1.8, and 12 units, respectively. The VHI3D model takes the same order of time as the 2D model, while significantly less time than the fully 3D model.

Chapter 6

PARTICLE TRAJECTORY MODELS AND POLLUTANT TRANSPORT MODELS

6.1 PARTICLE TRAJECTORY MODELS

Determining pollutant transport or powerless vessel movement in marine and lake environments is important in assessing damages, due to oil spills, or in vessel rescue emergencies. Experiments using drogue and satellite drifters have been employed recently to determine the current structure in lakes (Murthy et al., 1985; Wu et al., 1987; Tsanis, 1989; Boyce and Chiocchio, 1991). After the current field has been obtained by the circulation model, it is a straightforward process to formulate a particle trajectory prediction model.

6.1.1 A Second-Order Lagrangian Trajectory Model

The equations governing the current-induced particle motions are

$$\frac{\partial X}{\partial t} = U(x, y); \quad \frac{\partial Y}{\partial t} = V(x, y) \tag{6.1}$$

where X and Y are the particle's coordinates, and U and V are the depth-averaged velocity components. A second-order finite difference scheme can be applied to numerically integrate the preceding equations (Bennett and Clites, 1987). This scheme is given by

$$X^{n+1} = X^n + U(X^n, Y^n)\Delta t + \frac{1}{2}\frac{\partial U}{\partial x}(X^{n+1} - X^n)\Delta t + \frac{1}{2}\frac{\partial U}{\partial y}(Y^{n+1} - Y^n)\Delta t \tag{6.2}$$

$$Y^{n+1} = Y^n + V(X^n, Y^n)\Delta t + \frac{1}{2}\frac{\partial V}{\partial x}(X^{n+1} - X^n)\Delta t + \frac{1}{2}\frac{\partial V}{\partial y}(Y^{n+1} - Y^n)\Delta t \tag{6.3}$$

The values of the velocities U and V and their derivatives at time level n are computed using a bilinear interpolation scheme, as shown in Figure 6.1. They are computed in each mesh from the values of u and v at the four corner points of the grid square in which the particles begin the time step. The slipping factor (resistance

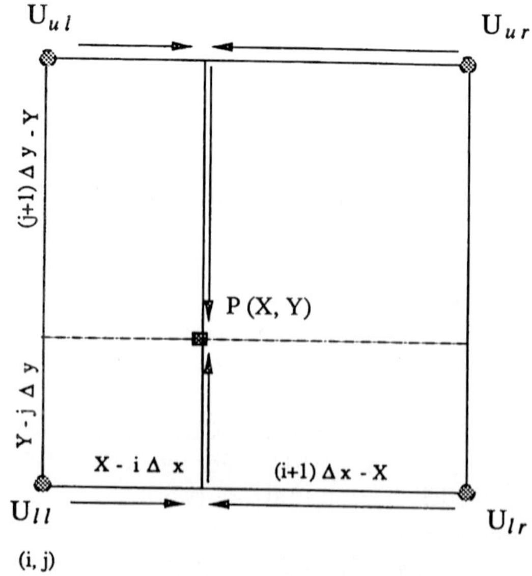

Fig. 6.1. A bilinear interpolation scheme for the particle trajectory prediction model (Wu, 1993).

to the water) and/or windage factor (resistance to the wind) may be considered in the individual application.

6.1.2 A Random-Walk Approach

It is assumed in Figure 6.1 trajectory prediction model that the particle follows the current and its trajectory is determined uniquely by the current velocity. However, the particle trajectories are not uniquely defined by the local velocity field calculated by the numerical model and may be affected by the turbulence and diffusion due to the sub-grid scale processes, which are not simulated in the circulation model. The simulated trajectories in the earlier section can be considered the centroid of a diffusive path or cloud. The dispersion can be simulated by a Monte-Carlo random-walk approach. The random displacement, evaluated by a random number and horizontal dispersion coefficients, is added to the movement computed from the flow field (Leendertse and Liu, 1977; Koutitas and Gousidou-Koutita, 1986). Assuming that, in lakes, the diffusive field is homogeneous, the trajectory prediction equations can be modified as

$$X^{n+1} = X^* + r_1\sqrt{6\Delta t D_x} \tag{6.4}$$

$$Y^{n+1} = Y^* + r_2\sqrt{6\Delta t D_y} \tag{6.5}$$

where X^* and Y^* are the predictions from the Lagrangian model (Eqs. (6.2) and (6.3); r_1 and r_2 are pseudo-Gaussian random numbers with mean zero, and a standard derivation of unity and; D_x and D_y are dispersion coefficients in the x and y directions, respectively. According to the drogue-cloud experiments in Lake Tai, the horizontal dispersion coefficients are approximately 0.1 to 1.0 m^2/s (Wu et al., 1987). The values in Lake Erie are about 1–10 m^2/s (Elzawahry, 1985).

6.2 NUMERICAL SCHEMES FOR THE 2D POLLUTANT TRANSPORT MODELS

Given a definition of the current field from the hydrodynamic model, and estimates of the source and sinks, an integration of the pollutant transport and water quality models can predict the spatial and temporal distributions of pollutant in the water column. The pollutant transport model is based on the conservation of the constituent mass equation, expressed in three dimensions as

$$\frac{\partial C}{\partial t} + u\frac{\partial C}{\partial x} + v\frac{\partial C}{\partial y} + w\frac{\partial C}{\partial z} = \frac{\partial}{\partial x}\left(D_x\frac{\partial C}{\partial x}\right) + \frac{\partial}{\partial y}\left(D_y\frac{\partial C}{\partial y}\right) + \frac{\partial}{\partial z}\left(D_z\frac{\partial C}{\partial z}\right) + S + [R]C \tag{6.6}$$

where C represents the concentration of the dissolved constituent; u, v, and w are velocity components in the x, y, and z directions, respectively; D_x, D_y, and D_z are turbulent dispersion coefficients (they are usually taken as constants); S represents sources and sinks of material; and $[R]$ is a reaction matrix. In most cases, a 2D transport model is used:

$$\frac{\partial C}{\partial t} + U\frac{\partial C}{\partial x} + V\frac{\partial C}{\partial y} = D_x\frac{\partial^2 C}{\partial x^2} + D_y\frac{\partial^2 C}{\partial y^2} + S + [R]C \tag{6.7}$$

The boundary conditions completing the model are (a) solid boundaries—zero normal flux, (b) free transmission boundaries—uniform flux, and (c) pollutant sources—concentration is known.

Equation (6.7) is solved by the finite difference method, and the same staggered grid as in the circulation model is used (the concentration is set at the centre of the mesh, see Fig. 6.2). The 2D finite difference representation of concentration can be written as

$$C(x, y, t) = C(i\Delta x, j\Delta y, n\Delta t) = C_{ij}^n \tag{6.8}$$

and define

$$CX_{ij} = U_{ij}\frac{\Delta t}{\Delta x}, \quad CY_{ij} = V_{ij}\frac{\Delta t}{\Delta y} \tag{6.9}$$

as the Courant number, and set

$$DX = \frac{D_x\Delta t}{\Delta x^2}, \quad DY = \frac{D_y\Delta t}{\Delta y^2} \tag{6.10}$$

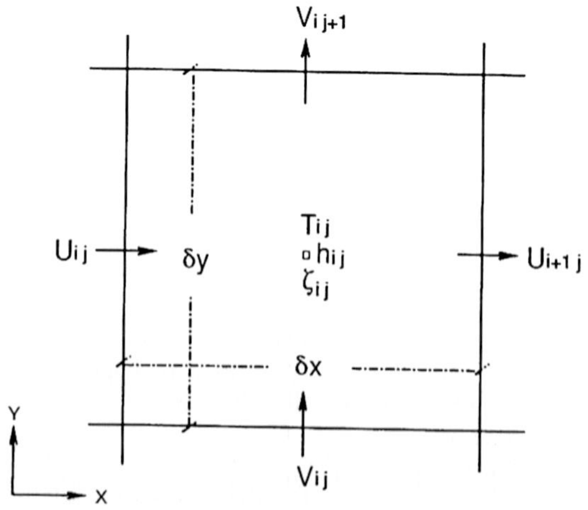

Fig. 6.2. Variable arrangements in a staggered finite difference grid system for the 30 hydrodynamic model; view of a horizontal grid (Wu, 1993).

The explicit central finite difference scheme

$$\delta D^n_{ij} = \left(D_x \frac{\partial^2 C}{\partial x^2} + D_y \frac{\partial^2 C}{\Delta y^2} \right) \Delta t = DX(C^n_{i+1,j} - 2C^n_{ij} + C^n_{i-1j}) + DY(C^n_{ij+1} - 2C^n_{ij} + C^n_{ij-1}) \quad (6.11)$$

is used for the diffusion terms. We will examine five numerical schemes which are typically applied for the advection terms. The source term or simple decay term is not included in the following scheme, since it can be directly included.

6.2.1 Explicit Scheme

$$C^{n+1}_{ij} = C^n_{ij} - CX_{ij}(C^n_{i+1j} - C^n_{i-1j}) - CY_{ij}(C^n_{ij+1} - C^n_{ij-1}) + \delta D^n_{ij} \quad (6.12)$$

6.2.2 First-Order Upwind Scheme

$$\begin{aligned} C^{n+1}_{ij} &= C^n_{ij} - CX_{ij}(C^n_{ij} - C^n_{i-1j}) - CY_{ij}(C^n_{ij} - C^n_{ij-1}) + \delta D^n_{ij} \\ &\quad \textit{for } U>0, V>0; \\ &= C^n_{ij} - CX_{ij}(C^n_{i+1j} - C^n_{ij}) - CY_{ij}(C^n_{ij+1} - C^n_{ij}) + \delta D^n_{ij} \\ &\quad \textit{for } U<0, V<0 \end{aligned} \quad (6.13)$$

6.2.3 McCormack Two-Step Scheme

$$C_{ij}^{*n+1} = C_{ij}^{n} - CX_{ij}(C_{i+1j}^{n} - C_{ij}^{n}) - CY_{ij}(C_{ij+1}^{n} - C_{ij}^{n}) + \delta D_{ij}^{n} \tag{6.14}$$

$$C_{ij}^{**n+1} = C_{ij}^{*n} - CX_{ij}(C_{ij}^{*n+1} - C_{i-1j}^{*n}) - CY_{ij}(C_{ij}^{*n+1} - C_{ij-1}^{*n}) + \delta D_{ij}^{*n} \tag{6.15}$$

$$C_{ij}^{n+1} = \frac{1}{2}(C_{ij}^{n} + C_{ij}^{**n+1}) \tag{6.16}$$

6.2.4 Quick Scheme

The quadratic upstream interpolation for convective kinematics (QUICK) is a third-order upwinding scheme (Leonard, 1979). Figure 6.3 shows a control volume centered at node (i, j), focusing on the convection across the left face for velocity $U_l > 0$. The downstream (D), central (C), upstream (U), and transverse (T, B) nodes in such cases are presented in the figure. For the basic QUICK scheme, the convected value averaged at the left face is

$$C_l = \frac{1}{2}(C_D + C_C) - \frac{1}{8}(C_D - 2C_C + C_U) + \frac{1}{24}(C_T - 2C_C + C_B) \tag{6.17}$$

which includes the linear interpolation term, the upstream normal curvature term, and a small term representing the effect of upstream-biased transverse curvature in computing the face average. A similar formula is used for calculating the convected

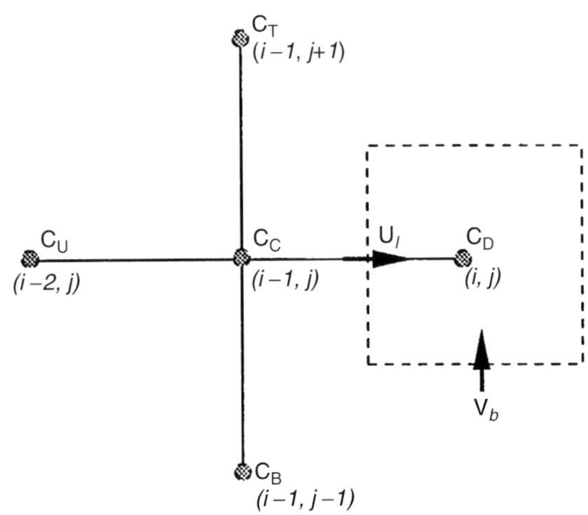

Fig. 6.3. 2D control volume showing nodes involved in estimating the left-face value for $U_l > 0$ in QUICK or SHARP schemes (Wu, 1993).

value C_b at the bottom faces. As a result, the left-face fluxes (convective plus diffusive) are

$$FLUXL_{ij} = CX_{ij}C_l - DX(C_{ij} - C_{i-1j}) \tag{6.18}$$

and bottom-face fluxes

$$FLUXB_{ij} = CY_{ij}C_b - DX(C_{ij} - C_{i-1j}) \tag{6.19}$$

The explicit algorithm is then written as:

$$C_{ij}^{n+1} = C_{ij}^{n} + FLUXL_{ij} - FLUXL_{i+1j} + FLUXB_{ij} - FLUXB_{ij+1} \tag{6.20}$$

6.2.5 Sharp Scheme

The same procedure is used in the Simple High-Accuracy Resolution Program (SHARP). The normal curvature factor (CF) is a constant in the QUICK algorithm (1/8 in Eq. (6.17)). According to the range of normalized variable $\bar{C}_C = (C_C - C_U)/(C_D - C_U)$, the different approximations (third-order upwinding, quadratic formula, exact exponential upwinding formula) for the factor *CF* are employed in the SHARP method (Leonard, 1988).

6.2.6 Comparison of Numerical Schemes

In order to examine the behavior of the already mentioned schemes, a one-dimensional (1D) advection–diffusion problem was used. The grid size was 1.0 m and the velocity u was 1.0 m/s. The horizontal dispersion coefficient D_x was taken as 0.002, 0.01, 0.1, and 1.0 m^2/s, which corresponds to the Peclet number P_e of 500, 100, 10, and 1, respectively. The simulations by the Upwind, McCormack, QUICK, and SHARP schemes were compared with the corresponding exact solutions. The conclusion can be clearly made from the comparisons in Figure 6.4. For the high advection case (Fig. 6.4(a), (b)), an oscillational solution is produced by the McCormack scheme, with slight overshoots by the QUICK scheme. The first upwind scheme does not produce overshoots, but the artificial (or numerical) diffusion distorts the results. The SHARP scheme produces good results without any oscillations or overshoots. When the diffusion is close to the advection, all schemes produce similar results, as shown in Figures 6.4(c) and (d).

6.3 NUMERICAL SCHEMES FOR THE 3D POLLUTANT TRANSPORT MODEL

The 3D pollutant transport model is a sub-model of the IDOR 3D hydrodynamic/pollutant transport model discussed in Chapter 4. In this pollutant

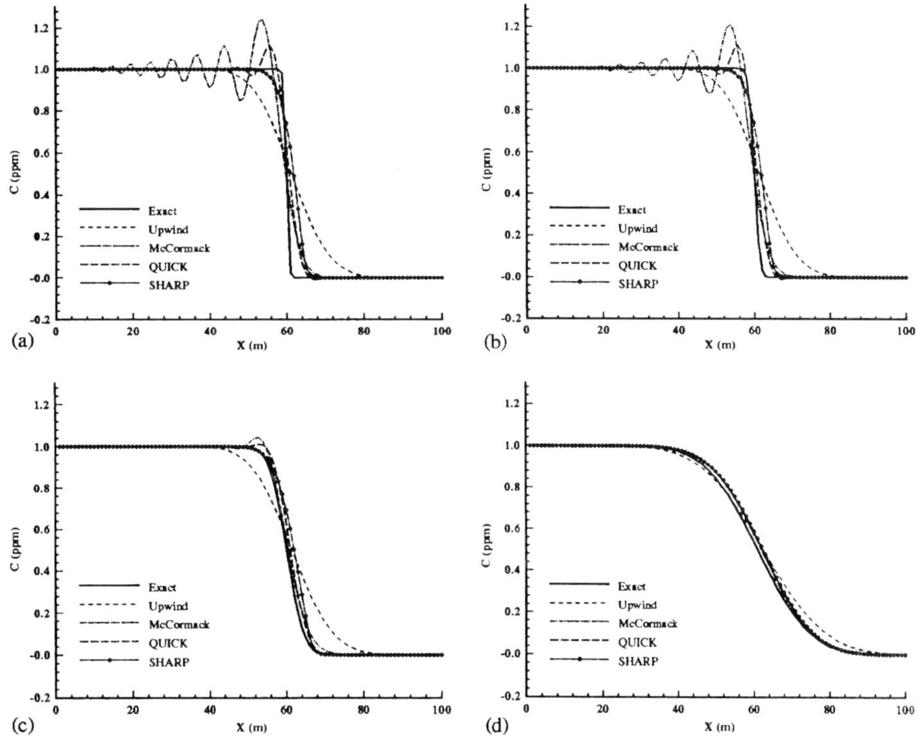

Fig. 6.4. Comparison of simulations by the *Upwind*, *McCormack*, *QUICK* and *SHARP* schemes with analytical solutions under various Peclet numbers ($P_e = u\Delta x/D_x$). (a) $P_e = 500$, (b) $P_e = 100$, (c) $P_e = 10$; (d) $P_e = 1$ (Wu, 1993).

transport model, a temporal second-order Adams–Bashforth scheme is used for the temporary term, a spatial second-order upwind scheme (Donor-Cell) is used for the advective terms, and a central scheme is used for the diffusive terms based on the control volume method.

(1) Temporal term

$$\frac{C_{i+1/2,j+1/2}^{k+1/2,n+1} - C_{i+1/2,j+1/2}^{k+1/2,n}}{\Delta t} \Delta V_{i+1/2,j+1/2}^{k+1/2,n} = 1.5Z_{i+1/2,j+1/2}^{k+1/2,n} - 0.5Z_{i+1/2,j+1/2}^{k+1/2,n-1} \tag{6.21}$$

where

$$\begin{aligned} Z_{i+1/2,j+1/2}^{k+1/2,n} &= Z_{\text{adv}}(i+1/2, j+1/2, k+1/2, n) \\ &+ Z_{\text{diff}}(i+1/2, j+1/2, k+1/2, n) \\ &+ Z_S(i+1/2, j+1/2, k+1/2, n) \end{aligned} \tag{6.22}$$

(2) Advective terms

$$
\begin{aligned}
&Z_{\text{adv}}(i+1/2, j+1/2, k+1/2, n) \\
&= -\left(u^{k+1/2,n}_{i+1,j+1/2}, C^{k+1/2,n}_{m,j+1/2}, \theta_h\right)^{m=i+3/2}_{m=i+1/2} \Delta y \Delta z^{k+1/2,n}_{i+1,j+1/2} \\
&\quad + \left(u^{k+1/2,n}_{i,j+1/2}, C^{k+1/2,n}_{m,j+1/2}, \theta_h\right)^{m=i+1/2}_{m=i-1/2} \Delta y \Delta z^{k+1/2,n}_{i,j+1/2} \\
&\quad - \left(v^{k+1/2,n}_{i+1/2,j+1}, C^{k+1/2,n}_{i+1/2,m}, \theta_h\right)^{m=j+3/2}_{m=j+1/2} \Delta y \Delta z^{k+1/2,n}_{i+1/2,j+1} \\
&\quad + \left(v^{k+1/2,n}_{i+1/2,j}, C^{k+1/2,n}_{i+1/2,n}, \theta_h\right)^{m=j+1/2}_{m=j-1/2} \Delta y \Delta z^{k+1/2,n}_{i+1/2,j} \\
&\quad - \left(w^{k,n}_{i+1/2,j+1/2}, C^{m,n}_{i+1/2,j+1/2}\right)^{m=k-1/2}_{m=k+1/2} \Delta x \Delta y \\
&\quad + \left(w^{k+1,n}_{i+1/2,j+1/2}, C^{m,n}_{i+1/2,j+1/2}, \theta_h\right)^{m=k+1/2}_{m=k+3/2} \Delta x \Delta y
\end{aligned}
\tag{6.23}
$$

(3) Diffusive terms

$$
\begin{aligned}
&Z_{\text{diff}}(i+1/2, j+1/2, k+1/2) \\
&= N_h \Bigg[\frac{1}{\Delta x} \Big\{ \left(C^{k+1/2,n}_{i+3/2,j+1/2} - C^{k+1/2,n}_{i+1/2,j+1/2} \right) \Delta y \Delta z^{k+1/2,n}_{i+1,j+1/2} \\
&\quad - \left(C^{k+1/2,n}_{i+1/2,j+1/2} - C^{k+1/2,n}_{i-1/2,j+1/2} \right) \Delta y z^{k+1/2,n}_{i,j+1/2} \Big\} \\
&\quad + \frac{1}{\Delta y} \Big\{ \left(C^{k+1/2,n}_{i+1/2,j+3/2} - C^{k+1/2,n}_{i+1/2,j+1/2} \right) \Delta x \Delta z^{k+1/2,n}_{i,j+1/2} \\
&\quad - \left(C^{k+1/2,n}_{i+1/2,j+1/2} - C^{k+1/2,n}_{i+1/2,j-1/2} \right) \Delta x \Delta z^{k+1/2,n}_{i+1/2,j} \Big\} \Bigg] \\
&\quad + N_v \Bigg\{ \frac{1}{\Delta^{k,n}_{i+1/2,j+1/2}} \left(C^{k-1/2,n}_{i+1/2,j+1/2} - C^{k+1/2,n}_{i+1/2,j+1/2} \right) \Delta x \Delta y \\
&\quad - \frac{1}{\Delta z^{k+1,n}_{i+1/2,j+1/2}} \left(C^{k+1/2,n}_{i+1/2,j+1/2} - C^{k+3/2,n}_{i+1/2,j+1/2} \right) - \Delta x \Delta y \Bigg\}
\end{aligned}
\tag{6.24}
$$

(4) Source terms

$$
Z_s(i+1/2, j+1/2, k+1/2) = S^{k+1/2,n}_{i+1/2,j+1/2} \Delta V^{k+1/2,n}_{i+1/2,j+1/2} \tag{6.25}
$$

Chapter 7

NUMERICAL ANALYSIS OF COARSE-FINE GRID MODEL

7.1 AN APPROACH TO MODELLING NEARSHORE AREAS[5]

Recently, a 2D circulation model was developed to simulate the Toronto Waterfront of Lake Ontario (MOE, 1989) and Collingwood Harbour of Lake Huron (MOE, 1992), using *in-situ* measured currents as open boundary conditions. There are a number of problems with this model. Field observations are expensive. A coarse current meter array fails to provide the detailed structure in the open boundary, which is complicated and dependent on wind conditions to the extent that no interpolation can follow such changes. Measurements cannot provide the water elevations at all grid points along the open boundary; this causes a violation of the mass conservation law within the calculation domain. Finally, this technique cannot predict circulation and transport, which is an important aspect of the application. This is because the model is driven by the measured current at the open boundary, and these data are not available.

Numerically, two approaches can be employed. The first approach is to formulate a radiation boundary condition, which will allow disturbances to "radiate" out of the modelled domain with a minimum of boundary reflection. Descriptions of such boundary conditions can be found in Roed and Cooper (1986), Blumberg and Kantha (1985), Miller and Thorpe (1981), and Krestenitis (1987). The most widely used type of condition is the Sommerfeld condition and its modified forms. Krestenitis (1987) presented five different open boundary conditions constructed from free radiation, *Neuman* and *Sommerfeld conditions*. They were developed for the open-sea boundary of a coastal area and were investigated through an application to the Thermaikos Gulf in the Aegray Sea. The results, in terms of free surface contours and current circulation patterns, were not always realistic, particularly the water elevations. It was also not possible to evaluate the model's results. The second approach is the so-called grid-nesting method, whereby a much larger domain, such as the whole lake or harbor, is first modelled using a coarser grid model, and the evolving fields are used to determine the sub-boundary conditions. This technique is well documented in the meteorological literature. Murthy et al. (1986) applied this method in the dynamic and transport modelling of the Niagara River plume in Lake Ontario. However, there were no detailed descriptions of schemes. For ocean coastal modelling, the first approach is necessary to determine the open boundary conditions for large

[5]Materials in Sections 7.1–7.3 are reproduced from Tsanis and Wu (1995), copyright with permission.

areas. The tide elevation is given by the observation data (Blumberg and Kantha, 1985). The second approach is appropriate for modelling a specific bay or smaller nearshore area (Crean et al., 1988). However, it is more feasible to employ a nested-grid modelling approach for the nearshore areas in lakes, since the whole lake can first be modelled with a coarse grid.

7.2 ALGORITHM FOR COUPLING THE COARSE AND FINE GRIDS

A nested-grid system consists of small, fine-grid domains embedded in larger coarse-grid domains. The depths for the fine-grid model can be re-schematized from hydrographic charts for a better representation of topography and various engineering designs. In order to remain unaffected by local changes in depth, the open boundaries for the coarse-grid model should be located away from the region of interest. Also, it is better to keep the same depth along the open boundaries in the fine-grid model to ensure conservation of transport.

Open boundary conditions can be obtained for the fine-grid model from the coarse-grid model. The water elevations and velocities at the open boundaries in the coarse-grid model are stored at certain time intervals, such as 15 min or 1 h. The time step used in the fine-grid model is usually smaller than the one used in the coarse-grid model. Therefore, a temporal interpolation scheme is needed to provide the corresponding values in the open boundary for the fine-grid model at each step. The velocities and elevations along the open boundaries for the fine-grid model are prescribed using spatial interpolation on the coarse-grid model. Simple interpolation can be used. The errors introduced are few because the interpolation is within one grid. Only the elevation and one velocity component need such an interpolation (u component at southern and northern boundaries, and v component at western and eastern boundaries), while another component can be directly prescribed from the coarse grid. A typical interpolation for a western open boundary, using a nested-grid system that is one-quarter of the size, is shown in Figure 7.1(a), along with the variable arrangement. The spatial interpolation schemes are

$$u_1 = u_2 = u_3 = u_4 = U_{ij} \tag{7.1}$$

$$v_1 = V_{ij} \tag{7.2}$$

$$v_2 = v_3 = v_4 = SI(V_{ij}, V_{ij+1}) \tag{7.3}$$

$$\zeta_1 = \zeta_2 = \zeta_3 = \zeta_4 = SI(\zeta_{i-1j}, \zeta_{ij}) \tag{7.4}$$

where $SI(V_1, V_2)$ refers to a linear interpolation between two variables V_1 and V_2 A more accurate scheme for Eqs. (7.3) and (7.4) may include values from additional neighboring grids.

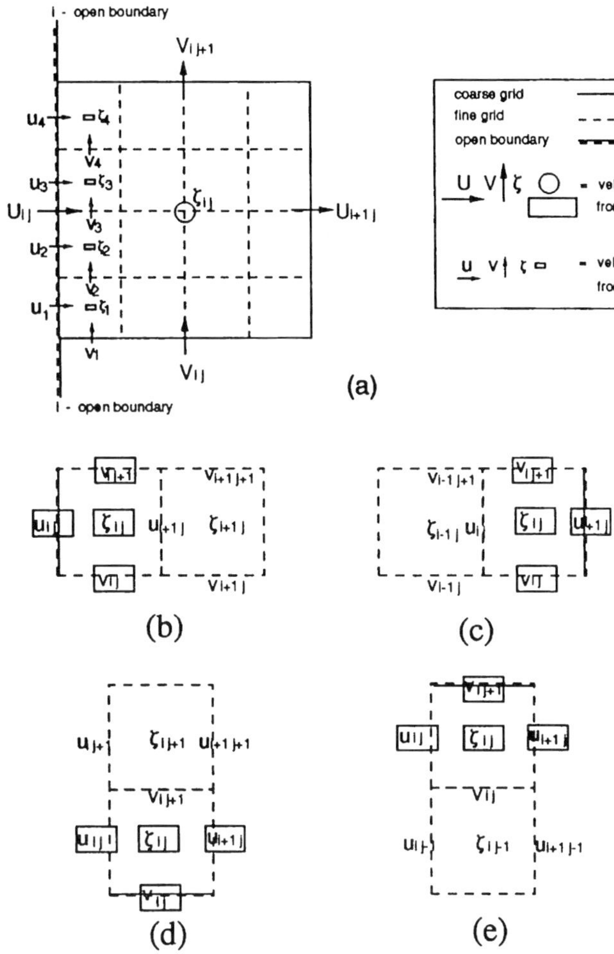

Fig. 7.1. Open boundary conditions in the nested-grid modelling: (a) the detailed variable arrangement and spatial interpolation diagram at a west open boundary mesh in both the coarse-grid and fine-grid system. The implementation of open boundary conditions for: (b) west; (c) east; (d) south; and (e) north open boundary in the fine grid system (Tsanis and Wu, 1995).

There are two methods to prescribe the open boundary conditions: (a) prescribe only the u and v velocity components, and (b) prescribe both the velocities and the elevations. Eventually, method (b) is used because both velocity and elevation along an open boundary can be derived from the coarse-grid model. The arrangement of velocity components for the different open boundaries, that is, north, south, east, and west, are different, as shown in Figures 7.1(b), (c), (d), and 7.1(e), respectively. This arrangement results in additional complexity in the calculation program, yet mass continuity is satisfied.

7.3 APPLICATIONS OF THE COARSE-FINE GRID MODEL TO HAMILTON HARBOUR

7.3.1 Introduction

There are 50 recommendations under the Remedial Action Plan (RAP) for Hamilton Harbour (MOE, 1992b). In order to protect the shoreline; promote fish and waterfowl habitat, public access, fishing and wildlife appreciation; new artificial islands will be built in three nearshore areas (see Fig. 7.2) in Hamilton Harbour by the Fish and Wildlife Restoration Committee (FWRC, 1992).

Before any of these projects can proceed, the federal government environmental review process will screen the implications of the projects on the natural and social environment. It is therefore necessary to know (a) the existing state of current movement within the Harbour under various wind and environmental conditions; (b) the effect of islands on currents; and (c) the effect of islands on pollutant transport and dispersion. It is very difficult to reproduce the field condition in the physical model for the above task, due to scale effects on pollutant transport (Tsanis et al., 1990). Therefore, there is a need to develop a numerical hydrodynamic/pollutant transport model as an evaluative tool for the assessment of various options. There are two circulation models applied to Hamilton Harbour in a coarse-grid of 300 m: a 2D model by Rasmussen and Badr (MOE, 1974) and a 3D model by James and Eid (1978). However, these models are only applicable to

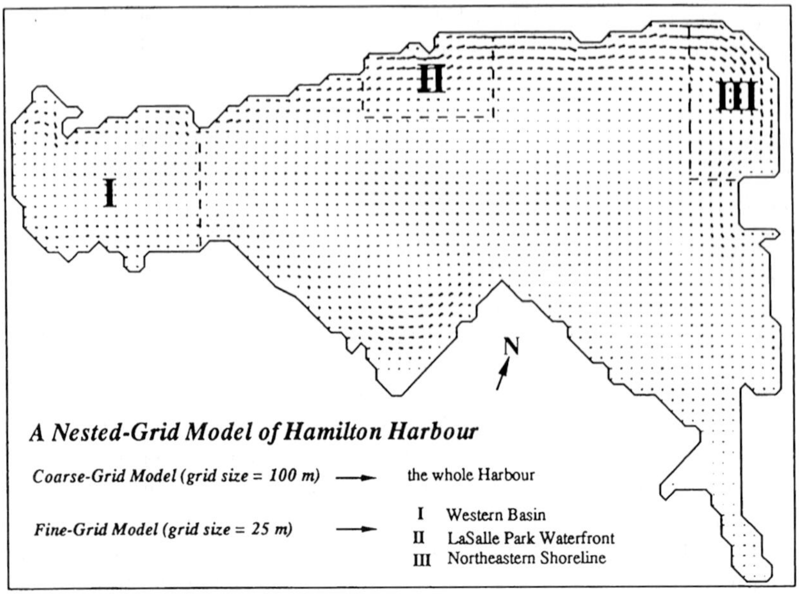

Fig. 7.2. A nested-grid modelling system for Hamilton Harbour and three nearshore areas, that is, Western Basin (I), LaSalle Park Waterfront (II), and Northeastern Shoreline (III). The grid size is 100 m for the coarse-grid model and 25 m for the fine-grid model (Tsanis and Wu, 1995).

the whole harbour. It is necessary to employ a high resolution, or fine-grid model, to better represent the complicated shoreline, in/outflows and sewage sources, and the details of engineering structures. It is impractical to employ such fine-grid models for the whole lake or harbor (a 25 m grid discretization for the whole harbor results in a larger matrix that requires very long computational time), thus an appropriate method is required.

The modified 2D depth-averaged circulation model Q3D (Koutitas 1988; Wu and Tsanis 1991), is employed to simulate depth-averaged circulation patterns and provide the current field for the pollutant transport model of Hamilton Harbour. Basically, the Q3D model is similar to the conventional 2D depth-averaged circulation model. The vertical velocity variation is included in the evaluation of bottom friction terms and depth-averaged advection terms. For a detailed model formulation, readers can refer to Koutitas (1988) and Wu (1993). The circulation model outputs are the depth-averaged velocity components (U in the east–west directions and V in the north–south directions), and water elevations (height above the mean water level). The vertical current structure which can be obtained from the model is not considered in this study. The model equations were solved in a staggered-grid system (see Fig. 7.1), where the water elevation is calculated at the center of the grid, U at the left and right faces, and V at the top and bottom faces of the grid.

The 2D transport equation (see Eq. (6.7)) for a conservative constituent is used. A central difference scheme is applied for the diffusion terms. For the advection terms, the third-order up-winding Simple High-Accuracy Resolution Program (SHARP) (Leonard, 1988) is applied. The numerical diffusion and dispersion are minimized, resulting in a very accurate scheme (Leonard, 1988; Wu, 1993).

First, the hydraulic flow was calculated, by the circulation model, to examine the impact of (a) exchange flow between Hamilton Harbour and Lake Ontario through the Burlington Canal, and (b) other inflows. The inflow from Cootes Paradise was taken as 3.8 and 2.2 m^3/s from CSO's, 2.4 and 3.6 m^3/s from the Burlington and Hamilton STP's respectively. The outflow through the Burlington Canal was taken as 12.0 m^3/s (C.O.A, 1992). The simulated hydraulic flows were very weak, only about 0.1–1.0 cm/s. This is due to the large size of the Harbour and small amount of in/outflows.

In the simulation of wind-induced circulation, the total kinetic energy under various wind speeds reaches a steady-state condition in a very short period because the lake is shallow, and, because of bottom friction, the damping is large. For example, it only takes 2 h to reach 90% of the steady-state value for a 10 m/s (west) wind. A test result for a variable wind event indicates that the harbor's water responds very quickly to wind changes. To obtain the wind-induced circulation pattern, an idealized 10 m/s wind with direction from four compass points (at 90° intervals) was applied to Hamilton Harbour. The simulated results for east, south, west, and north winds are presented in Figures 7.3(a)–7.3(d). The depth-averaged currents follow the wind close to the shore (aligned in a way that is parallel to the shore) and are against the wind in the deeper middle part of the Harbour. Three-to-four smaller eddies are the major features of the depth-averaged current patterns. The first large eddy occupies the northeastern Harbour and has clockwise rotation

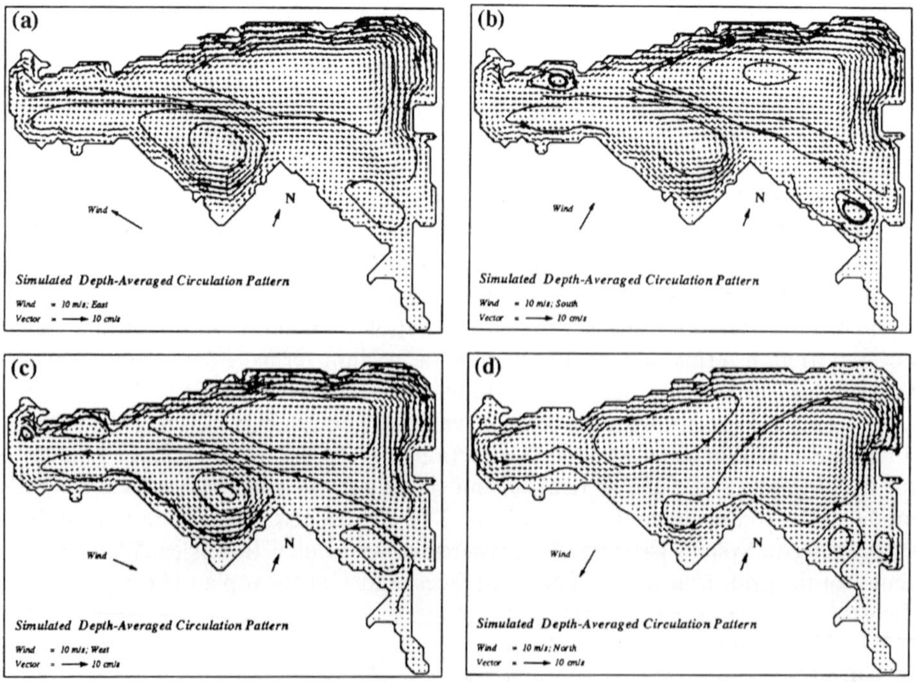

Fig. 7.3. The simulated depth-averaged circulation patterns under a 10 m/s wind with the direction of (a) east, (b) south, (c) west, and (d) north (Tsanis and Wu, 1995).

for north and west winds, and a counterclockwise rotation for south and east winds. For the cases of northwest and southwest winds, the large eddy splits into two large counterclockwise eddies (Wu, 1993). The second eddy occupies the south central Harbour and has a clockwise rotation for east and north winds, and a counterclockwise rotation for west and south winds. The size of these eddies changes with the wind direction. For example, the western end of the Harbour is dominated by southerly and south-westerly winds. Also, the size of these eddies changes with the wind direction. The western end of the Harbour is dominated by a counterclockwise eddy under north winds. It has a clockwise eddy for a south wind, and two counter-rotating eddies for west and east winds. Tests were performed using different wind speeds: 7, 15, and 20 m/s. The circulation features are very similar for all wind speeds, with the only difference being the current speed (the depth-averaged current is higher in the shallower parts of the Harbour when compared to the current in the deeper parts for increasing winds).

7.3.2 Horizontal Mixing Times in Hamilton Harbour

In the study of the interlocking of physical and bio-geochemical processes, an important time scale is the horizontal mixing time of the basin. It can be defined as the time taken, on average, for an amount of a conservative contaminant, introduced at a point within the basin to be mixed uniformly throughout the basin. A test can then be designed for Hamilton Harbour to determine this horizontal mixing time, with the aid of the circulation/pollutant transport model used in this section.

A 100 ppm pollutant was introduced at four typical locations: *A* (western basin), *B* (south shore), *C* (southeastern corner), and *D* (northeastern corner) as identified in Figure 7.4(a). A 10 m/s wind from four directions (N, S, E, W) was applied uniformly to the Harbour and produced different flow patterns. The maximum concentration (C_{max}) was recorded and the simulation was stopped when C_{max} reached 1 ppm. At this time, the initial plug of the pollutant was considered to be fully mixed under the ambient current. As an example, Figure 7.4 shows the concentration contour of a pollutant at a fully mixed state at four locations under the west wind condition. It clearly presents the extent and outline of the pollutant released at these locations. For locations *B* and *D*, the current was highly advective. Consequently, the pollutant mainly stayed around *A* and *C* after being released, due to the existing weak current and well-mixed eddy around these locations. Figures 7.5(a), (b), (c), and (d) show the curves of C_{max} as a function of time, and Figure 7.5(d) also shows the mixing time at *A*, *B*, *C* and *D* under the four wind

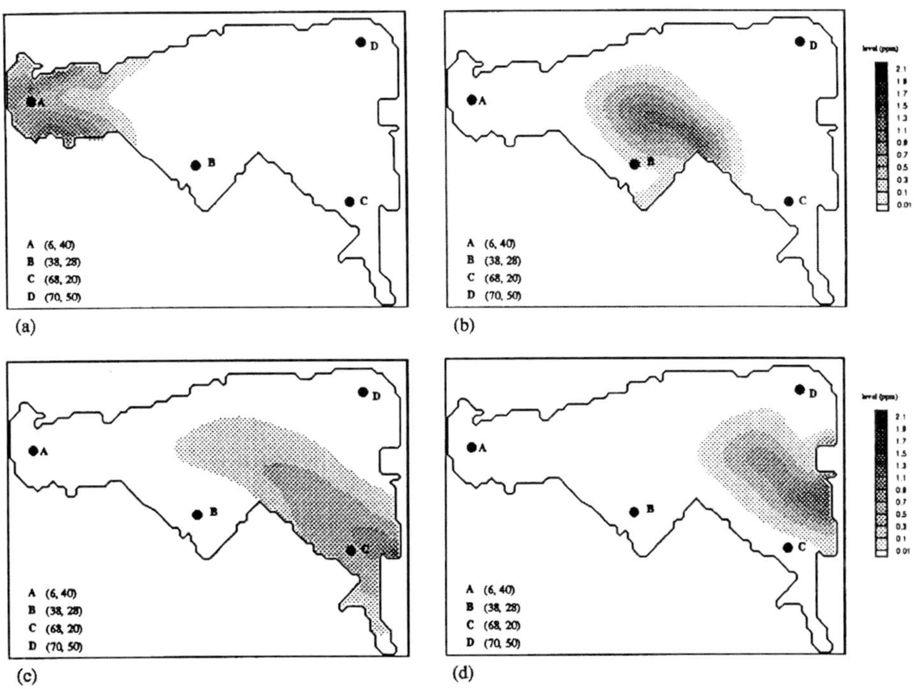

Fig. 7.4. Numerical experiment on the horizontal mixing time scale in Hamilton Harbour: concentration contours when blocks of 100 ppm pollutant released from various locations are fully mixed under a 10 m/s west wind at (a) location *A* (west basin), (b) location *B* (south basin), (c) location *C* (southeastern shore), and (d) location *D* (northeastern corner) (Tsanis and Wu, 1995).

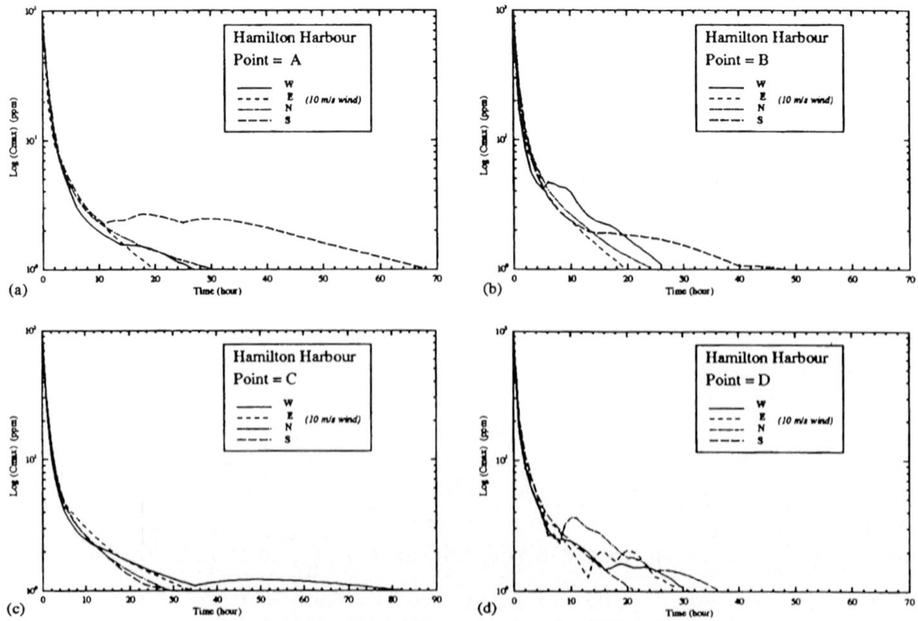

(e)

point \ time (h) \ wind	W (10 m/s)	E (10 m/s)	N (10 m/s)	S (10 m/s)
A	27.01	24.09	30.55	68.64
B	27.00	20.19	24.75	48.11
C	82.97	34.07	29.26	29.93
D	30.61	31.28	36.92	21.60

Fig. 7.5. Numerical experiment on the horizontal mixing time scale in Hamilton Harbour: time series of maximum concentration under a 10 m/s wind with four typical four wind directions (W, E, N, S) at (a) location A, (b) location B, (c) location C, (d) location D, (e) horizontal mixing time at A, B, C, and D under the four wind conditions. The time taken for a plug of 100 ppm pollutant to be fully mixed to level of 1 ppm (Tsanis and Wu, 1995).

conditions. In general, the horizontal mixing time in Hamilton Harbour is about one-to-two days, and, in some wind directions and locations (west wind at *C* and south wind at *A*), over three days. It should be pointed out that the mixing time may become longer for lower wind speeds. It also depends on the horizontal dispersion coefficient. In the simulation, a typical value of $0.5\,m^2/s$ (McCorquodale et al., 1986) was used.

7.3.3 A Nested-Grid Model of Hamilton Harbour

A nested-grid system for Hamilton Harbour and three nearshore areas—including Western Basin, LaSalle Park Waterfront, and Northeastern Shorelines—is

Table 7.1. Parameters in the nested-grid model of Hamilton Harbour.

Name	Area	Model grid	Grid size (m)	IMAX	JMAX	Open B.C.
Hamilton		Coarse	100	79	57	
Harbour	I	Fine	25	75	79	1
Western Basin	II	Fine	25	63	43	3
LaSalle Park	III	Fine	25	35	67	2
Waterfront						
Northeastern						
Shoreline						

presented in Figure 7.2 and Table 7.1. A coarse grid of 100 m × 100 m is used for the whole Harbour, and a fine grid of 25 m × 25 m for the three areas. The coarse-grid model for the Harbour is rotated 23.4° clockwise from the north for better representation of the eastern and northern shorelines. Likewise, the fine-grid model for the three nearshore areas is also rotated in the same manner. The shorelines and topography in the fine-grid models are modified to include detailed variations which cannot be represented in the coarse-grid model for the whole Harbour. It is expected that such modifications will have little effect on the simulation.

7.3.4 Circulation Modelling in the Three Nearshore Areas

To justify the nested-grid modelling technique, the circulation and water elevations from the fine-grid model should generally be similar to those from the coarse-grid model for the area. The simulated circulation patterns from the coarse-grid model and from the fine-grid model under a 10 m/s west wind for LaSalle Park Waterfront are presented in Figure 7.6. Because the shorelines are mostly straight, strong flow belts parallel to the shorelines have always existed close to the shorelines: west to east for the western and southern winds, and east to west for the eastern and northern winds. In the case of the west and east winds, the belts are stronger because the wind direction is parallel to the shorelines. Weak eddies exist in the open water areas. Compared to the result from the coarse grid, the fine-grid model keeps conformity and, at the same time, reveals a more detailed resolution of the velocity field. As in the case of north and west winds, the weak gyre in the central area becomes more detailed in the fine-grid model. In the case of a south wind, two gyres are evident in the fine-grid model which are not clear in the coarse-grid model. Detailed currents are also provided close to the shoreline, where the coarse grids do not exhibit detailed features.

For comparative purposes, a total kinetic energy (KE) in the simulation at each time step is computed from the sum

$$KE^n = \sum_{i=1}^{i=i\max}\sum_{j=1}^{j=j\max}\left((U^n_{ij} + U^n_{i+1j})^2 + (V^n_{ij} + V^n_{ij+1})^2\right)(h_{ij} + \zeta_{ij})\Delta x\Delta y/8 \tag{7.5}$$

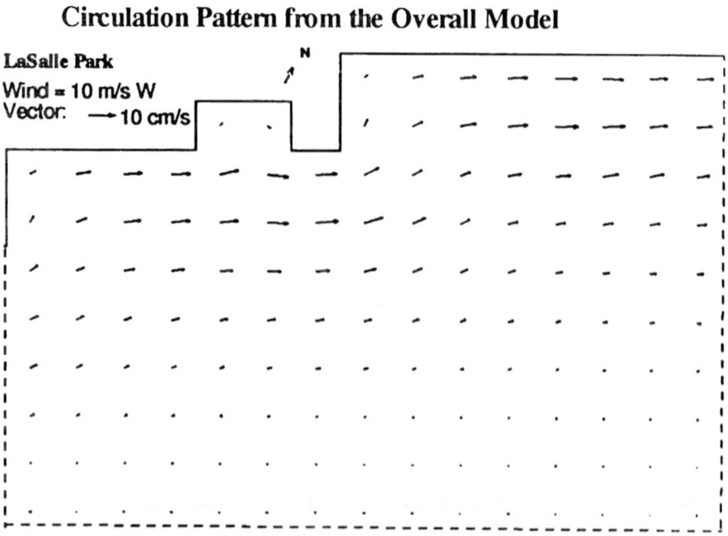

(a)

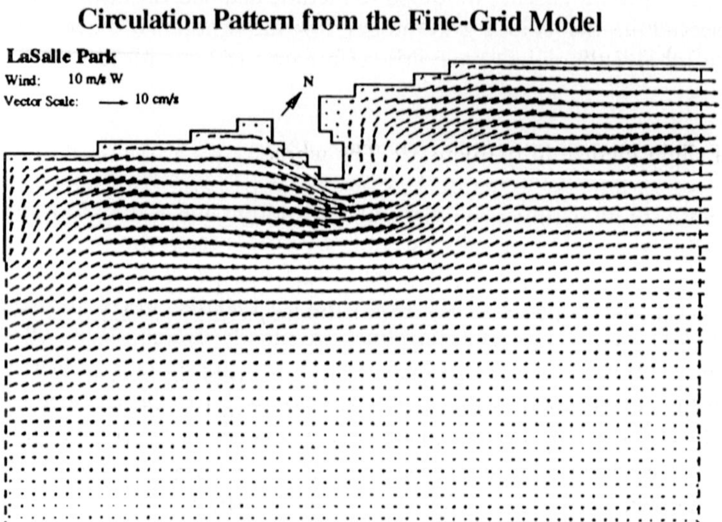

(b)

Fig. 7.6. The simulated depth-averaged circulation pattern in LaSalle Park Waterfront for a 10 m/s west wind from (a) the overall (coarse-grid) model, and (b) the fine-grid model (Tsanis and Wu, 1995).

It can also be used as a steady criterion in the simulation, that is, the steady state is reached when the ratio $|KE^{n+1} - KE^{n}|/KE^{n+1}$ becomes less than a test convergence value (10^{-5}). The kinetic energy for a 10 m/s west wind in both the overall grid and fine-grid models for the LaSalle Park waterfront is illustrated in Figure 7.7. The curves are very close, indicating the fine-grid model's success. The small difference is

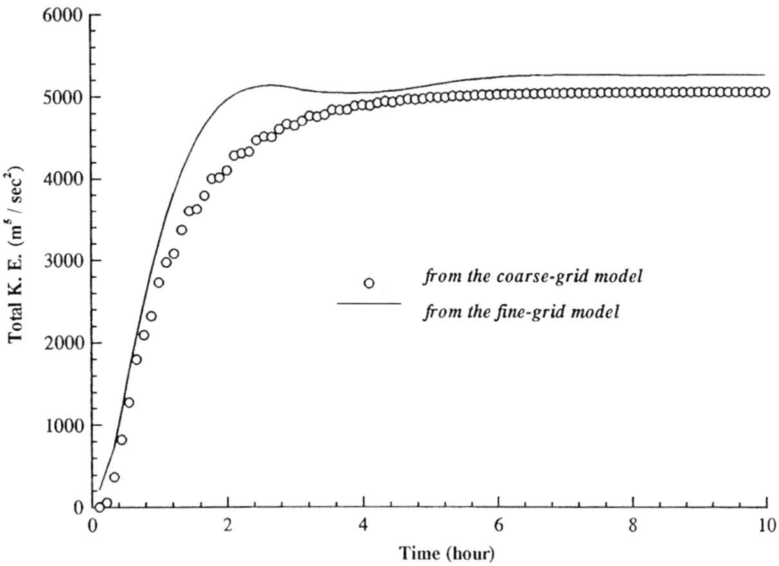

Fig. 7.7. The kinematic energy process in the simulations from both the overall and fine-grid model of LaSalle Park Waterfront, for a 10 m/s east wind (Tsanis and Wu, 1995).

expected, due to a slight change in bathymetry. The whole lake seiche (oscillation) may have some effect on this local area in the coarse-grid model. Because of that, the realistic water elevation field in a nearshore area largely depends on the relative location in the whole lake.

The model using direct open boundary conditions, such as free radiation or the Sommerfeld open boundary conditions, has difficulty reflecting the effect of the relative location. Therefore, it is not surprising that the water elevation field is very difficult to simulate when using the direct open boundary conditions as tested by Krestenitis (1987). As a result, mass continuity is not satisfied. However, using the nested-grid modelling technique proposed in this study, the water elevation field is dramatically well recovered. The contours clearly display the distribution of water elevations for various wind conditions. Figure 7.8 presents the contours of wind-induced elevation at the LaSalle Park Waterfront under a 10 m/s west wind. The elevations from the fine-grid model agree well with the ones from the overall grid model. Contours for other conditions, again, are in good agreement.

7.3.5 Remedial Measures

A coupled hydrodynamic/pollutant transport and dispersion model is used to examine the effects of islands on current and pollutant transport. It was decided, for economic and other reasons, that no islands will be in the immediate plan for the western area. Here, only the feasibility studies in the Northeastern Shorelines is presented.

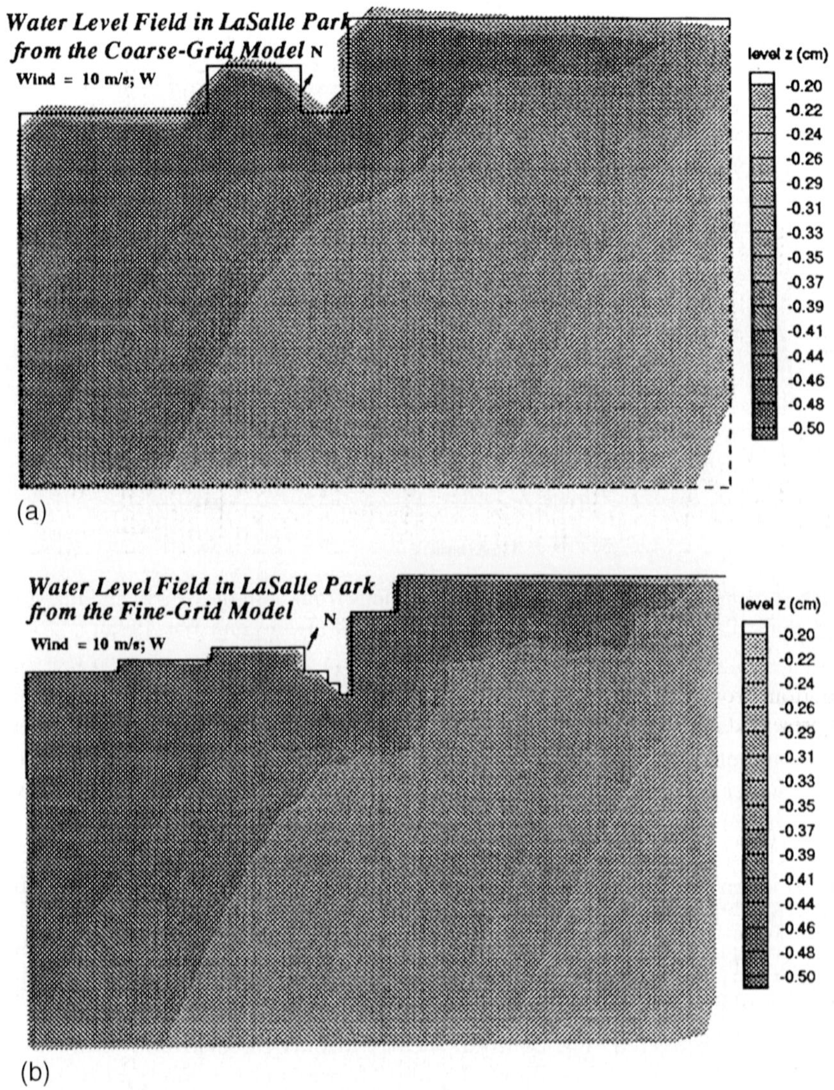

Fig. 7.8. The water elevation fields in LaSalle Park Waterfront for a 10 m/s west wind from (a) the overall (coarse-grid) model, and (b) the fine-grid model (Tsanis and Wu, 1995).

In the Northeastern Shoreline, three alternative island configurations, including about eight small islands along the eastern shorelines, have been preliminarily proposed and numerically studied. As a result, the final proposal is shown in Figure 7.9(b). Figure 7.9(a) presents the simulated depth-averaged circulation pattern from the fine-grid model in the Northeastern Shoreline: (a) without islands, and (b) with islands (a 10 m/s north wind was applied). It can be seen that a clockwise circulation, parallel to the shoreline, exists in the case of north-to-west wind conditions, while counterclockwise patterns exist for south-to-east wind

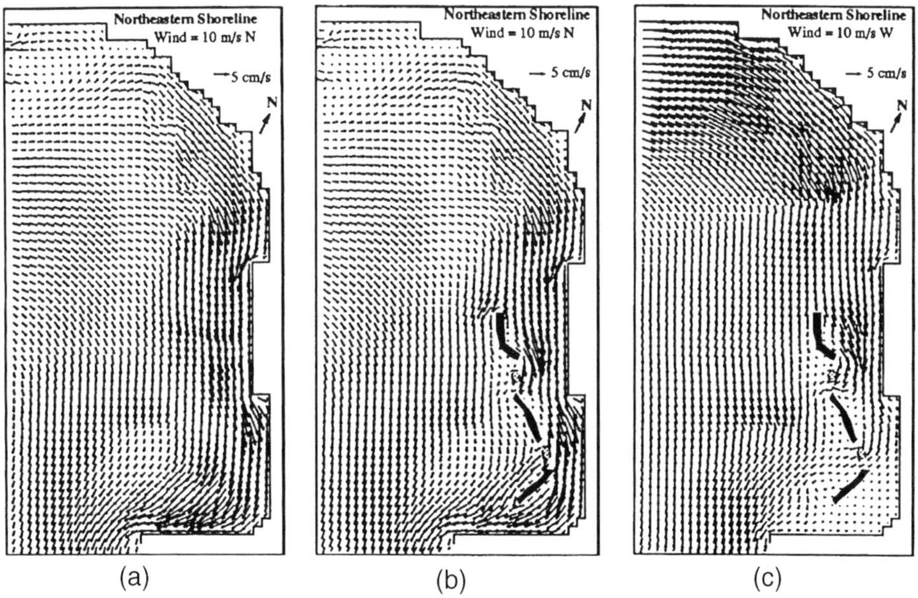

Fig. 7.9. The simulated depth-averaged circulation pattern from the fine-grid model in Northeastern Shoreline. (a) Without islands, (b) With islands, and (c) With islands and rougher bottom around islands. A 10 m/s west wind was applied. The final design of artificial islands is presented in (b), where dashed marks stand for half-depth islands (Tsanis and Wu, 1995).

conditions. The construction of new islands may increase the bottom roughness around the islands. If so, the current pattern will be changed, and in turn affect the pollution transport. Figure 7.9(c) gives the pattern with islands and a rougher bottom (the bottom friction is reduced by half) around the islands. The reduction in current magnitude is obvious. There are two major pollutant sources (COA, 1992): one from Indian Creek, and another from the outfall of Burlington's STP, in the middle of the basin (see Fig. 7.10). The most significant impact of the islands is expected under north-west wind. As an example, the north wind case is presented. Figure 7.10 shows the equal concentration of pollutants at 2 and 5 h, with and without islands. The concentration curves at 10 locations of concern, for cases with and without islands, as well as the case with the islands and a rougher bottom, are presented in Figure 7.11. The features of concentration time history under a 10 m/s north wind is listed in Table 7.2. It is observed that the impact of the islands is small. The rougher bottom may reduce the peak concentration, while having longer retention time around the islands.

The general features of depth-averaged circulation patterns under various wind conditions have been revealed from the simulation results by the hydrodynamic model. The depth-averaged currents follow the wind close to the shore (aligned parallel to the shore), and are against the wind in the middle and deeper part of the harbor. The major features of the depth-averaged current patterns are: (a) two large

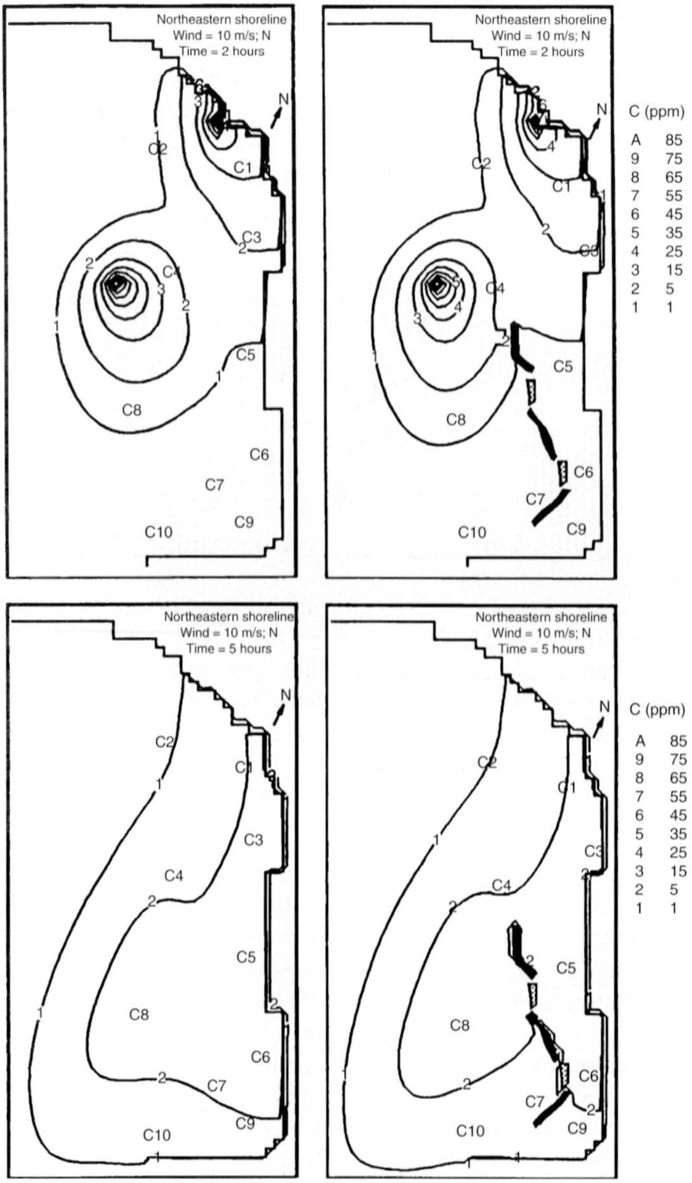

Fig. 7.10. The contours of pollutant concentration in Northeastern Shoreline from the fine-grid model without, and with islands, under a 10 m/s north wind, 2 and 5 h after the cessation of 2 h emission of pollutant from two sources (Tsanis and Wu, 1995).

eddies with opposite rotation covering almost half of the harbor's area, and (b) three-to-four smaller eddies. The model reaches steady state in less than 2 h, which indicates that, under any wind change, the circulation will be established in a shorter time. The hydraulically induced current, due to the inflow/outflow, is much

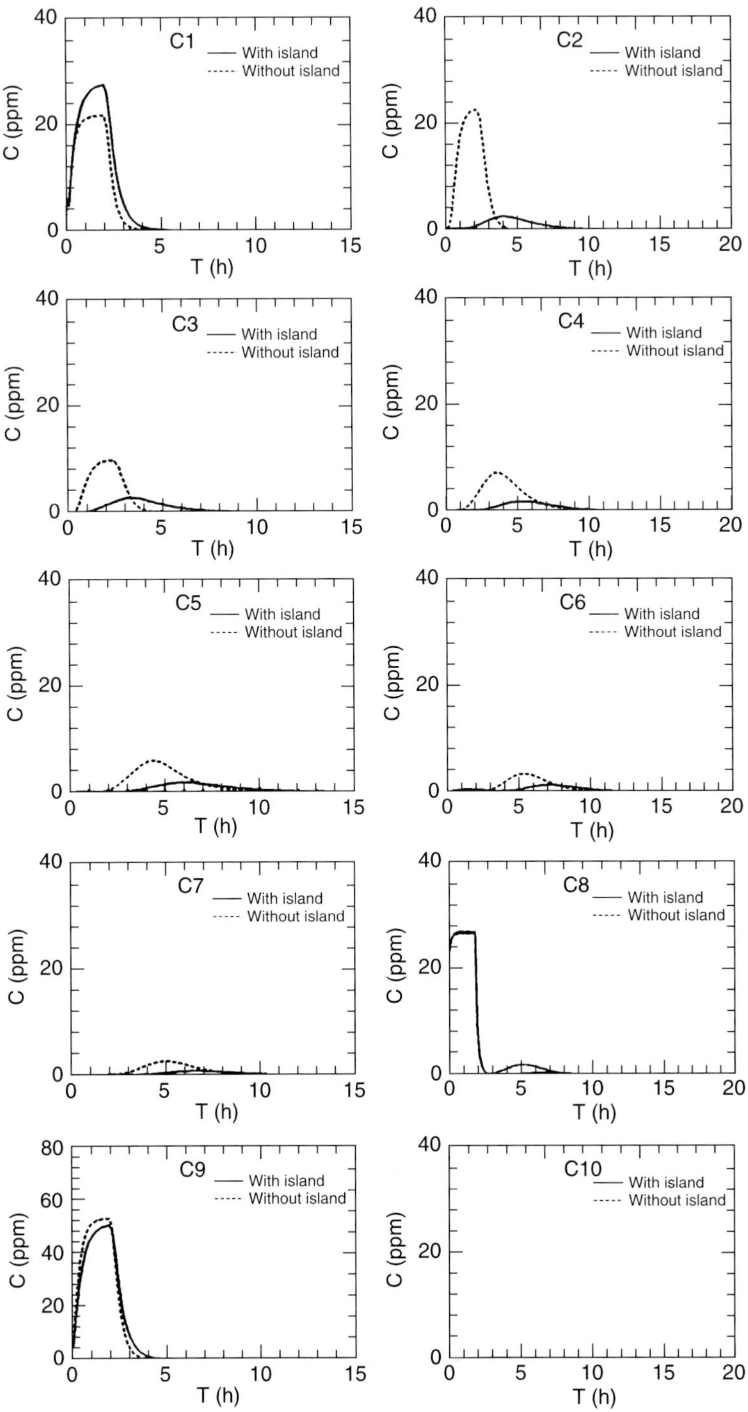

Fig. 7.11. Time series of pollutant concentration at 10 locations of concern in Northeastern Shoreline islands, under a 10 m/s north wind (Tsanis and Wu, 1995).

Table 7.2. Features of concentration time histories (10 m/s N; Northeastern Shoreline).

Point	$C_{i\max}$ (ppm)	$C_{\max}$ (ppm)	T_i (h)	T (h)	A_i/A
C1	16.35	16.28	23.83	23.17	1.00
C2	1.92	1.94	21.50	20.83	1.01
C3	10.99	10.86	23.50	23.33	1.02
C4	6.15	7.43	22.50	20.83	0.96
C5	8.71	8.86	22.83	23.00	1.00
C6	8.10	8.20	22.00	22.33	0.98
C7	6.80	7.74	22.00	22.17	1.06
C8	9.47	9.41	23.33	21.67	1.07
C9	7.64	7.87	21.50	21.67	0.97
C10	6.55	7.47	21.50	21.50	1.05

$C_{\max}$, maximum concentration (without islands); $C_{i\max}$, maximum concentration (with islands); T, duration of pollutants with concentration above zero (without islands); T_i, duration of pollutants with concentration above zero (with islands); A, area under the concentration curve (without islands); and A_i, area under the concentration curve (with islands).
Source: Tsanis and Wu (1995).

weaker than the wind-induced current, because of the large size of the harbor and small inflow/outflow.

A case simulation clearly demonstrates that the pollutant transport model, coupled with the circulation model, is capable of simulating the temporal and spatial distribution of pollutants emitted from multiple sources. With the aid of this modelling system, the so-called horizontal mixing time in Hamilton Harbour is revealed for the first time. In general, the time scale under a 10 m/s wind is about one-to-two days, depending on the location and wind condition.

A nested-grid modelling technique has been developed in which a coarse-grid model is first applied for the whole Harbour, and the velocities and water elevations along the open boundaries are provided for the fine-grid model in the nearshore area. The model simulation in three nearshore areas in Hamilton Harbour clearly indicate that the model can predict, in detail, the current structure, while mass conservation is satisfied. The impact of artificial islands was examined by studying changes in current patterns, pollutant peaks, exposure, and flushing time in different locations of concern. The final design of the proposed islands will require model verifications by field studies, and detailed analysis of sediment transport and wave effect on the islands.

7.4 TORONTO WATERFRONT[6]

A nested model was developed based on the 3D IDOR hydrodynamic model to simulate the currents along the Metropolitan Toronto waterfront. The nested model consists of a coarse-grid model and a fine-grid model. First, the coarse-grid

[6]Material in Section 7.4 is reproduced from Shen et al. (1995), copyright with permission from the International Association of Great Lakes Research.

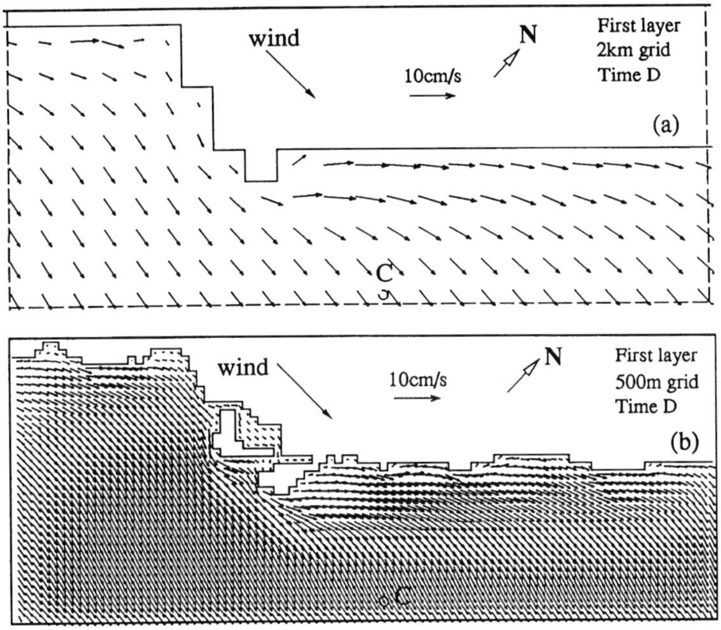

Fig. 7.12. Currents in Toronto waterfront obtained by coarse–fine grid models (Shen et al., 1995).

model was run over the whole of Lake Ontario, and the results (including velocity and water surface elevation) at the open boundaries were stored. The fine-grid model was then run over the nearshore area by interpolating the calculated results from the coarse-grid model at the open boundaries. Since the same vertical spatial grid and time step were used for both the coarse-grid and fine-grid models, the velocities and water elevations at the open boundaries only needed to be interpolated in the horizontal plane. For details on the interpolation of the open boundary, refer to Tsanis and Shen (1994c).

The IDOR 3D nested model was used to predict circulation patterns across the Metropolitan Toronto waterfront (see Fig. 7.12). A coarse ($2\,km^2$)-grid model was used for the whole of Lake Ontario, and a fine ($500\,m^2$)-grid model was used for the Metropolitan Toronto waterfront. The grid used for discretizing Lake Ontario was rotated 45° clockwise, in order to align it with the Toronto waterfront shoreline. The grid extended west of Etobicoke Creek and to the east of the Rouge River. A 60 s time step was used for both models. Horizontal eddy viscosities of 40 and $10\,m^2/s$ were assumed for Lake Ontario and the Metropolitan Toronto waterfront, respectively.

7.4.1 Coarse-Grid and Fine-Grid Model Comparisons

An error analysis applied to the nested model included comparisons of the total kinetic energies and the water elevations at a given point, between the coarse-grid

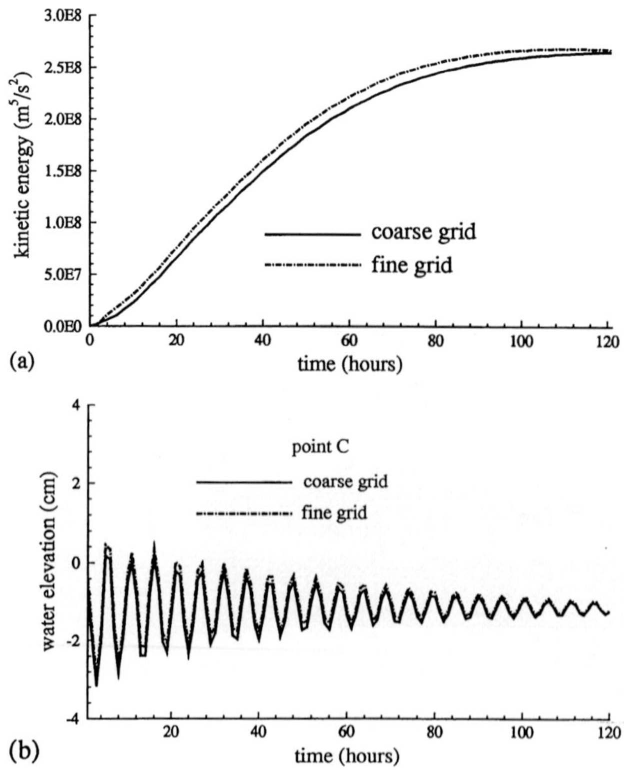

Fig. 7.13. (a) Total kinetic energy of Toronto waterfront area. (b) Water elevation at point C (Shen et al., 1995).

and fine-grid models. For the convenience of comparison, a constant 10 m/s west wind was applied evenly over the lake surface under isothermal conditions. Figure 7.13(a) presents time series of the total kinetic energy for the two models, and the total kinetic energy in the two models is similar. However, the slightly higher kinetic energy measured in the fine-grid model can be explained by the modification made for the higher resolution of the shore geometry and bottom topography. A comparison of the water elevation time series at point *C* (identified point in Fig. 7.12), obtained by the two models, is presented in Figure 7.13(b). The water elevation time series obtained by the two models are almost identical. It can therefore be concluded that the nested-model approach does not introduce any instability problems, and the truncation error along the open boundary is suppressed if the values at the open boundary are interpolated correctly.

7.5 LAKE BIWA

Lake Biwa is the largest lake in Japan, with 674 km^2 in area, $27.5 \times 10^9 \, m^3$ in storage capacity, and 3848 km^2 in catchment area. As shown in Figure 7.14, the lake

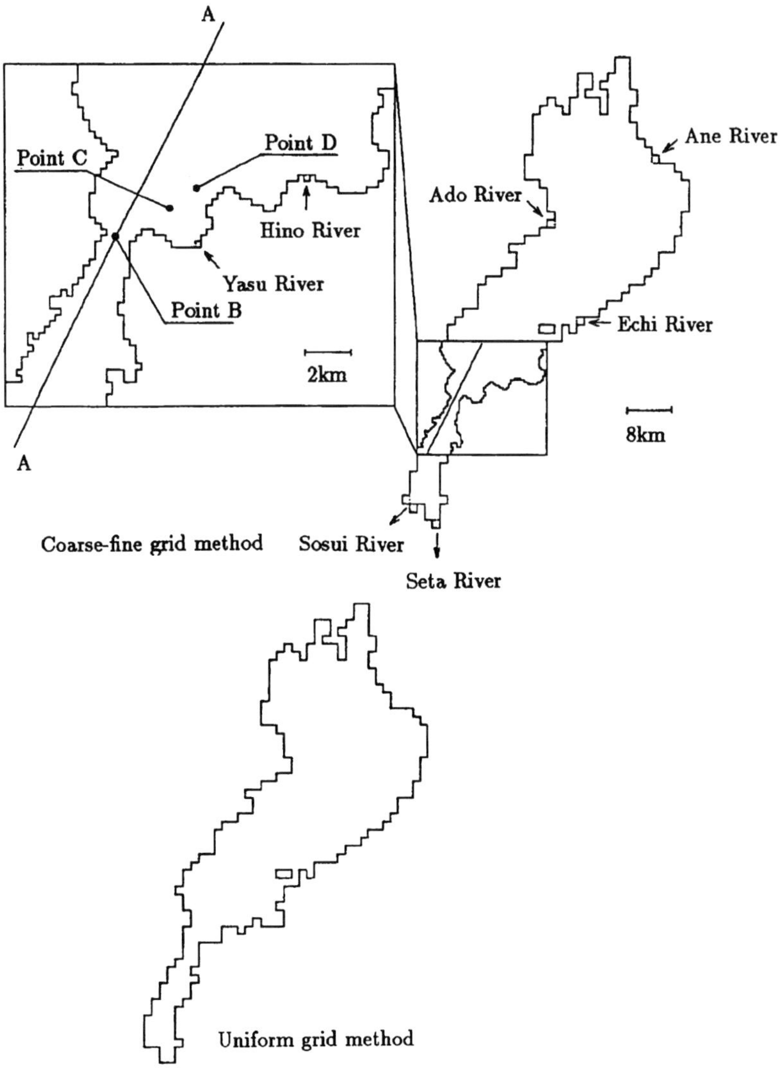

Fig. 7.14. Combined use of coarse grid and fine grid (Shen, 1991).

consists of the north and south basins, located on the north and south sides of Lake Biwa Bridge, respectively. Having 620 km^2 in area, the north basin makes up more than 90% of the whole lake. The water is deeper in the west side, with an average depth of 43 m and a maximum depth of 104 m. On the other hand, the south basin has only 54 km^2 in area and its average depth is ~4 m. There are 120 main rivers pouring into the lake, and only 2 rivers flowing out of the lake: the Seta River and Sosui River. In this example, several local dominant flows are simulated by the coarse–fine grid method. The first computational example is the density current occurring near Lake Biwa Bridge during the cooling period from autumn to winter.

Table 7.3. Comparison of the numbers of column and cell.

Domain	Column			Cell		
	Coarse	Fine	Total	Coarse	Fine	Total
Uniform grid method	662	–	662	9975	–	9975
Nesting method (Biwa Bridge)	567	1569	2136	8717	9811	18,528
Nesting method (Ane River)	614	725	1339	8778	8467	17,245

The second example is the vertical mixing and diffusion phenomena at the inflow estuaries, because of the difference in water temperature between the lake and inflow rivers. The mesh sizes of the coarse and fine grids should be determined not only according to the scale of the concerned flows and topography conditions, but also by the memory capacity of the computer and CPU time. The grid system is divided as shown in Figure 7.14, where the coarse-grid intervals are $\Delta x_1 = \Delta y_1 = 1\,\mathrm{km}$ and the fine grid intervals are $\Delta x_2 = \Delta y_2 = 250\,\mathrm{m}$. In the vertical direction, a uniform grid system is used, which is $\Delta z_1 = \Delta z_2 = 2.5\,\mathrm{m}$ in the area shallower than 50 m, and $\Delta z_1 = \Delta z_2 = 10\,\mathrm{m}$ in the area deeper than 50 m. The cell and column numbers of the coarse-fine grid system are listed in Table 7.3 (Hirose, 1990).

7.5.1 A Comparison with a Uniform Grid Model

In order to check whether any numerical oscillation due to the employment of the coarse-fine grid model occurs, a uniform grid model, which covers the whole domain only by coarse-grid net, is also used. In the coarse–fine grid model, the area near Lake Biwa Bridge is subdivided as shown in Figure 7.14. In the case of a uniform grid system with $\Delta x = \Delta y = 1\,\mathrm{km}$, it has been learnt from experience that stable solutions can be obtained if the time step Δt is chosen as 300 s. Because the mesh sizes of the fine grid used in this study are $\Delta x_2 = \Delta y_2 = 250\,\mathrm{m}$, Δt is chosen as 60 s, in consideration of the earlier mentioned.

For the initial conditions, it is assumed that the water surface of the whole lake coincides with the reference water surface in the still state, and the initial distribution of temperature is given with the thermocline existing at 20 m below the water surface. The southwestern wind, with a velocity of 2.5 m/s, is continuously blowing for 4 h.

Figure 7.15 illustrates the velocity distributions in the surface layer at 4 h and 8 h after the computation started, and Figure 7.16 shows the velocity distributions in the eighth layer, near the metalimnion, at the same times. Although no significant difference between the results of the two models can be observed, the velocity at the coarse–fine grid boundary, in the coarse–fine grid model, appears to be a little smaller.

Figure 7.17 shows the temporal variations of the water stages at various points. The solid line represents the results of the coarse–fine grid model and the dotted line expresses the results of the uniform grid model. Figures 7.17(a) and (b) show the

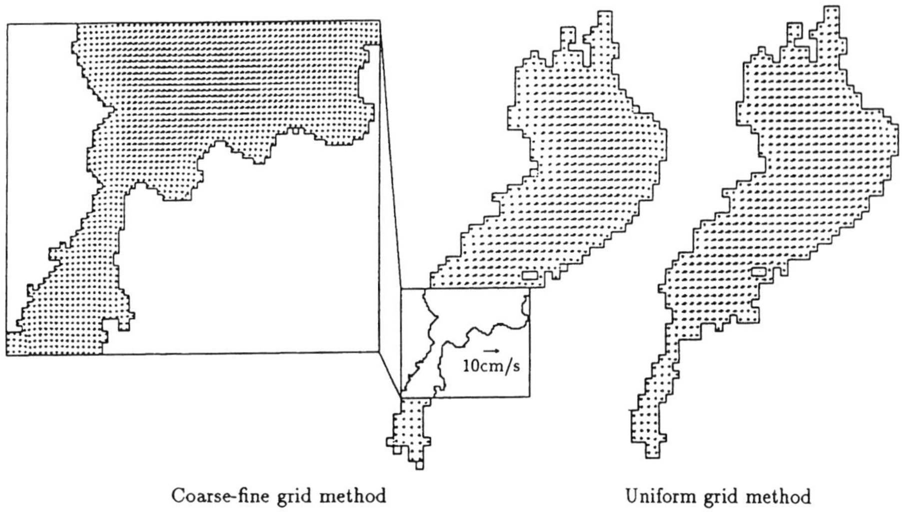

Fig. 7.15. Velocity distributions in the surface layer (Shen, 1991).

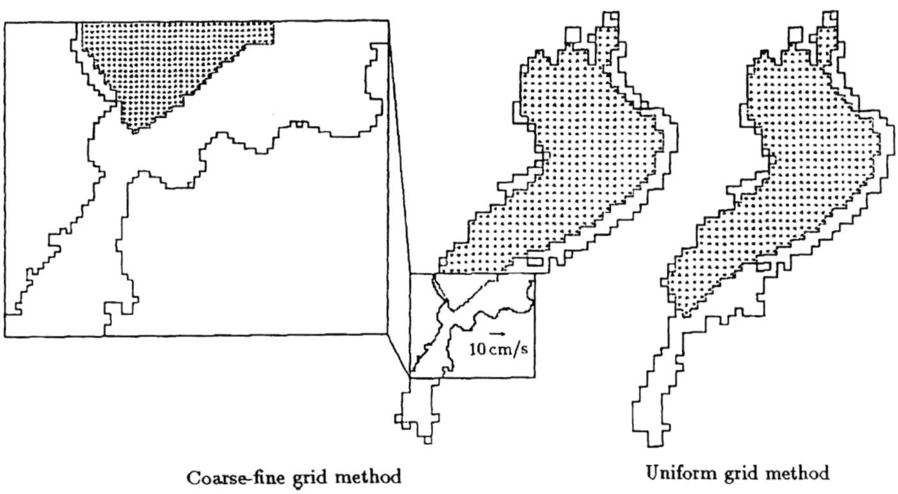

Fig. 7.16. Velocity distributions in the eighth layer (Shen, 1991).

water stages at the fine-grid side and coarse-grid side of the coarse–fine grid boundary, respectively. Figure 7.17(c) shows the water stages near Lake Biwa Bridge which are located at the inner area of the fine grid sub-domain. As seen in these figures, the coarse–fine grid model gives rise to a small oscillation in the fine grid sub-domain. Though this oscillation does not coincide with the results of the uniform grid model in both amplitude and phase, the response to the wind is similar. In the view of mathematics, the oscillation of water stage or discontinuity of velocity results from the different truncation errors of the coarse and fine grids.

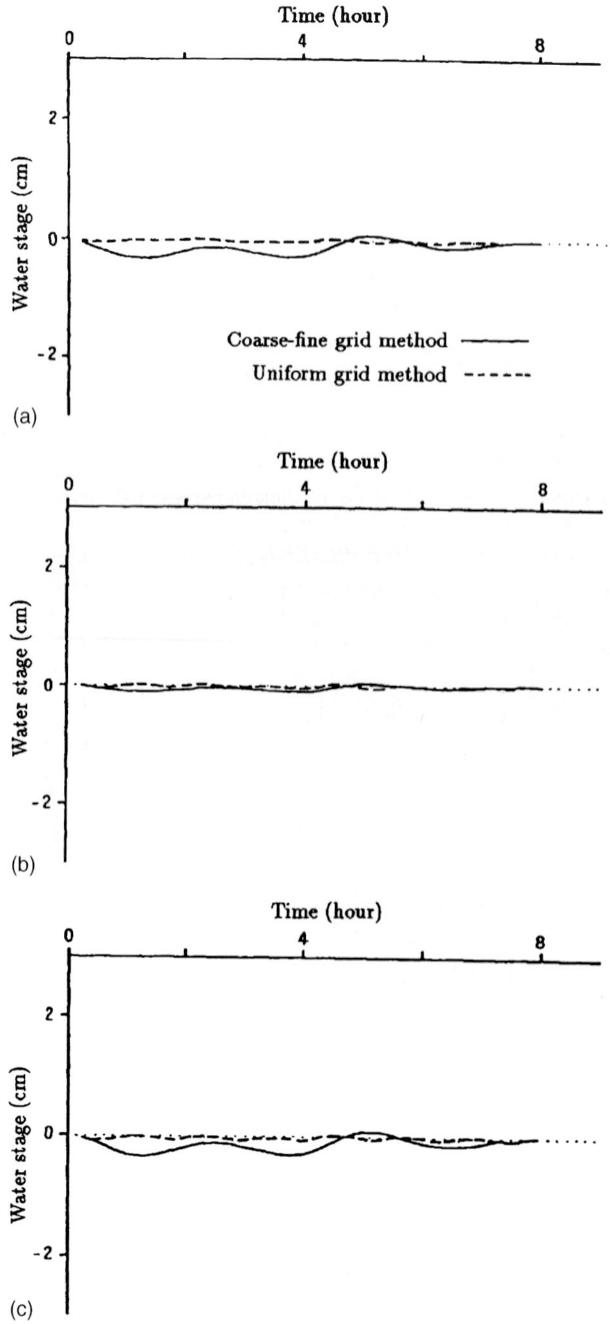

Fig. 7.17. Temporal variations of water stages at various points: (a) uniform-grid side of the boundary; (b) coarse-grid side of the boundary; (c) boundary near Lake Biwa Bridge (Shen, 1991).

Since the truncation error of a difference scheme with second-order accuracy is $O(\Delta x^3)$ and $\Delta x_2/\Delta x_1 = 1/4$ is chosen in this study, the difference of truncation errors between the coarse and fine grids becomes $O(\Delta x_2^3)/O(\Delta x_1^3) = 1/64$. Physically, the staggered scheme used in this study may also give rise to momentum discontinuity at the coarse–fine grid boundary. For example, the x-wise momentum is evaluated at the open circle O point for the control volume of the coarse grid piled up on the fine grid, but this momentum is not continuous with that seen at the control volume of the fine grid.

Another problem in calculation is the evaluation of water stage. When the water stage is calculated from the continuity equation and integrated over a column by using the Successive Over Relaxation (SOR) method, the water stages of neighboring coarse grids are adopted to express those of the fine grids. Therefore, the Poisson equation for the water stage is not solved correctly in the strictest sense, and it is necessary to treat the boundary of a coarse–fine grid more strictly in order to simulate the variation of water surface in detail. In this study, the most important thing is not to simulate rigorously the movements of water surface, but rather the flows induced by wind or density difference. Moreover, the numerical oscillation of water stage produced by the coarse–fine grid model is very small in amplitude compared to actual water stage and therefore, will not greatly influence the flows. Figure 7.18 illustrates the temporal variation of a water temperature at the boundary of a coarse–fine grid model near the metalimnion. The solid line expresses the results of a coarse–fine grid model and the dotted line, the results of the uniform grid model. The temperature difference between the two models is nearly 0.1°C at 8 h after the computation starts, and no numerical oscillation of internal model can be observed in either of the model.

Since the control volumes of the coarse grid are stacked on the fine grid side on the boundary of the coarse–fine grid, the water depths of all fine grids which one coarse grid covers (16 fine grids in this study) should be the same. In this example, though the same time step ($\Delta t = 60$ s) is used for the fine and coarse grids, the time step of the coarse grid can be taken as longer than that of the fine grid in terms of the grid intervals from the stability condition. However, if different time steps are chosen for coarse and fine grids, the combination of different time steps must be

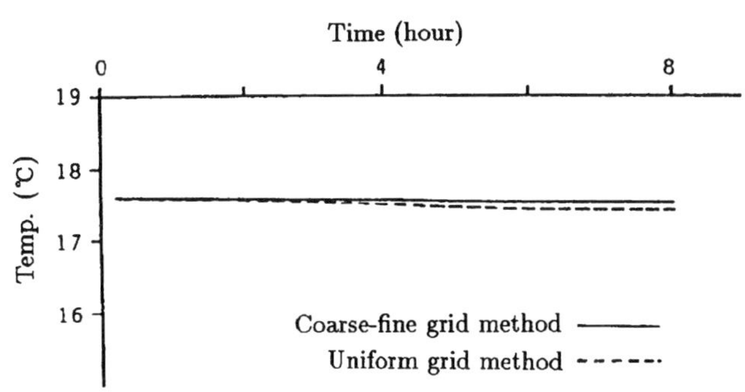

Fig. 7.18. Temporal variation of temperature at the boundary of coarse–fine grid near the metalimnion (Shen, 1991).

considered, as well as the combination of coarse and fine grids. Therefore, the treatments of boundary will become much more complicated. Moreover, since the column ratio of coarse to fine grids is 1:3 and the cell ratio is approximately 1:1, near Lake Biwa Bridge (Table 7.3), there is little advantage in choosing different time steps. Despite the fact that the number of control volumes in the coarse–fine grid model is only twice that seen in the uniform grid model, more than three times the CPU time is required for the coarse–fine grid model. This probably results from the treatment of the coarse–fine grid boundary and the large number of iterative calculations by the SOR method.

Chapter 8

MODEL VERIFICATIONS WITH ANALYTICAL SOLUTIONS AND LABORATORY DATA

8.1 POLLUTANT TRANSPORT AND RESIDENCE TIME IN A MODEL BASIN[7]

Shallow basins are often used for treatment processes such as sedimentation, waste stabilization and cooling (Thackston et al., 1987). In order to study the dynamics of these basins, data collection from the prototype basin is most desirable. However, this is expensive, time consuming, and sometimes impossible (in the case of new basin designs). As a result, a physical model or numerical model is often employed, aside from direct measurements in the prototype.

A physical model is a scaled-down version of the prototype; on the other hand, a numerical model is a set of equations which are thought to represent the characteristics of the process being studied. Physical models have a long tradition, and are appealing because the results can easily be visualized. The models may be expensive to build, but, once constructed, may be used for a variety of studies. Numerical models became more popular with the development of modern computers. A numerical model has several advantages over a physical model. For instance, it may be quicker to apply and less costly, if less expensive computers are used. Coriolis forces and heat exchanges may be more easily incorporated in the numerical model than in the physical model (Rumer and Hooper, 1970; Li et al., 1975; Nydegger, 1979). Extreme conditions, such as very large wind, inflow, and impacts of various engineering designs on the water body can be studied at a lower cost. However, for a numerical model, certain numerical methods have to be employed which may introduce numerical errors. Boundary conditions and model parameters must be properly specified in order to provide acceptable simulations. As a result, measurements from the prototype or physical model are required and model calibration must be performed.

A number of numerical models have been developed in the last decade to simulate large-scale circulation. Calibrating numerical models is imperative and good calibration with physical models is rare, mainly due to a lack of well-designed physical experiments and the failure to integrate both models. The flow visualization and pollutant transport experiments in the physical model of the Windermere Basin provide a unique opportunity to bridge such a gap. This section will focus on calibrating a hydrodynamic model and will study pollutant transport

[7]Materials in Section 8.1 are reproduced from Wu and Tsanis (1994), copyright with kind permission from the IAHR, Delft, The Netherlands.

and residence time in the physical model, with the aid of a pollutant transport model.

8.1.1 A Physical Model of Windermere Basin

The Windermere Basin lies at the southeast corner of Hamilton Harbour, which is at the western end of Lake Ontario. It receives inflows from creeks, sewage treatment plants, and several sewer overflows. Contaminations in this basin have become serious, and a cleanup of the basin is to be undertaken by dredging a portion of the basin. The physical model study was conducted at the National Water Research Institute (NWRI) of the Canadian Centre for Inland Waters (CCIW) to determine the hydraulic properties of the proposed basin design. Many basin configurations and flowrates have been tested (Engel, 1988) and only the standard case is presented in this study. The physical model, as shown in Figure 8.1(a), is a fixed bed model (horizontal scale ratio of 1:60 and a vertical scale ratio of 1:15). The depth is 175 mm, with a central trap of 350 mm and a flowrate of 11.4 L/s. In this distorted scale model, it is necessary to make the model bed surface rougher in order to simulate the flow conditions in the prototype. In the prototype, the Manning's n is 0.009 and according to the laws of similitude it is 0.0115 for the physical model. A value of $n = 0.0115$ represents the surface roughness of mortar or concrete (Chow, 1959). Therefore, a natural mortar surface was chosen as the bed surface of the model (Engel, 1988). Flow patterns were visualized by using potassium permanganate. In addition to the pollutant, 22 weighted ping-pong balls were released at the entrance of the basin at various times as an additional aid in visualizing flow paths, and in calculating the surface velocities in the model basin. Several flow conditioning schemes were examined, including the effects of spur dikes, a weir at the outflow of the basin, a flow deflector at the entrance of the basin, and a sediment trap.

8.1.2 Numerical Modelling of Circulation Pattern

Numerical models were applied to model the circulation in the physical model of the Windermere Basin (Blaisdell, 1990; Tsanis et al., 1990; Blaisdell et al., 1991). It was demonstrated that the non-linear advective terms and horizontal viscosity terms in the 2D circulation model are necessary to predict the zones of re-circulation observed in the physical model. These studies were limited to the modelling of hydraulically induced currents. Therefore, further studies are required on the effect of flow patterns (wind and/or hydraulically induced) on residence time and pollutant transport.

The pollutant transport is accomplished by two distinct physical mechanisms—advection and diffusion. The first mechanism re-distributes mass through the displacement of fluid parcels that follow the flow. The second mechanism re-distributes mass through actual exchange between adjacent parcels. Mass exchange is set in a way that tends to smooth out existing gradients.

The 2D depth-averaged hydrodynamic model (2DH) is used to simulate the hydraulic circulation in the physical model of the Windermere Basin. A 44×47

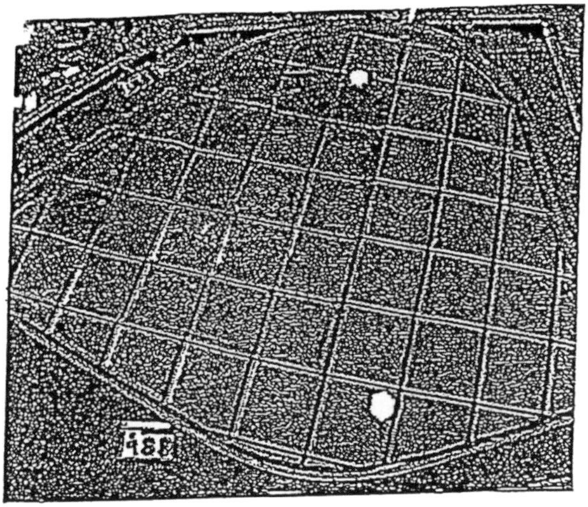

(a)

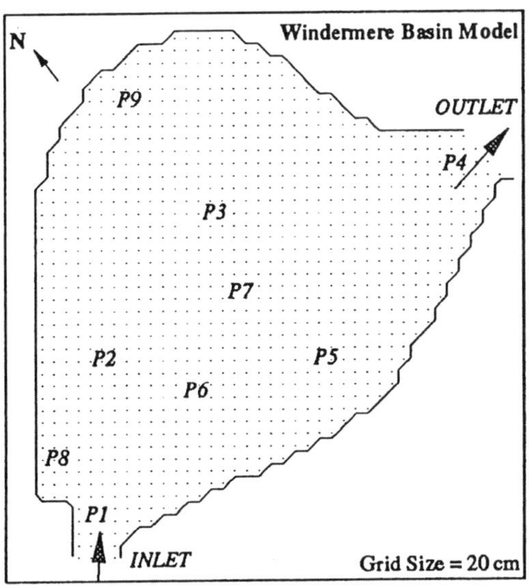

(b)

Fig. 8.1. Model of Windermere Basin: (a) the physical model at the NWRI of the Canadian Centre for Inland Waters (CCIW) during testing, as shown by VHS tape (1 m squared grid is shown, in which the white spots are ping-pong balls), (b) the mathematical model (the grid is square with sides of 20 cm and is rotated counterclockwise by 45°) (Reprinted from Wu and Tsanis (1994), copyright with kind permission from the IAHR, Delft, The Netherlands).

square grid system with the grid size of $\Delta x = 0.2\,\text{m}$ is used, as shown in Figure 8.1(b). The grid is rotated 45° counterclockwise from the north, to represent the north shoreline better. Points P1–P9 indicate the locations where pollutant concentration will be displayed. Using an inflow $Q = 0.0114\,\text{m}^3/\text{s}$, and a Manning coefficient $n = 0.0115$, as in the standard case of the physical model of the Windermere Basin, the equations of motion are solved for a time step of $\Delta t = 0.04\,\text{s}$ (the Courant–Friedriches–Lewy (CFL) stability condition is satisfied), and a value of smoothing factor $t_h = 0.995$ (eddy viscosity $K_h = 0.00125\,\text{m}^2/\text{s}$). The inlet Reynolds number is $R_j = U_j/v = 1.415 \times 10^4$, where U_j is the velocity at the inlet, h the depth at the inlet, and v the kinematic viscosity of water. The inlet Froude number is given as $Fr_j = U_j/\sqrt{gh} = 6.21 \times 10^{-2}$. The resulting hydraulic circulation of this simulation is shown in Figure 8.2, where the black arrows show the velocity vectors for the standard case derived from the drifter results in the experiments (Blaisdell et al., 1991). The magnitude and direction of the velocity vectors and the size of the recirculation zones from the simulation are in good agreement with those determined experimentally.

The simulation clearly indicates three zones of recirculation, including the left-hand eddy at the inlet, the central eddy, and the dead zone at the top of the basin. The advective flow is deflected toward the left upon leaving the inlet areas. This is due to the squeezing of the advective flow between the two opposing eddies. As the central eddy develops, it forces the flow to the left, which results in deflection and squeezing of the flow. At the top left side of the basin, a weak counterclockwise eddy is present. Although none of the ping-pong balls were recirculated in this eddy, some of them showed signs of being affected by it. For instance, the velocity of the

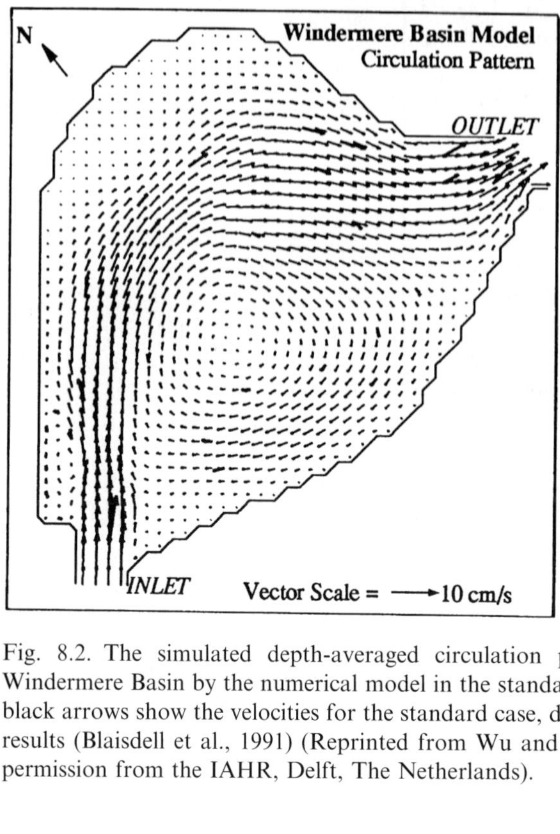

Fig. 8.2. The simulated depth-averaged circulation pattern in the physical model of Windermere Basin by the numerical model in the standard case (Wu and Tsanis, 1994). The black arrows show the velocities for the standard case, derived from the drifter experimental results (Blaisdell et al., 1991) (Reprinted from Wu and Tsanis (1994), copyright with kind permission from the IAHR, Delft, The Netherlands).

ping-pong ball would drop suddenly in this region, or would be deflected to the right or left. In addition, the advective path avoided the top of the basin, as if it were deflected by a weak eddy.

The bottom friction coefficient, or Manning's coefficient, and the horizontal eddy viscosity coefficient, or smoothing factor are the two important parameters in the numerical model. Horizontal eddy viscosity terms have been ignored in most numerical hydrodynamic models, based on the assumption that they are very small compared with other terms. However, recent studies of the Windermere Basin (Blaisdell, 1990) and Lake Ontario (Wu and Tsanis, 1991) have revealed that the horizontal eddy viscosity terms cannot be ignored, especially for the hydraulically induced flow cases. Their effects on the flow pattern are significant and there have been no direct measurements on the horizontal eddy viscosity coefficient. In the present 2DH model, they are included through a spatial smoothing scheme. The eddy viscosity coefficient is then calculated from the corresponding smoothing factor, along with the time step and grid size. Figure 8.3 shows the circulation patterns for eddy viscosity coefficients of 0.005, 0.0025, and 0.00125 m^2/s. It is clearly indicated that a smaller size of eddy is obtained for larger coefficients: a plug flow (no eddies) forms for the larger eddy viscosity. The simulation with a value of 0.00125 m^2/s (smoothing factor $t_h = 0.995$) gives the closest result to the one measured. The Manning's coefficient n has a similar effect on the flow simulation. Figure 8.4 shows the circulation patterns for Manning's coefficient $n = 0.1150$ ($C_b = 0.01680$, $h = 0.175$ m), 0.0345 ($C_b = 0.01512$), and 0.0115 ($C_b = 0.00168$). The smaller size of eddy is presented for a larger Manning's coefficient, which corresponds to a rougher bottom and a higher resistance to the flow. A plug type flow appears for a value of 0.115. A value of $n = 0.0115$ gives the closest result to the one measured.

8.1.3 Numerical Modelling of Dye Transport

The hydraulically induced circulation pattern (see Fig. 8.2) was used as an input in the 2D advection–diffusion model to simulate the transport of a continuously injected pollutant in the basin. The location of the pollutant injection was in a channel, a few meters upstream from the inlet. The pollutant was fully mixed, both in the vertical and horizontal planes, as it was entering the basin.

A test was performed for a continuous pollutant released into the basin, using different values of the dispersion coefficient, that is, 0.1, 0.01, 0.001, and 0.0005 m^2/s. The inlet Peclet numbers had values of 0.16, 1.6, 16, and 80 respectively. Figure 8.5 shows the equal pollutant concentration contours at 6 min after the pollutant release. In Figure 8.5(a) the pollutant entering the basin is diffused faster than it is advected, resulting in the pollutant front covering the whole basin while moving toward the basin's outlet. In Figures 8.5(c) and (d), the advection dominates the diffusion. The effects of the left-side eddy at the inlet, the central eddy, and the dead zone are evident. The diffusion in both cases is insignificant (advection dominated flow). As we know, the dispersion coefficient varies with the plume width, as denoted by L (m). A formula with the power 4/3 of L has been recommended by Lam et al. (1984). To avoid the non-linear problem, as

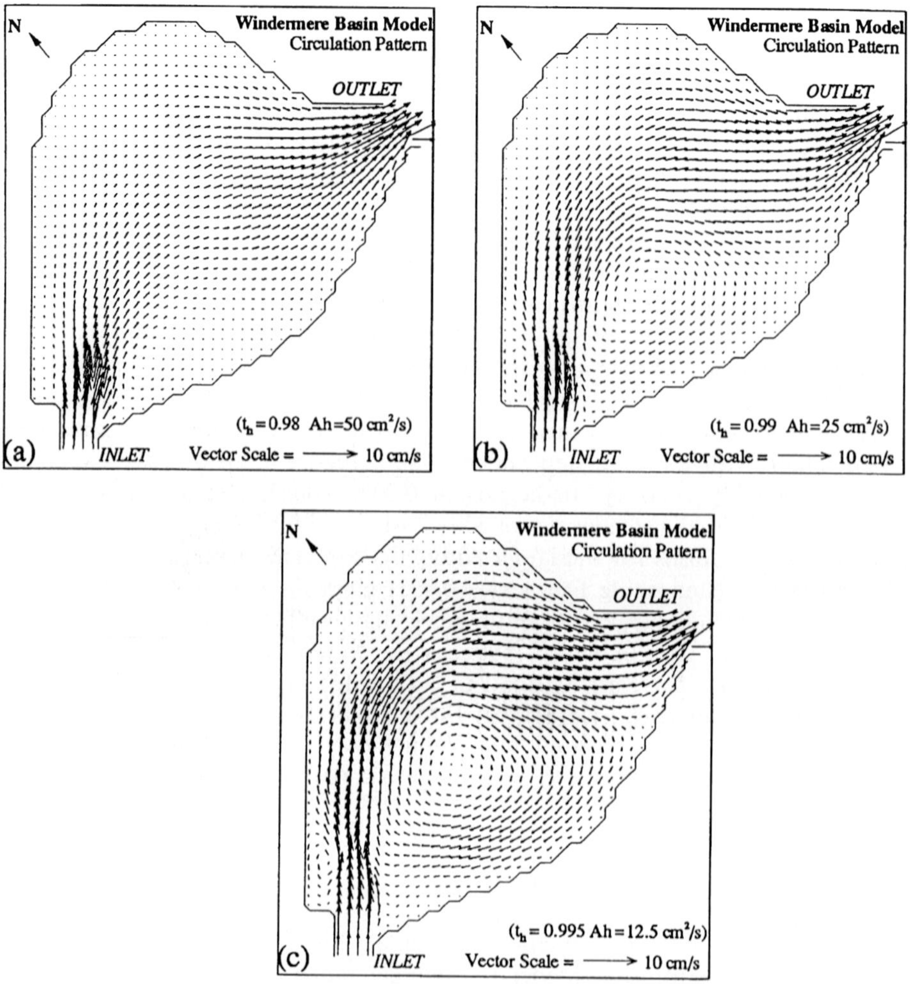

Fig. 8.3. Effect of horizontal eddy viscosity K_h on the simulated circulation pattern in the physical model of the Windermere Basin: (a) $K_h = 50.0\,\text{cm}^2/\text{s}$ ($t_h = 0.98$), (b) $K_h = 25.0\,\text{cm}^2/\text{s}$ ($t_h = 0.99$), and (c) $K_h = 12.5\,\text{cm}^2/\text{s}$ ($t_h = 0.995$) (Reprinted from Wu and Tsanis (1994), copyright with kind permission from the IAHR, Delft, The Netherlands).

well as the difficulty in implicitly determining the dispersion coefficient, a variable but explicit coefficient is used, that is, $D = 0.001\ L^{(4/3)}$ (m^2/s), where L is measured from the simulation results in Figure 8.5(c). The concentration field for the variable coefficient is shown in Figure 8.5(e). As observed from Figures 8.5(c) and (e), the concentration fields are similar, with no significant differences. When the pollutant tends to be fully mixed in the basin, it is reasonable to use a constant coefficient in the numerical model.

Comparison of the foregoing results with those of the experimental, reveals that the best agreement between the simulations and the experiments will be

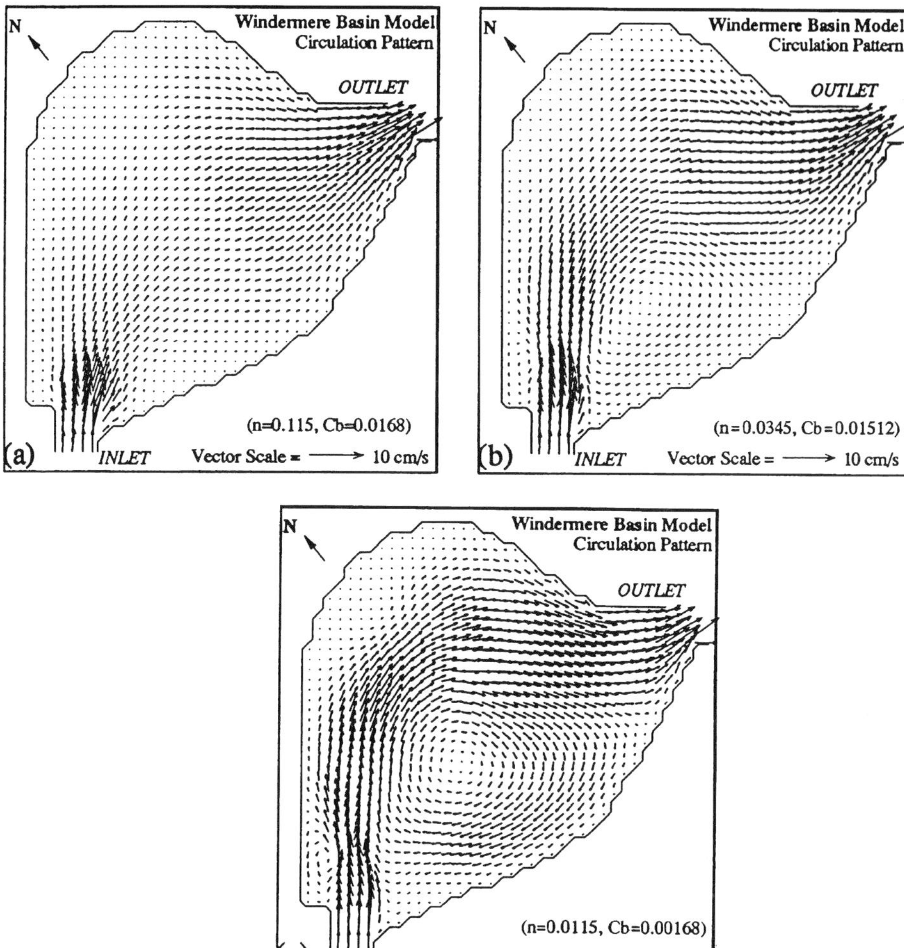

Fig. 8.4. Effect of Manning's coefficient n or bottom friction coefficient C_b on the simulated circulation pattern in the Windermere Basin: (a) $n = 0.1150$ ($C_b = 0.01680$), (b) $n = 0.0345$ ($C_b = 0.01512$), and (c) $n = 0.0115$ ($C_b = 0.00168$) (Reprinted from Wu and Tsanis (1994), copyright with kind permission from the IAHR, Delft, The Netherlands).

achieved by using a value of dispersion coefficient around $0.0015\,m^2/s$, which is of the same order as the eddy viscosity ($K_h = 0.00125\,m^2/s$) used in the hydrodynamic model. This conclusion is consistent with the assumption by Blumberg (1986); that is, the ratio of kinematic eddy viscosity to turbulent matter diffusivity is ~0.7–1.0.

Using the dispersion coefficient of $0.0015\,m^2/s$, a simulation in the standard case for a continuous pollutant release was undertaken. These simulations were then compared with the corresponding images reproduced from the VHS video tape of the experiment (Engel, 1988). Figures 8.6(a)–(c) present the simulated contours for

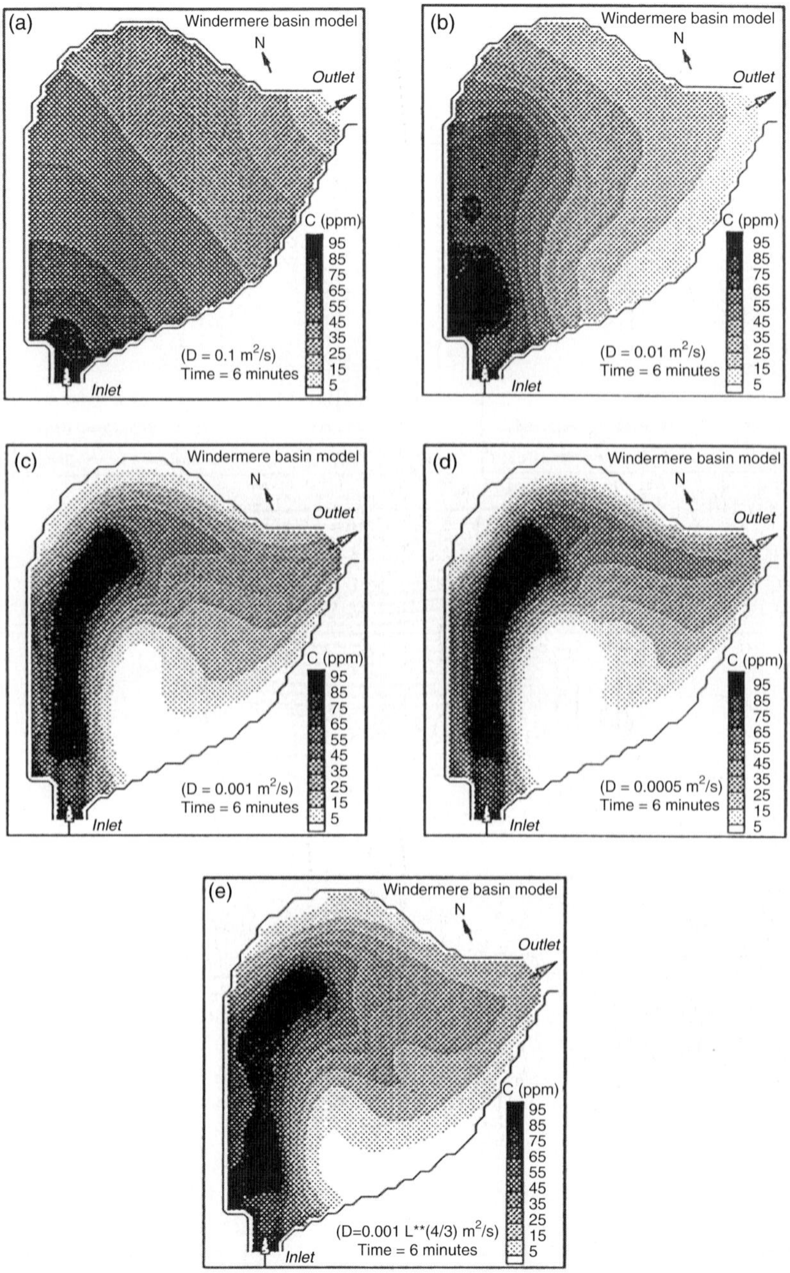

Fig. 8.5. Simulated pollutant concentration flooded contours 6 min after the pollutant release, with dispersion coefficients of (a) $D = 0.1\,m^2/s$, (b) $D = 0.001\,m^2/s$, (c) $D = 0.001\,m^2/s$, (d) $D = 0.0005\,m^2/s$, and (e) $D = 0.001\,L^{(4/3)}\,m^2/s$ (continuous pollutant release case) (Reprinted from Wu and Tsanis (1994), copyright with kind permission from the IAHR, Delft, The Netherlands).

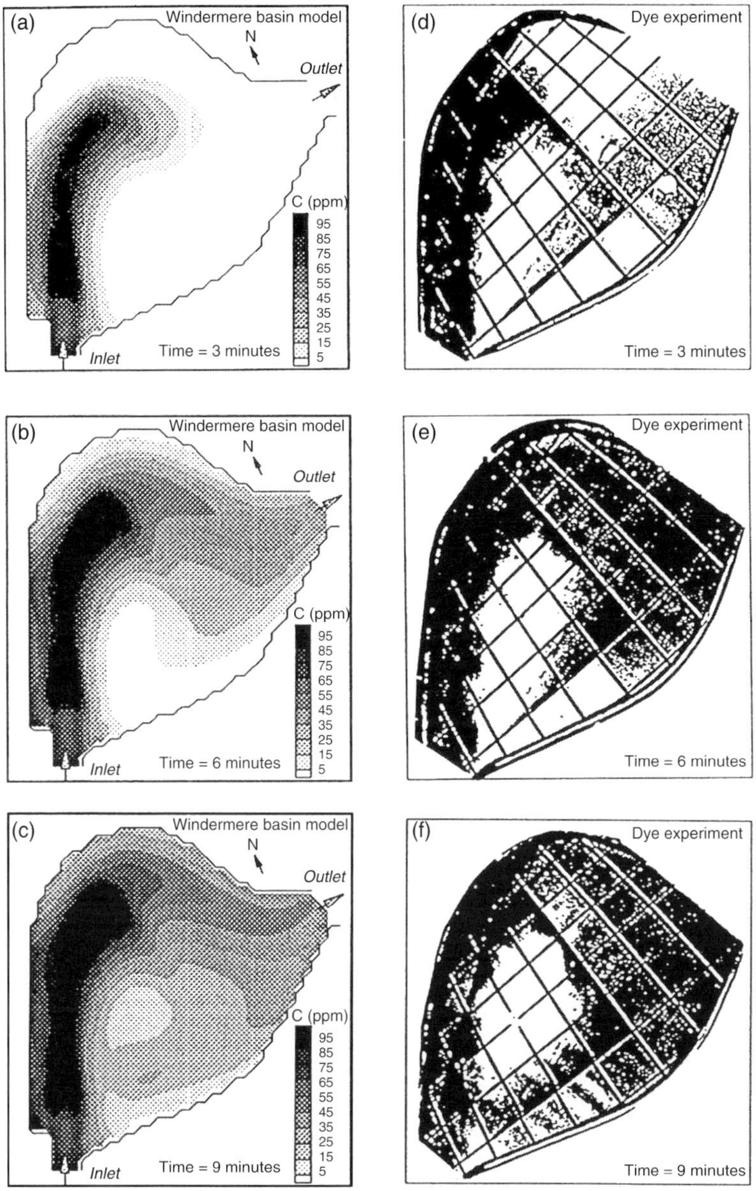

Fig. 8.6. Simulated equal pollutant concentration contours in the physical model of the Windermere Basin for a 1.5 min release of pollutant at: (a) 3 min, (b) 6 min, and (c) 9 min. Images from the pollutant experiment in the laboratory model at: (d) 3 min, (e) 6 min, and (f) 9 min after a continuous emission of pollutant into the basin (continuous pollutant release case) (Reprinted from Wu and Tsanis (1994), copyright with kind permission from the IAHR, Delft, The Netherlands).

the cases under consideration, that 3, 6, and 9 min after the pollutant release. Figures 8.6(d)–(f) show the images from the experiment at the same time intervals. It is observed that the fronts and outlines in the simulation are close to those in the experiment (the images from the experiment are slightly distorted in brightness when they are copied). No attempt was made to compare pollutant concentration distribution inside the visual outline, because of the lack of laboratory measurements.

Typical flow patterns that occur in a shallow sedimentation basin, as given by Thackston et al. (1987), include the advective, mixed, and dead zones. The advective zone is the regime where currents are highly advective and advance directly from inlet to outlet. The dead zone is the regime where velocities are very small, and very weak re-circulation currents exist. The mixed zone consists of large strong gyres, where the exchange between the mixed zone and the main advective flow zone is higher than the dead zone. The effect of zone distribution on the trace movement has been discussed (Thackston et al., 1987; Engel, 1988). However, no examples are available so far to clearly demonstrate such an effect.

By using the numerical model, an excellent numerical experiment can be designed to demonstrate the effect of zones of circulation on the pollutant transport and distribution in a large, shallow sedimentation basin such as the Windermere Basin. In this test, a pollutant with a concentration of 100 ppm was continuously released for 1.5 min. The standard circulation case (see Fig. 8.2) with the calibrated set of parameters ($K_h = 12.5\,\mathrm{m}^2/\mathrm{s}$, $n = 0.0115$, and $0.0015\,\mathrm{m}^2/\mathrm{s}$) was employed in the experiment. Figures 8.7(a) through (f) show the simulated flooding contours of pollutant concentration at 1, 3, 5, 9, 30, and 90 min after the initial emission of the pollutant, respectively.

The pollutant reaches the outlet at about 4.5 min, but before arriving, the pollutant follows the strong current in the main advective flow zone. Part of the pollutant enters the small counterclockwise eddy on the left–hand side of the inlet. No pollutant enters the dead zone at the top of the basin and the central large eddy. After the pollutant front arrives at the outlet (5–9 min), part of the pollutant leaves the basin, part of the pollutant follows the return flow entering the central large clockwise eddy (mixed zone), and part of the pollutant enters the dead zone at the top. The pollutant does not reach the center of the central mixed zone at this stage. Then, in the advective zone, the strong currents cause the pollutant to leave this zone very quickly. The pollutant that entered the mixed zone on the left-hand side of the inlet eventually leaves this area because the size of this eddy is small and the exchange between this zone and the main advective flow zone is high (see Fig. 8.2). The pollutant stays in the dead zone in the top of the basin for a much longer time than in other areas, due to the slow exchange with the main advective flow zone (1 ppm pollutant remains in the zone even at 1 h). It is interesting to observe the pollutant distribution in the central mixed zone. The pollutant accumulates gradually in the central area of the eddy in less than 30 min, then leaves this zone from the outer regime to the inner regime through the exchange with the main advective flow. Even after 1.5 h, about 0.1 ppm pollutant remains at the center of the zone (see Fig. 8.7(f)) and finally, all the pollutant leaves the basin within 2 h.

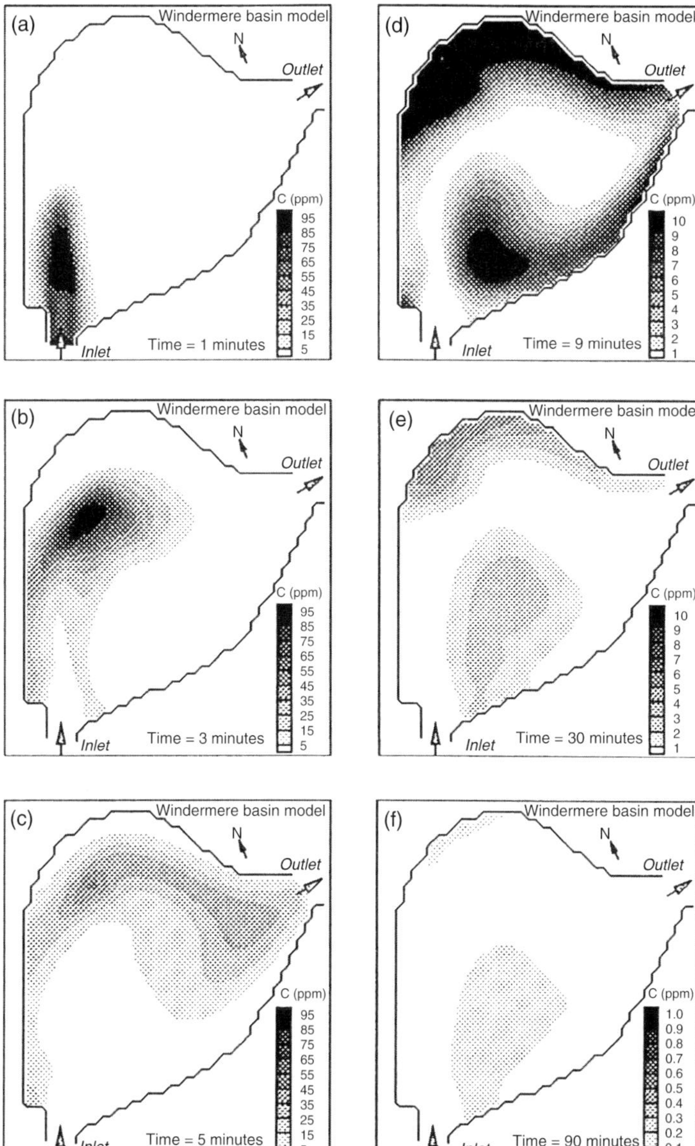

Fig. 8.7. Simulated pollutant concentration flooded contours in the physical model of the Windermere Basin for a 1.5 min release of pollutant at: (a) 1, (b) 3, (c) 5, (d) 9, (e) 30, and (f) 90 min after the pollutant is introduced to the basin (Reprinted from Wu and Tsanis (1994), copyright with kind permission from the IAHR, Delft, The Netherlands).

8.1.4 Numerical Study of Residence Time

In an ideal sedimentation basin, every part of the fluid entering the basin is evenly distributed in the vertical cross-section of the entrance, and the flow advances at a uniform and constant velocity to the outlet. In such a case, the flow regime consists of a plug flow only, and each part resides in the system for a period equal to the theoretical residence time, given as $T_r = V_o/Q$, where T_r is the theoretical residence time, V_o the volume of water in the basin, and Q the inflow. However, in a real basin, the flow pattern deviates from the plug flow due to the existence of dead zones, lateral dispersion, and mixing. The more a basin deviates from the ideal, the lower the trapping efficiency will be (Thackston et al., 1987). The hydraulic efficiency E_h is defined as $E_h = \bar{t}/T_r$, where $\bar{t}$ is the actual mean residence time. For the standard case of the Windermere Basin model, the theoretical residence time is 0.2414 h ($Q = 0.0114\,m^2/s$, $V_o = 9.907\,m^3$). The effect of various zones such as the dead, mixed, and advective flow zones on the mass transport and on the residence time has been clearly demonstrated in the preceding section. For a sedimentation basin such as the Windermere Basin model, the degree of treatment depends in some ways on the actual residence time and as a result, the estimation of mean residence time and its distribution is important.

8.1.5 Numerical Tracer Test

The flow regime and residence time for a sedimentation basin are usually determined by a tracer test. The flow curve (trace concentration at outlet as a function of time) can then be used to determine the hydraulic efficiency. However, as pointed out by Thackston et al. (1987), the literature contains little information about the hydraulics of large shallow basins due to the difficulty of performing tracer tests in remote locations on large facilities with long residence times. It is also not an easy task to analyze the experimental results from a physical model. Furthermore, factors cannot be isolated when performing such tests, both in the field and in physical models. It is feasible and convenient to perform such tests using a numerical model.

The parameters used in this test are the same as those used in the test described in Figure 8.7 when the pollutant is released for 4 min, which is the time needed for the pollutant to first reach the outlet. Figure 8.8 shows the simulated results on the residence time tracer (pollutant) test. The time series of pollutant concentrations at nine locations, as labeled in Figure 8.1(b), are presented in Figures 8.8(a) and (b), where P1–P4 are located in the main advective flow zone, and P5–P9 in the mixed and dead zones. The pollutant arrives at points in the advective zone in a very short time, for example, at P3 in $0.15T_r$ and at P4 in $0.28T_r$ ($T_r = 0.2414$ h is the theoretical residence time for the basin).

The delay of the arrival of peak concentration is obvious, and the peak concentrations are decreased with the increase in distance from the inlet (95 ppm at P2, 45 ppm at P3, and 40 ppm at P4). The *short-circuiting* exists for the advective flow zone, in which the large portion of pollutant exits the basin in much less time than T_r ($< 0.5T_r$). In contrast, the pollutant leaves these points at mixed zones and dead zones after a much longer than T_r ($\sim 2T_r$ to $4T_r$). A useful index A (%) can be constructed according to the pollutant occupying (>1 ppm concentration) volume relative to the whole volume. As shown in Figure 8.8(c), about 55% of the whole

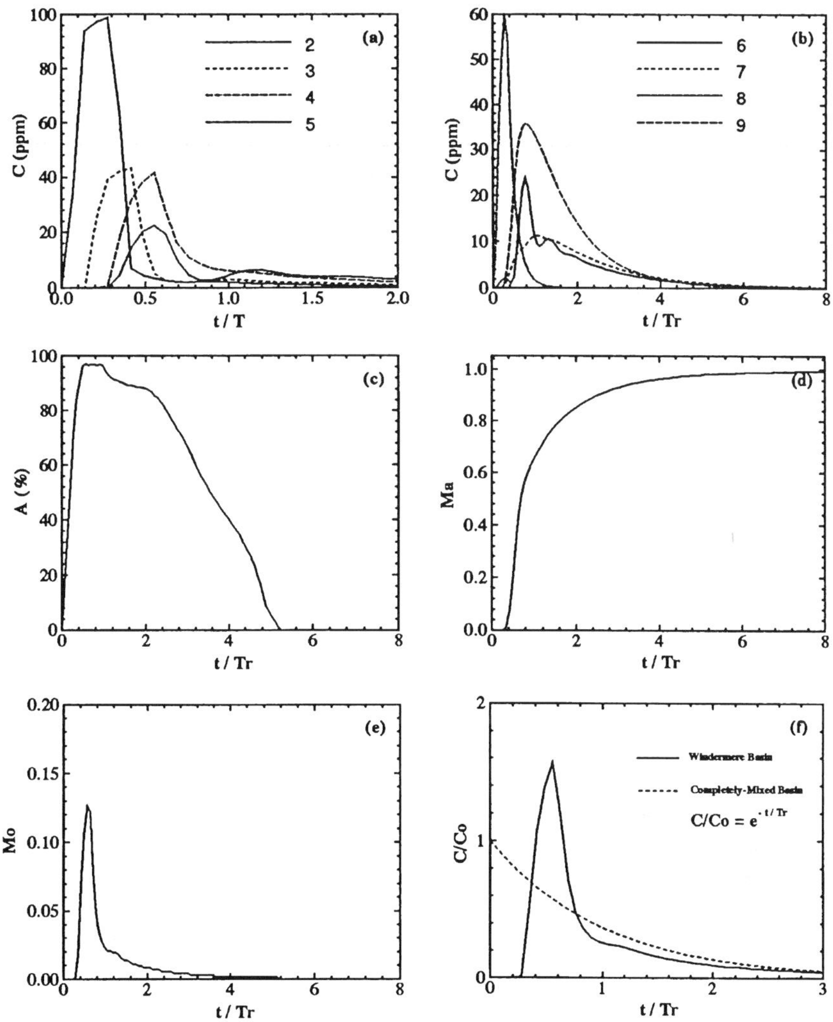

Fig. 8.8. Study of the residence time in the physical model of the Windermere Basin: (a) the pollutant concentrations at points in advective flow zone, (b) the pollutant concentrations at points in mixing and dead zone, (c) the effective volume as a function of relative time, (d) the ratio of accumulated pollutant quantity at outlet to the total emissioned pollutant quantity, and (e) the pollutant curve from the physical model of Windermere Basin, and (f) the theoretical curve from a completely mixed basin as a function of the time normalized by the theoretical residence time (Reprinted from Wu and Tsanis (1994), copyright with kind permission from the IAHR, Delft, The Netherlands).

volume was occupied when the pollutant first arrived at the outlet, and 95% was occupied within the residence time T_r.

The features of zones can also be reflected when one examines the pollutant history at the outlet. Figure 8.8(d) shows the accumulated pollutant mass at the outlet, normalized by the total mass emitted from the inlet, as the function of time normalized by the theoretical residence time. Figure 8.8(e) shows the transient pollutant mass at the outlet. It is observed that about 70% of the pollutant leaves the basin within T_r or 0.2414 h 99% of the pollutant leaves the basin in $6T_r$. Figure 8.8(f) shows the pollutant curve in the above test (solid line) and the theoretical curve from a completely mixed basin (dashed line). The theoretical curve is obtained from $C/C_o = e^{-t/Tr}$ (C_o is the inlet concentration). The shape of the curve in the Windermere Basin resembles that of a completely mixed basin, except that there is a lag time of $0.2T_r$ before the pollutant occurs at the outlet, a slight peak ($C/C_o = 1.6$) in pollutant concentration at $0.6T_r$, and a decay limb slightly below the completely mixed curve after $0.8T_r$. In summary, in the physical model of the Windermere Basin, the advective flow zone results in short-circuiting for a large portion of the pollutant, and the mixed zone and dead zones result in a very long flow through time for a small portion of the pollutant.

8.1.6 Methods of Increasing Residence Time

Hydraulic residence time is directly influenced by such physical variables as depth, shape, side slope, and shoaling, as well as hydraulic characteristics such as the degree of mixing, stratification, and flow-zone distribution. Increasing the residence time implies an increase in the hydraulic efficiency of the settling basin. No attempt is made in this study to examine all factors affecting the residence time. It is assumed that the depth and basin configuration remain unchangeable. For the Windermere Basin, the standard case is implied. The objective of increasing residence time is to increase the path length of the advective flow, and the area of the dead zone or weak mixed zone.

The first test is to examine the wind effect on the residence time distribution, which is very difficult to study in the field and with the physical model. In the test, the pollutant is continuously emitted for 4 min. The depth-averaged circulation patterns under a 10 m/s NE (blowing from outlet to inlet) and a SW wind (blowing from inlet to outlet) are shown in Figure 8.9, respectively. The northeast (NE) wind causes a decrease in size in the central eddy, and the dead zone at the top of basin, and the left-hand side eddy at the inlet. The advective flow path is deflected more to the right when compared with the hydraulic case. The southwest wind causes the disappearance of the dead zone and the decrease in size of the left side eddy at the inlet, and splits the central eddy into two counterclockwise and clockwise eddies. The advective flow path is deflected slightly to the right, parallel to the top shore of the basin to the outlet. The pollutant quantities at the outlet for these two wind conditions and for the hydraulic case are also presented in Figure 8.9. The time the pollutant takes to reach the outlet from the inlet is very close in the three cases, indicating that the wind does not affect the advective flow path significantly in terms of pollutant transport. The curve under the NE wind is flatter than the curve of the southeast (SE) wind and the curve without wind, indicating a larger mean residence time in the NE wind than in the other two cases. The peak in the NE wind condition is lower than other cases. These findings are attributable to the two large central eddies in the NE wind condition. Overall, the wind effect on the residence

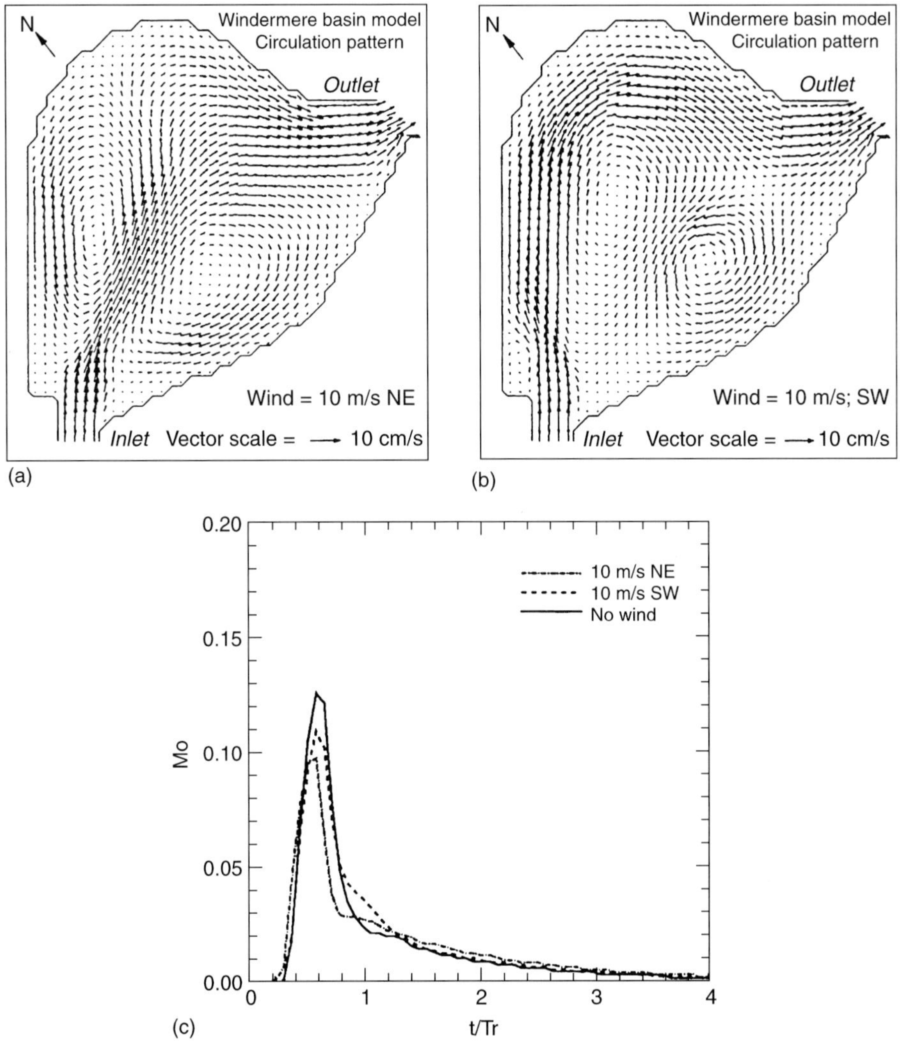

Fig. 8.9. Effect of wind conditions in the physical model of the Windermere Basin. Pollutant quantity at outlet to the total released pollutant quantity as function of the time normalized by the theoretical residence time (Reprinted from Wu and Tsanis (1994), copyright with kind permission from the IAHR, Delft, The Netherlands).

time is small, but it increases the mixing in the basin, resulting in a lower concentration peak at the outlet. This depends on the shape, depth and, in turn, the circulation pattern in the basin.

Site conditions, which promote unevenly distributed flow, tend to degrade the residence time away from ideal plug flow. Spur dikes are by far the most effective and reliable devices which can be used to increase residence time (Thackston et al., 1987). Three dike configurations, and other devices, were tested in the physical model study of the Windermere Basin. The configuration DIKE ♯3 was considered the best case (Engel, 1988). In this configuration, as shown in Figure 8.10(a), two parallel spur dikes are constructed in the basin, one at the right-hand side of the inlet, and another at the central basin. The same test condition is used as in the test of residence time. Figures 8.10(a)–(c) present the simulated circulation patterns under this dike configuration, equal pollutant concentration contours at 9 min, and

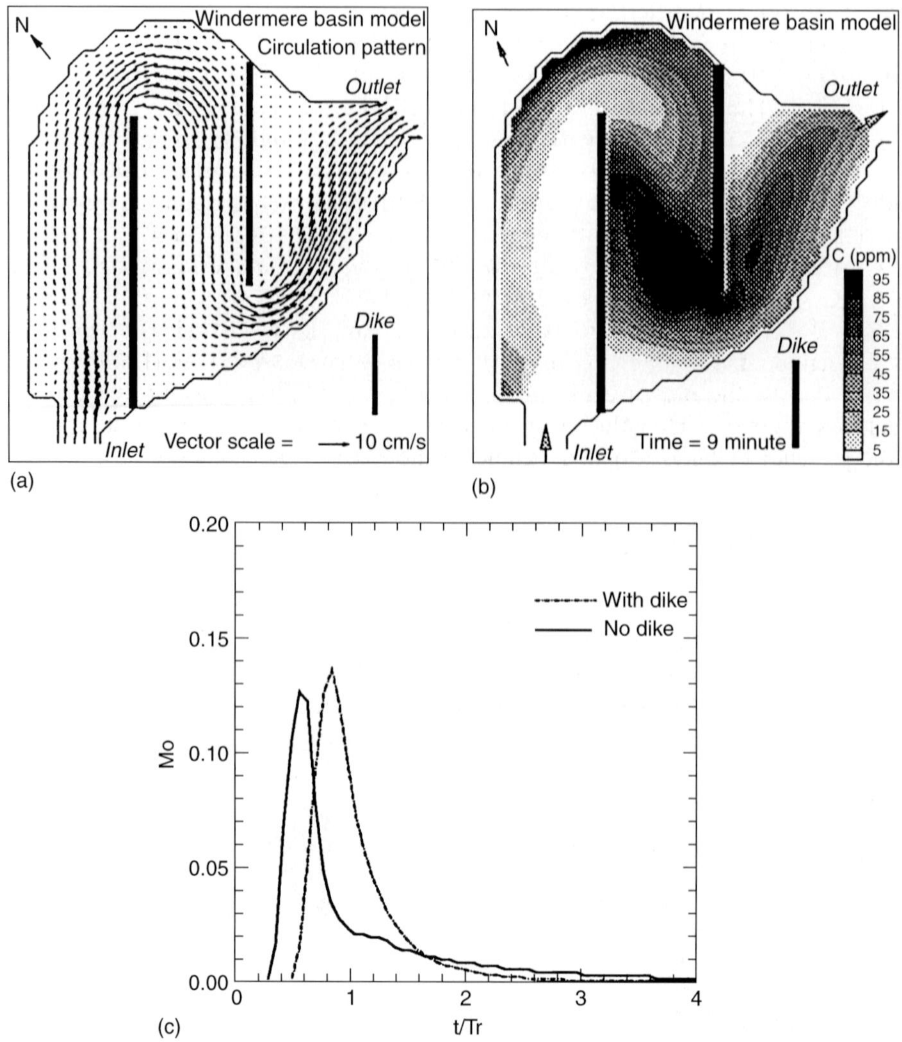

Fig. 8.10. Effect of dikes in the physical model of the Windermere Basin: (a) circulation pattern with a dike, (b) equal pollutant concentration flooded contours at 9 min after the pollutant was introduced to the basin, and (c) pollutant quantity at the outlet to the total released pollutant quantity as a function of the time normalized by the theoretical residence time (Reprinted from Wu and Tsanis (1994), copyright with kind permission from the IAHR, Delft, The Netherlands).

the pollutant quantities at the outlet, with and without dike, respectively. It is observed from Figure 8.10(a) that the dikes cause the advective flow to meander through the whole basin, along the dikes and dead zones behind the dikes, which is also evident in the physical model study (Engel, 1988). This has the effect of increasing the advective flow path length and breaking up the large eddies. As expected under such flow patterns, the arriving time is longer and the peak is slightly higher (reduced mixing), compared with the case without the dike. As a result, the mean residence time is significantly increased.

In summary, the following conclusions can be drawn from this example: (a) conventional 2DH was successful in simulating the depth-averaged hydraulically induced flow. The simulation was calibrated in a physical model of the Windermere Basin using the experimentally determined circulation pattern derived from drifter trajectories; (b) the 2D advection–diffusion model was successful in modelling the transport and dispersion of a pollutant emitted into the basin, and a numerical experiment was executed to demonstrate how the various zones such as advective flow, mixed, and dead zones affect the pollutant transport and distribution; (c) the numerical study of residence time indicated that, for a large portion of flow, a very short residence time exists, due to the advective flow (short-circuiting). Also, for a small portion of flow, a much longer time than that for theoretical residence is needed, due to the presence of eddies and dead zones. Simulations revealed no significant effect of wind-induced currents on the residence time. Finally, simulation in the presence of dikes revealed significant increases in the residence time.

8.2 STUDY OF WIND-INDUCED FLOWS[8]

Wind-induced flow is the most important type of flow in lakes and oceans, and the countercurrent flow in a closed channel includes this important mixing mechanism for geophysical wind-induced flow. As a result, it has been widely studied by laboratory experiments and numerical models. As shown in Figure 8.11,

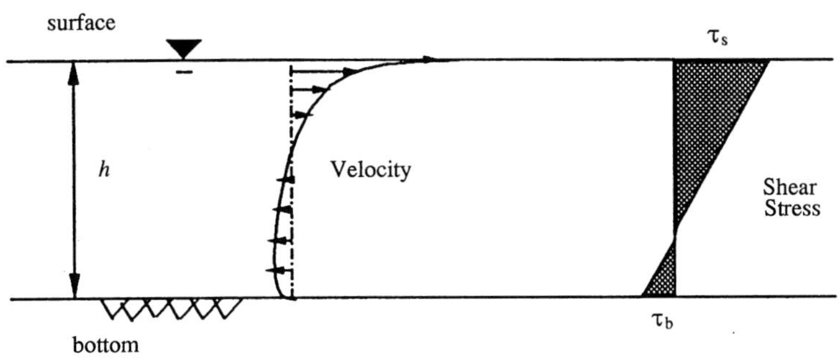

Fig. 8.11. Schematic representation of a fully developed countercurrent flow (Reprinted from Wu and Tsanis (1995a), copyright with permission from ASCE).

[8]Material in Section 8.2 is reproduced from Wu and Tsanis (1995a), copyright with permission from ASCE.

for fully developed conditions and 1D shearing, mass conservation requires zero net flow over the depth, while the shear stress distribution over the depth is linear (Tsanis and Leutheusser, 1988).

A series of experiments have been conducted for the study of wind-induced flow by using conventional laboratory air–water tunnels (Baines and Knapp, 1965; Wu, 1975; Tsuruya et al., 1985), and a volume of air in a moving bottomless box (Tsanis, 1986). The mean flow and turbulence properties were measured and analyzed. However, because of the limited length of the channel used (longitudinal aspect ratio was usually lower than 50) and the effect of side walls, a precise countercurrent flow with zero net flux was difficult to obtain in the majority of experiments. This was confirmed from the non-zero mass flux and the non-linearity of shear stress distributions found in extant experimental data (Tsanis, 1986, 1989).

The numerical modelling approach shows great potential in the study of shear-induced countercurrent flows. It is less costly and can be easily controlled and executed. Some features, such as the set-up and seiche in the channel, are readily predicted in the numerical model, which is rarely reported in the experiments. A number of wind-induced circulation models have been developed for oceans, lakes, and estuaries over the past years, as summarized by Cheng and Smith (1990). Before applying a model to natural water bodies, one has to examine the model's response to the expected forcing mechanism in a numerical sense. One of the most difficult tasks for wind-induced circulation models is to simulate the vertical current structure. Model evaluation, that is, comparing simulations with analytical and experimental results, should be performed. The countercurrent flows have been employed to calibrate the wind-induced coastal circulation models (Pearce and Cooper, 1981; Sheng, 1983; Swanson, 1986); however, the features of surface drift current and return flow near the bottom were not well simulated in these calibrations.

In this section, a general wind-induced circulation model is employed to study the wind-induced countercurrent flows. The simulations are first compared with a newly developed analytical solution, then verified using the experimental results obtained by Baines and Knapp (1965) and Tsuruya et al. (1985). The sensitivity study of model parameters, such as eddy viscosity and vertical layer discretization, is performed.

8.2.1 Model Comparison with Analytical Solutions

The wind-induced flow in a closed basin may be either laminar or turbulent. Exact solutions exist for steady cases.

Analytical solution for steady laminar flow

Assume that the convective acceleration, Coriolis forces, horizontal diffusivity, and cross (y-direction) terms are negligible when compared to the vertical shear gradient. The solution given by Heaps (1984) for the non-slip bottom condition for constant vertical eddy viscosity is

$$u(z) = \frac{\tau_s}{4\,\rho\,K_v\,h} z(3z - 2h) \tag{8.1}$$

where $u(z)$ is the velocity at any depth z; h the water depth; K_v the vertical eddy viscosity; τ_s the wind stress acting on the surface ($\tau_s = \rho_a C_D W^2$, in which ρ_a is the air density; C_D the wind drag coefficient; W the wind velocity, and ρ the water density). The origin of the z coordinate ($z = 0$) is located at the bottom and the direction is upward. This coordinate will also be applied to the analytical solution of turbulent flow and notably with such an application, the solution is in a simpler form.

A new analytical solution for steady shear-induced turbulent flow
Tsuruya et al. (1985) derived a double-logarithmic solution which depends on z_s, z_b (the characteristic lengths determined at the surface and at a bottom, respectively) and the ratio of bottom to surface shear velocities $m = u_b^*/u_s^*$. However, z_b and m are inter-dependent and difficult to solve. Tsanis (1989) derived an analytical solution which depends on z_s, z_b, and the ratio of bottom to surface shear stress $\eta = \tau_b^*/\tau_s^* = m^2$. The solution was obtained under an assumption on the relation between the characteristic lengths.

A new analytical solution which is independent of m and η could be derived as follows. The velocity profile takes a double-logarithmic profile. The following conditions are used to determine the coefficients: (a) at the free surface, the shear-stress is evaluated by the wind velocity; (b) at the bottom, the velocity is equal to zero; and (c) the depth-averaged velocity (U) equals zero. Therefore, the velocities at any depth, normalized by the surface shear velocity, that is, $\bar{u}_0(z) = u(z)/u_s^*$, can be obtained by

$$\bar{u}_0(z) = a\,\ln\left(1 + \frac{z}{z_s}\right) + b\,\ln\left(1 - \frac{z}{z_b + h}\right) \tag{8.2}$$

where (a) and (b) are calculated from Eq. (5.14) and (5.15) with the zero depth-averaged velocity U.

A constant viscosity leads to a parabolic velocity profile, as in the case of laminar flow. The simulated result from the 3D model using the same constant vertical viscosity coefficient is in good agreement with the analytical solution, as shown in Figure 8.12. The zero velocity occurs at one-third depth, below the water surface. The horizontal eddy viscosity has some effect on the profile. In the case of turbulent flow, a parabolic eddy viscosity distribution ($z_{sh} = 0.00022$, $z_{bh} = 0.00014$, and $\lambda = 0.35$) is used. Figure 8.13 shows the predictions by the 3D model, along with the analytical solution, and the model predictions are in excellent agreement with this.

8.2.2 Model Verification with Experimental Data

Baines and Knapp's experimental apparatus (1965) was a wind channel with a uniform depth of 0.3048 m. As shown in Table 8.1, two wind conditions were

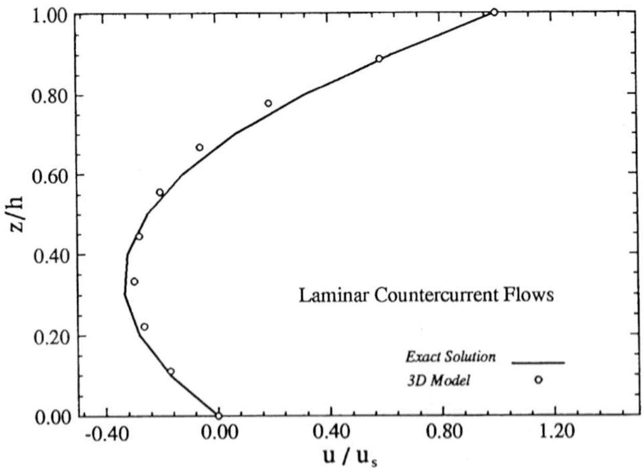

Fig. 8.12. Comparisons of the model simulation with the analytical solution for the steady laminar countercurrent flows (Reprinted from Wu and Tsanis (1995a), copyright with permission from ASCE).

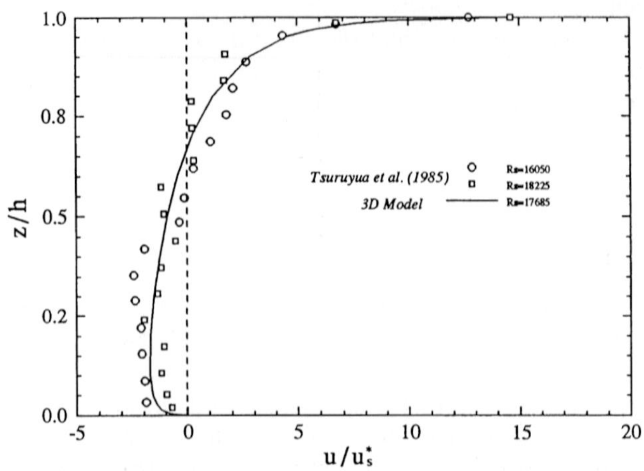

Fig. 8.13. Comparison of model simulation with the analytical solution for the steady turbulent countercurrent flow (Reprinted from Wu and Tsanis (1995a), copyright with permission from ASCE).

employed, that is, 3.901 and 6.096 m/s. Thirty-one velocity measurements were made at various depths, including the surface, in both cases. The surface shear velocities were evaluated from the applied wind velocities (Baines and Knapp, 1965). The Reynolds numbers $R = u_s h/\nu$, where u_s is the surface velocity, were 32,700 and 46,500, respectively and the same conditions were used in the 3D model. The parameters in the vertical eddy viscosity distribution are $z_{sh} = 0.00022$, $z_{bh} = 0.00014$, and $\lambda = 0.35$. The velocity profiles, normalized by the surface shear velocities obtained by the model, are presented in Figure 8.14, together with the

Table 8.1. Experimental results (Baines and Knapp, 1965) and simulations by the 3D model.

Quantity		Test 1	3D	Test 2	3D
Depth	h (m)	0.3048	0.3048	0.3048	0.3048
Wind velocity	W (m/s)	3.901	3.901	6.096	6.096
Surface shear velocity	u^*_s (cm/s)	0.6233	0.6233	0.9416	0.9416
Surface velocity	u_s (cm/s)	10.72	10.67	15.25	15.14
Reynolds number	$R_s = u_s h/\nu$	32700	32522	46500	46147
Normalized surface velocity	u_s/u^*_s	17.20	17.11	16.20	16.08
Absolute roughness	Z_{0w} (mm)	0.3521	0.3521	0.4795	0.4795
Zero velocity height	Z_0	0.69	0.68	0.75	0.68

Source: Reprinted from Wu and Tsanis (1995a), copyright with permission from ASCE.

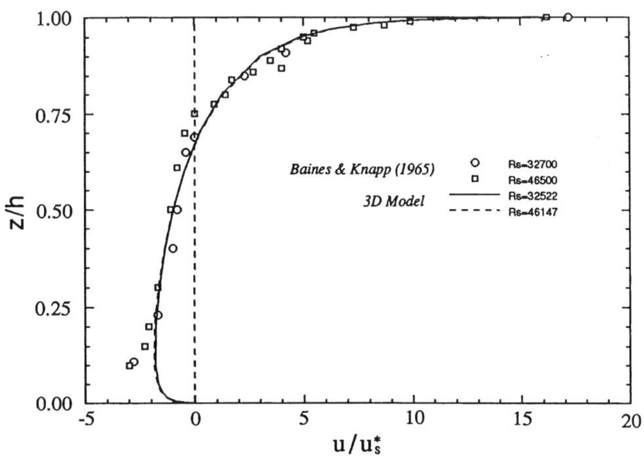

Fig. 8.14. Normalized velocity profiles—verification of the 3D model with the experimental data (Baines and Knapp, 1965) (Reprinted from Wu and Tsanis (1995a), copyright with permission from ASCE).

experimental results. The normalized surface velocities from the model are 17.11 and 16.08 for cases 1 and 2, respectively. These normalized velocities are close to the experimental data, 17.2 and 16.2, respectively and the velocity was logarithmic near the surface and the bottom. The return flow in the experiment exhibited a distinct peak near the bottom, due to the short length of the channel and the presence of the sloping beach (Baines and Knapp, 1965). The detailed currents had not been measured close to the bottom (lower than 10% of the whole depth).

A better presentation of the velocity profile is in terms of the coordinates of the "smooth" inner law of the velocity distribution, related to either the surface or the bottom (Tsanis, 1986; Andreasson, 1991). In Figure 8.15(a), curve (A) represents the viscous sub-layer portion of the velocity profile,

$$\frac{u_s - u}{u_s^*} = \frac{(h - z)u_s^*}{\nu} \tag{8.3}$$

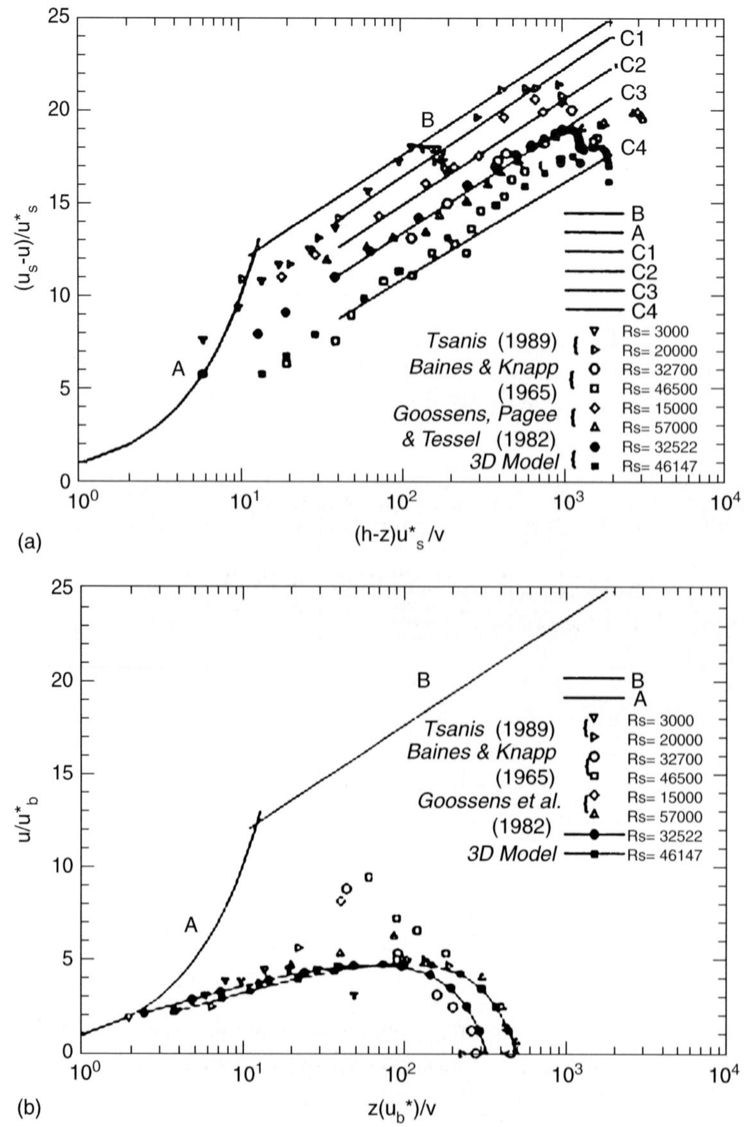

Fig. 8.15. "Smooth" inner-law of velocity distributions of the experimental data (Baines and Knapp, 1965) and the model simulations, relative to (a) the surface interface, and (b) the bottom interface (Reprinted from Wu and Tsanis (1995a), copyright with permission from ASCE).

and curve (B), corresponding to the logarithmic portion of the velocity profile for the hydrodynamically smooth conditions, is described by:

$$d\frac{u_s - u}{u_s^*} = 5.75\ \log\ \frac{(h-z)u_s^*}{\nu} + 6.0 \tag{8.4}$$

The curves labeled (C_i) $(i = 1, 2, 3, 4)$ represent the curves for experimental data with various Reynolds number. These curves,

$$\frac{u_s - u}{u_s^*} = 5.75 \log \frac{(h - z)u_s^*}{\nu} + 8.5 - 5.75 \log R_{kw} \tag{8.5}$$

are parallel to, but below, the curve (B), by an amount related to the roughness of the water side of the air–water interface. This value is proportional to the roughness Reynolds number, $R_{kw} = u^*_s Z_{0w}/\nu$, where Z_{0w} is the absolute roughness of the interface (Schlichting, 1968). The roughness length Z_{0w} can then be calculated from R_{kw}. The experimental data from Baines and Knapp (1965) and simulations by the 3D model are presented in Figure 8.15, together with experimental data from Tsanis (1989) and Goossens et al. (1982). Both data and simulations are seen to be in excellent agreement with the experimental data under both wind conditions.

Figure 8.15(b) shows the same model and experimental data plotted in inner-law coordinates relative to the bottom. It is expected from this figure that, in contrast to the conditions prevailing in the surface, the turbulence in the bottom return flow is quite undeveloped in the experiments. However, a good linearity in the semi-log coordinates is observed from the simulation, showing that a steady return flow was more easily achieved in the numerical model.

These conclusions could be supported from the vertical distribution of shear stresses; Figure 8.16 presents the shear stress distributions for existing experimental data, and theoretically linear distributions for laminar flow (Tsanis and Leutheusser, 1987) and high-Reynolds number turbulent flow ($\eta = \tau_b/\tau = -0.10$). It can be seen that all of the experimentally deduced Reynolds stresses differ significantly from the linear distribution. As pointed out by Tsanis (1986), the reason is the limited length

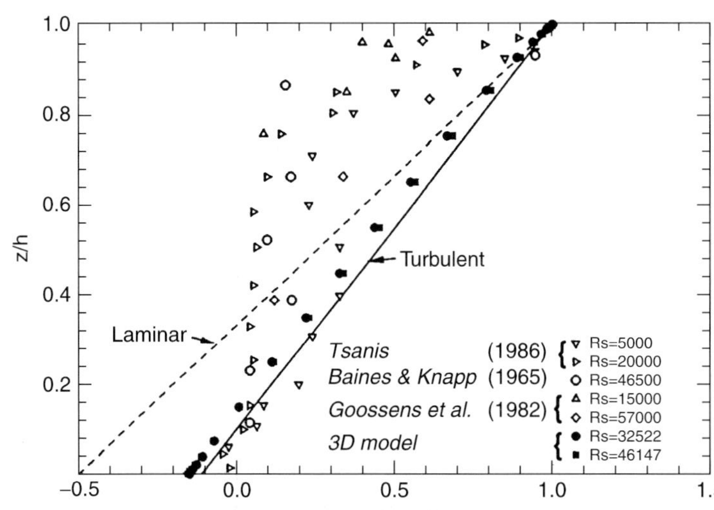

Fig. 8.16. Reynolds stresses as a function of transverse position (Reprinted from Wu and Tsanis (1995a), copyright with permission from ASCE).

Table 8.2. Experimental results (Tsuruya et al., 1985) and model simulations.

Quantity		Test 1	Test 2	3D
Depth	h (m)	0.15	0.15	0.15
Wind velocity	W (m/s)	6.73	6.90	6.80
Surface shear velocity	u^*_s (cm/s)	0.815	0.835	0.823
Surface velocity	u_s (cm/s)	10.70	12.15	11.79
Reynolds number	$R_s = \mathrm{u\ h}/\nu$	16500	18225	17685
Normalized surface velocity	u_s/u^*_s	12.74	14.55	14.32
Absolute roughness	Z_{0w} (mm)	0.21	0.36	0.31

Source: Reprinted from Wu and Tsanis (1995a), copyright with permission from ASCE.

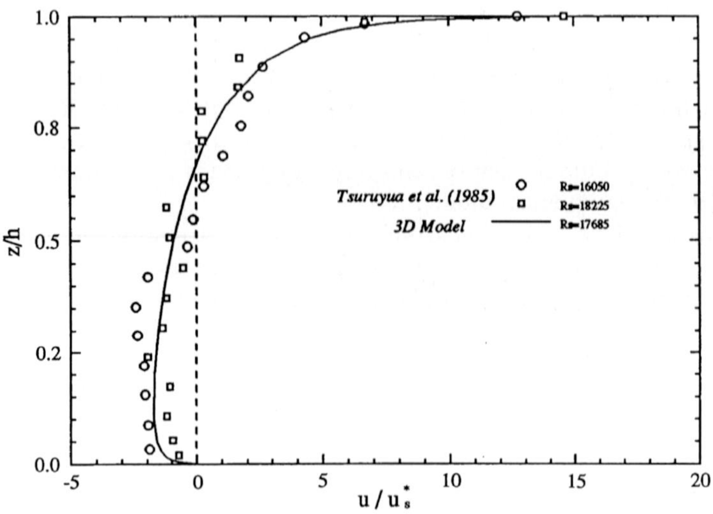

Fig. 8.17. Normalized velocity profiles—verification of the 3D model with the experimental data (Tsuruya et al., 1985) (Reprinted from Wu and Tsanis (1995a), copyright with permission from ASCE).

of the channel and the immature state of the structure of turbulence at a relatively low Reynolds number. In the numerical model, the channel is long enough; the mass conservation law is satisfied; and the simulation is controlled by the total kinematic energy (*KE*), that is, the steady state is reached when the ratio $|KE^{n+1} - KE^n|/KE^{n+1}$ becomes less than a test convergence value (10^{-8}). Therefore, as shown in Figure 8.16, a good linearity is observed for the simulation.

Another experiment carried out by Tsuruya et al. (1985) is used for verification. A wind-wave tank was used in the experiments. The flow visualizations were undertaken under two wind conditions, 6.73 and 6.90 m/s, as shown in Table 8.2. In the numerical 3D model the same depth was used and the wind velocity was taken as 6.8 m/s, a value between the two wind conditions used in the experiments. It is observed from Figure 8.17 and Table 8.2, that the model simulations are in good agreement with the experimental data. The surface velocity is well predicted, and the simulations fall between the two sets of experimental data. Comparisons are made in terms of the coordinates of the "smooth" inner law of the velocity distribution, relative to both the surface and bottom, as shown in Figures 8.18(a) and (b), respectively. It is observed that this experimental data keeps a better linearity than that of Baines and Knapp (1965), because the return flow is well measured through the use of a false bottom in this experiment.

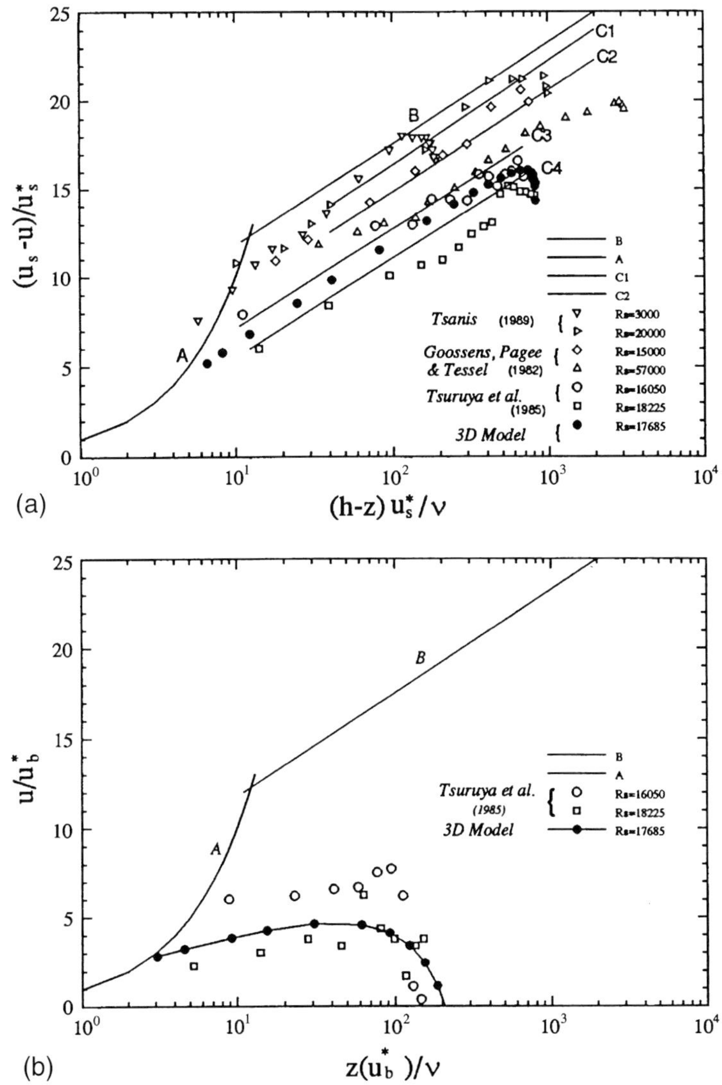

Fig. 8.18. "Smooth" inner-law of velocity distributions of the experimental data (Tsuruya et al., 1985) and model predictions, relative to (a) the surface interface and (b) the bottom interface (Reprinted from Wu and Tsanis (1995a), copyright with permission from ASCE).

8.2.3 Numerical Experiments

Effect of eddy viscosity

Physically, when the eddy viscosity is increased the internal shear stress and, consequently, the turbulence is increased. Therefore, the drift and return flow are decreased. Figure 8.19(a) shows a test using various horizontal eddy viscosity coefficients, ranging from 125 to 2500 cm^2/s. It is noticeable that the effect of horizontal eddy viscosity is insignificant, because the horizontal dimension is several orders of magnitude larger than the vertical dimension. As expected, the

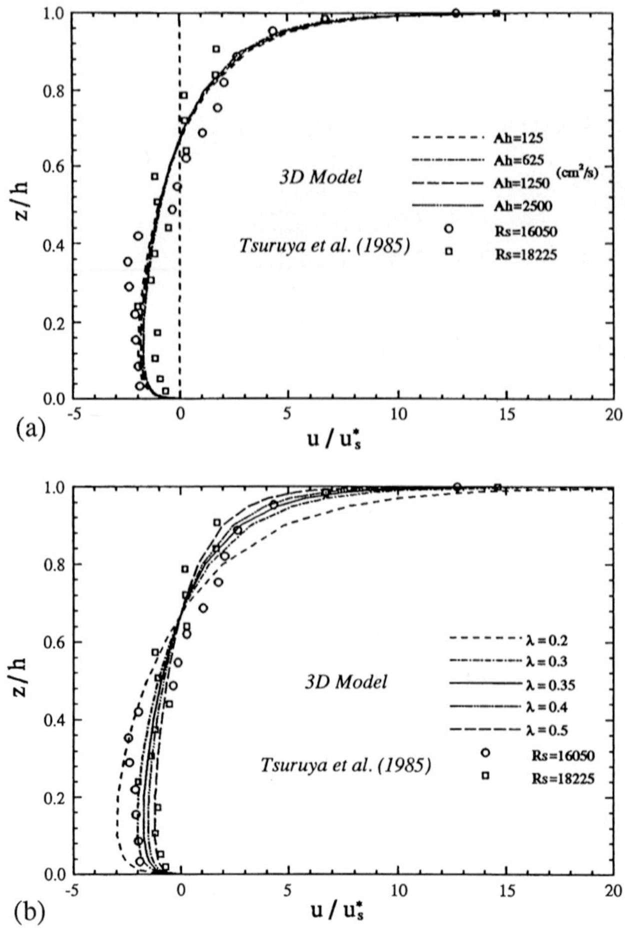

Fig. 8.19. Sensitivities of model parameters: (a) the effect of the horizontal eddy viscosity on the simulated velocity profile; symbols refer to the experimental data from Tsuruya et al. (1985); (b) effect of λ (constant in the vertical eddy viscosity distribution) on the simulated velocity profile; symbols refer to the experimental data from Tsuruya et al. (1985) (Reprinted from Wu and Tsanis (1995a), copyright with permission from ASCE).

vertical eddy viscosity significantly affects the predictions. Figure 8.19(b) shows a test for different values of λ in the parabolic vertical eddy viscosity distribution, where two sets of experimental data from Tsuruya et al. (1985) are presented for comparison. Strong currents are obtained for small λ: for example, normalized surface velocity reaches as high as 24.9 when $\lambda = 0.2$, while only 10.05 when $\lambda = 0.5$. Overall, a value of λ around 0.35 gives good simulation results which compare well with experimental data.

Effect of vertical discretization

Determining the surface velocity is very important for the prediction of surface particle or oil spill movement, and for the study of wave–current interaction. Determining the velocity field near the bottom is critical for the determination of the resuspension of bottom sediments; this is one of the difficult tasks for numerical models. It is noticed from tests that, except for the formulation of boundary conditions, the vertical layer discretization has significant effect on the prediction of surface drift current and near-bottom return current. Figure 8.20 shows the model predictions with different vertical layer discretizations, that is, equal layers (5, 15, 30 layers), unequal layers (7, 9 layers), and unequal layers enhanced both near the surface and the bottom (30 layers). By inspecting this figure, it is found that the layout has an insignificant effect on the velocities in the middle range between 0.2 and 0.8 of whole depth. However, high resolutions are needed to predict the surface and bottom velocity fields. Using the same model parameters, if vertical grid size $\Delta\sigma$ near the surface is greater than 0.06, the surface velocity normalized by surface shear velocity is lower than 8. However, when grid size is about 0.001, a

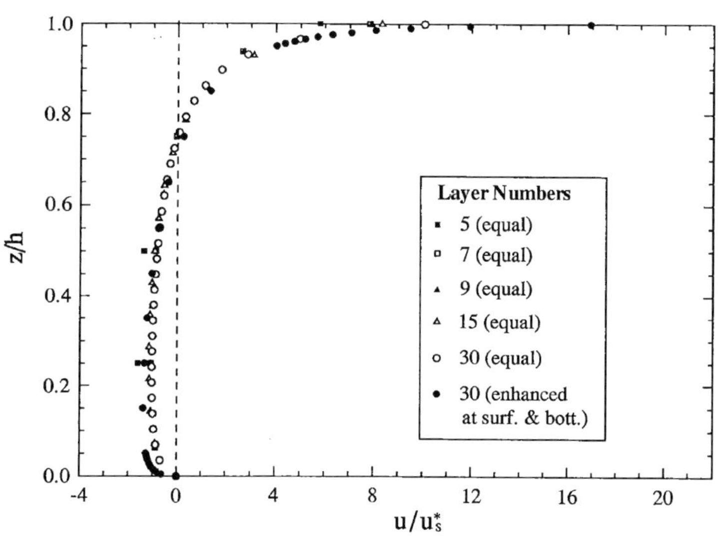

Fig. 8.20. Effect of the vertical layer discretizations on the simulated velocity profile (Reprinted from Wu and Tsanis (1995a), copyright with permission from ASCE).

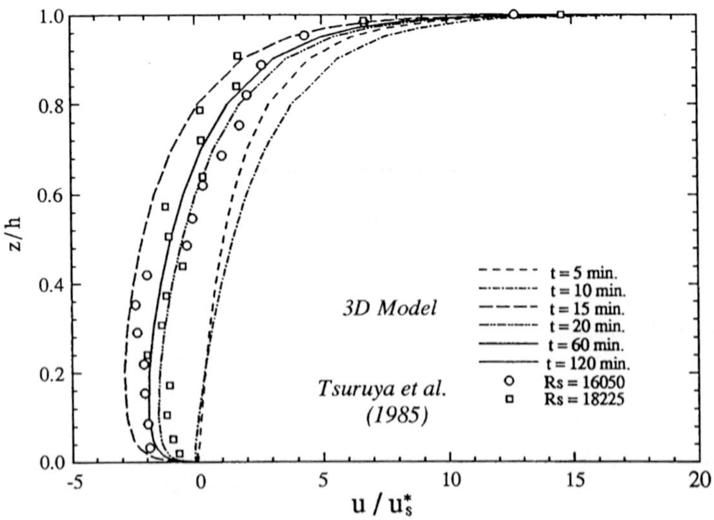

Fig. 8.21. Development of a shear-induced turbulent flow (Reprinted from Wu and Tsanis (1995a), copyright with permission from ASCE).

normalized surface velocity can reach 16–20, values which compare well to experimental results.

Developing shear-induced turbulent flow

The development of shear-induced turbulent flow can be examined from the evolution of the velocity distribution, as depicted in Figure 8.21. At the beginning (see profiles at 5 and 10 min), currents at all layers follow the wind direction, and there is no return flow near the bottom. A strong return flow is then extended to most of the region, except near the surface (see profile at 15 min). Afterward (120 min, in this test), the surface drift flow is balanced with the return flow, and a steady state is reached.

Wind-induced set-up

The wind-induced set-up is a by-product of the simulation. Figure 8.22 shows a typical wind-induced set-up, in which the total set-up is about 0.6 cm. Equal area for the water gain region (leeward) and the water loss region (upwind) confirms a mass conservation in the simulation. Theoretically, the total set-up can be evaluated from the momentum equation as:

$$\Delta\zeta = \frac{L}{gh} u_s^{*2}(1 - \eta) \tag{8.6}$$

where η is the ratio of the bottom to surface shear stress. The calculated set-up in this test is 0.59 cm and the simulated set-up is close to the calculated one.

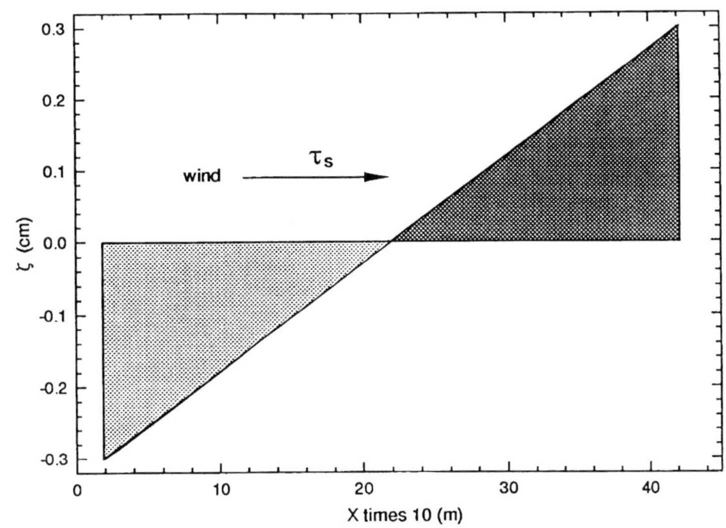

Fig. 8.22. Wind-induced set-up in a closed channel (Reprinted from Wu and Tsanis (1995a), copyright with permission from ASCE).

Seiche

Owing to the boundedness of the channel, free gravity waves combine and form a standing wave, or seiche. The theoretical period of seiche T is calculated by

$$T = 2\,L/\sqrt{gh} \tag{8.7}$$

where L is the length of the channel. A 5 m/s west wind is applied in a 9100 m long by 10 m deep channel, and with a corresponding period of 1836 s. From the time series of water elevation at leeward end, as shown in Figure 8.23(a), a period of 1826 s is observed. The relative error to the theoretical period is 0.4%. It is observed, from the total kinetic energy of the simulation (Fig. 8.23(b)), that there is an oscillation with a period of 925 s, which has a larger amplitude at the beginning, and zero amplitude when approaching the steady state. The period in the kinetic energy is half of that in the standing wave, and therefore the simulated period of seiche is 1850 s and the relative error is 0.7%. It is also seen from the time series of total kinetic energy, that at zero, net mass flux in the vertical plane is preserved; therefore, precise countercurrent flow is achieved.

Wind-induced current in a closed channel has been numerically studied by a multi-layered 3D circulation model. A new analytical solution, which is independent of the ratio of bottom to surface shear velocity or shear stress, has been derived for the steady shear-induced turbulent flow. The simulations for steady countercurrent flows are calibrated with the corresponding analytical solutions and verified using the experimental results obtained in air–water channels. It is found that the velocity distributions in the simulations follow the universal law of the wall, both in the drift current and in the return portion of the flow, while the

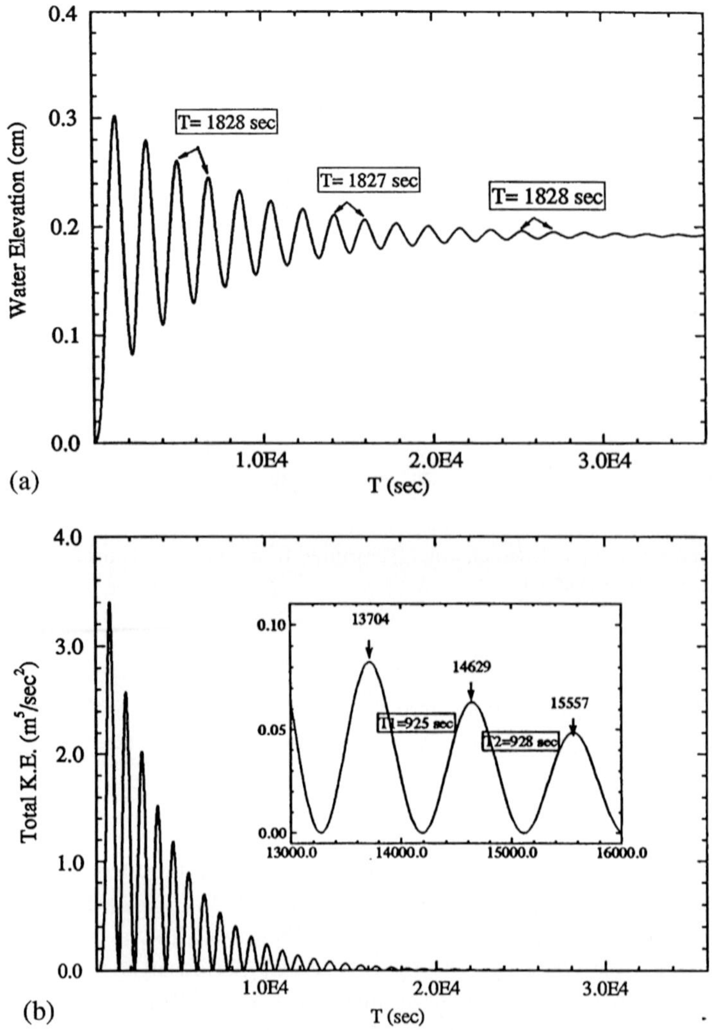

Fig. 8.23. Time series of (a) the water elevation at the leeward end of the channel; and (b) the total kinematic energy evaluated by depth-averaged velocity in the model simulation (Reprinted from Wu and Tsanis (1995a), copyright with permission from ASCE).

latter state is hard to achieve in most experiments due to the side wall friction and limited length of channel. Numerical experiments reveal that the horizontal eddy viscosity is insignificant, while the magnitude and distribution of the vertical eddy viscosity is very important in simulating the vertical velocity distribution. In order to obtain the detailed and accurate current structure, a high-resolution vertical grid must be used, especially near the surface bottom regions where the velocity gradients are high.

It is concluded that the 3D circulation model is able to simulate the features of wind-induced countercurrent flow, wind set-up, and seiche. The precise, steady, wind-induced countercurrent flow is easily obtained in this model; the numerical model then provides an alternative to the experimental approach, in the study of countercurrent flow. On the other hand, this study makes it possible to examine the credibility of existing 3D lake circulation models, and gives us insights into the possibilities in modelling drift and near-bottom currents.

Chapter 9

MODEL APPLICATIONS TO THE GREAT LAKES

9.1 GENERAL CIRCULATION IN THE GREAT LAKES

The Great Lakes of the St. Lawrence River system (Fig. 9.1), shared by Canada and the United States, constitute one of the largest masses of fresh water on Earth. The lakes themselves—Superior, Michigan, Huron, Erie, Ontario, and the relatively small St. Clair—have a total water area of 246,000 km^2, and a volume of 23,888 km^3. Their maximum depths are 406, 281, 229, 64, 224, and 6 m, and the average depths are 149, 85, 50, 10, 86, and 3 m, respectively. They provide good transportation, relatively inexpensive hydroelectric power, valuable fisheries, an abundance of water for industrial purposes, and increasingly popular recreation facilities. In addition, the lakes are a fresh water supply, yet, at the same time, they act as a sewage disposal system for the population. In the event of a marine oil spill emergency in the Great Lakes, knowledge of the current structure under different conditions is required, in order to provide real-time predictions of spill movement. In addition to winds, oil spill movement is directly affected by water currents.

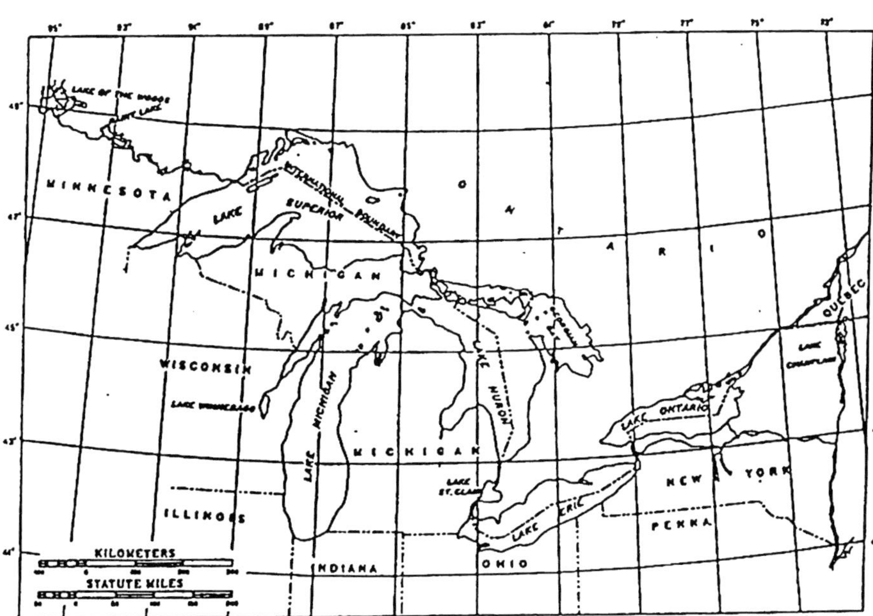

Fig. 9.1. Geographic location of the Great Lakes, including Lake Superior, Lake Michigan, Lake Huron, Lake St. Clair, Lake Erie, and Lake Ontario (Wu, 1993).

Table 9.1. Parameters used in the circulation modelling of the Great Lakes (Wu, 1993).

Parameters/Lakes	Superior	Michigan	Huron	Erie	Ontario	St. Clair
dx (m)	4000	4000	4000	4000	4000	1200
Imax	152	66	102	104	77	38
Jmax	71	127	94	30	30	39
C_D	0.000002	0.000002	0.000002	0.000002	0.000002	0.000002
t_h	0.9991	0.9991	0.9991	0.9991	0.9991	0.9996
K_h (m^2/s)	60	60	60	60	60	15
dt (s)	60	60	60	60	60	60
Q (m^3/s)	0.0	0.0	0.0	5500	6500	5700

Notes: *imax*, *jmax*, maximum value of i, j grid index, respectively; K_h, t_h, horizontal eddy viscosity and smooth factor; C_D, wind drag coefficient (relative to 10 m wind); dx, dt, spatial grid size and time step; and Q, average total inflow or outflow.

Therefore, the availability of a database of water currents in the Great Lakes is required. This database should be easily accessible and usable in an emergency environment.

The Great Lakes are interconnected through a system of channels and rivers. As a result, the circulation patterns in these lakes are both wind- and hydraulically driven. With the exception of Lake St. Clair, hydraulically induced current is negligible compared to wind induced. Thermal stratification during the summer season must be considered, as well as the spatial distribution of wind field and radiation. However, these factors are complicated and need a considerable amount of measured or forecasted meteorological and hydrological data, along with a complicated numerical circulation model. For the purpose of establishing an easily accessible database, it is economical to exclude these factors. A depth-averaged circulation model can only simulate depth-averaged currents and obviously cannot provide currents in different layers. A fully 3D circulation model requires a large amount of computational time, because the database should involve not only various wind speeds (3, 5, 7, 10, 15, 20 m/s) and directions (at intervals of 45°), but also different layers for all six lakes. Therefore, an efficient circulation model should be helpful and feasible for such circumstances; the Q3D and VHI3D are the appropriate models to be used for this purpose. The advantage over the multi-layered 3D model is that the currents at any depths can be obtained at almost no cost, using the depth-averaged currents and the related wind conditions (in the fully 3D model, vertical layers are discretized before any simulations, and the number of layers is limited). The Q3D and VHI3D models are applied in a 4 km grid for the Great Lakes, except for Lake St. Clair, where a 1.2 km grid is used. The parameters used in the model for each lake are listed in Table 9.1.

9.1.1 *Wind-Induced Circulation Patterns*

The VHI3D model is applied to Lake Superior, Lake Michigan, Lake Huron, and Lake Erie, with a grid size of 4 km, and the circulation pattern under a 10 m/s west wind is presented as a demonstration. The depth-averaged circulation patterns for

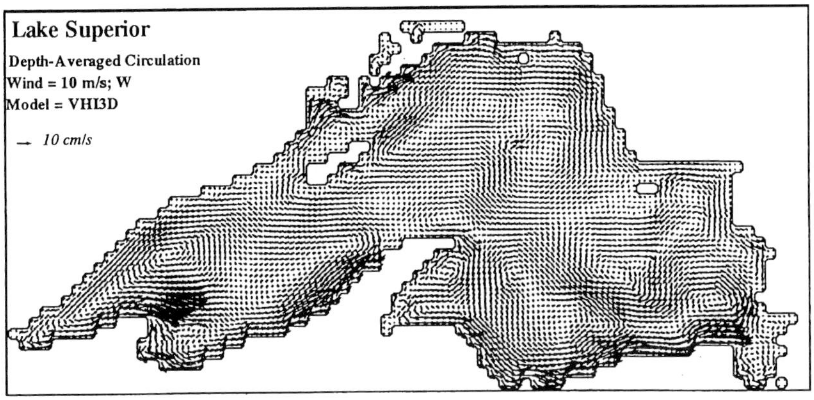

Fig. 9.2. Simulated depth-averaged circulation pattern under a 10 m/s west wind in Lake Superior (Wu, 1993).

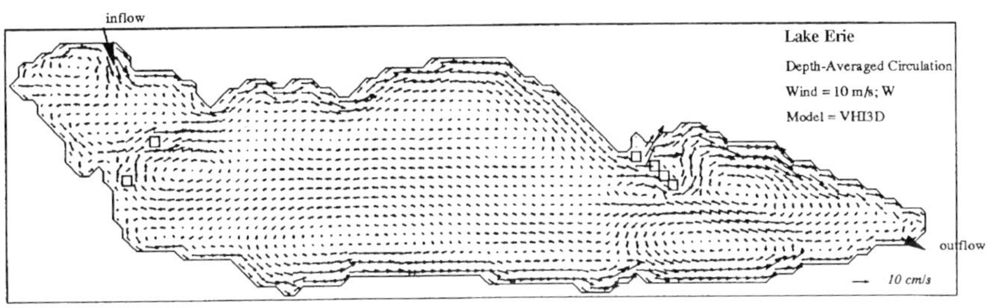

Fig. 9.3. Simulated depth-averaged circulation pattern under a 10 m/s west wind in Lake Erie (Wu, 1993).

these lakes are shown in Figures 9.2, 9.3, 9.4, and 9.5, respectively. In general, the patterns in Lake Superior, Michigan, and Huron are very complicated; the currents follow the wind in most near-shore areas, and many gyres can be found in the central areas. For these large lakes, spatial variation of wind speed and direction must be considered in order to provide a more realistic circulation pattern: this requires a large data set and is not considered in this study.

The Q3D and VHI3D models are applied in the 4 km grid of Lake Ontario to efficiently predict the 3D current structure in case of non-stratified seasons. Figure 9.6 presents the depth-averaged circulation patterns under a 10 m/s west wind by the Q3D and VHI3D models, as well as by the 2D model without horizontal viscosity and advective terms (Simons and Lam, 1986), and the 2DH model. It is observed that the circulation patterns are similar in all models; the current follows the wind in the shallow, north and south shorelines, and the return flow is formed in the deep central basin, resulting in two lake-wide, counterclockwise, and clockwise gyres. The magnitudes of currents predicted by the 2D and 2DH models are very close to each other. The horizontal eddy viscosity has an insignificant effect on the current when this coefficient is taken below a certain value; they are slightly larger than those of the Q3D and VHI3D models. The constants in the vertical eddy viscosity distribution in

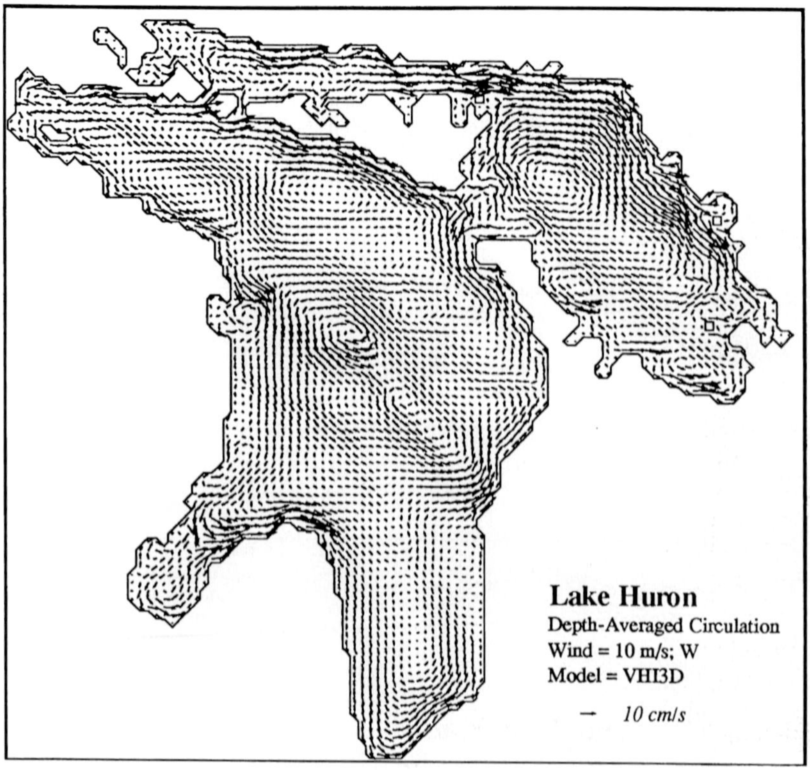

Fig. 9.4. Simulated depth-averaged circulation pattern under a 10 m/s west wind in Lake Huron (Wu, 1993).

the Q3D and VHI3D models are values that are calibrated in the numerical study of counter-current flow, as described in Chapter 5, and the magnitude of simulated current can be adjusted by changing this constant. The simulated circulation patterns from the VHI3D model, under a 10 m/s west wind and at four typical depths (equal σ layer), that is, surface, 5th layer (1/5 depth below water surface), 4th layer (2/5 depth), and 2nd layer (4/5 depth), are presented in Figure 9.7. The patterns indicate that the wind-induced flow in the shallow shorelines follows the wind over the whole depth, while, in the deep part of the lake, except close to the surface, a return flow against the wind dominates.

9.1.2 Wind-Induced Set-Up in Lake Ontario

Figure 9.8 shows the water elevation field under a uniform 10 m/s west wind in Lake Ontario. The water-level difference between the upwind and downwind ends of the basin is ~8 cm. Field data on the water-level difference between Burlington and Oswego, from the International Field Year for the Great Lakes (IFYGL) (Donelan et al., 1974), is used to calibrate the model simulation. In Figure 9.9, the small solid dots represent the field data and the straight line represents the empirical

Fig. 9.5. Simulated depth-averaged circulation pattern under a 10 m/s west wind in Lake Michigan (Wu, 1993).

relationship between wind shear stress (dyne/cm^2) and the water-level difference (cm). The open circles represent model simulations under four wind conditions (see Table 9.2). The comparison indicates a good agreement between field data and model simulations. According to Donelan et al. (1974), the best estimate for the overall 4 m wind drag coefficient is 1.35×10^{-3}; for unstable conditions, 1.5×10^{-3}; and for neutral condition, is 1.3×10^{-3} (see Table 9.3).

9.2 BACTERIAL TRANSPORT OF THE ST. CLAIR RIVER IN SARNIA[9]

Field surveys of indicator bacteria (*fecal coliform*, *fecal streptococci*, *Escherichia coli*, and *Pseudomonas aeruginosa*) show that the near shore waters of the St. Clair

[9]Materials in Section 9.2 is reproduced from Tsanis et al. (1995), copyright with permission from the International Association of Great Lakes Research.

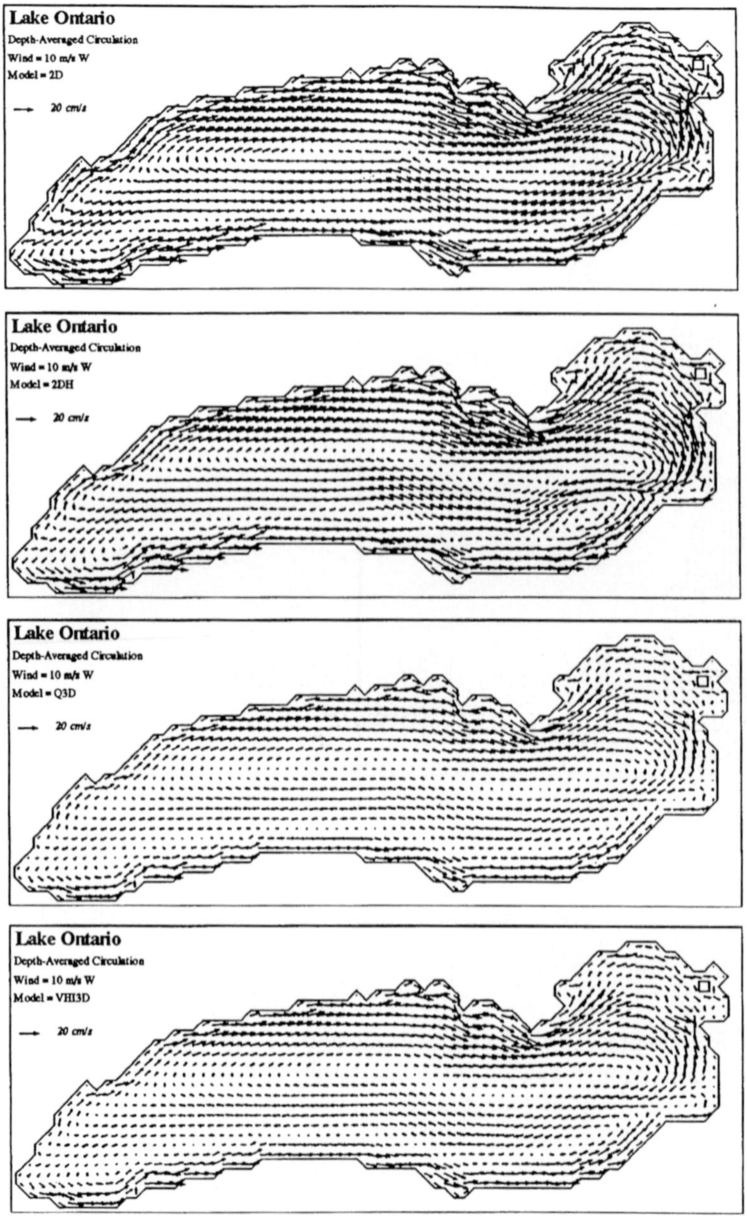

Fig. 9.6. Simulated depth-averaged circulation pattern under a 10 m/s west wind in Lake Ontario. Four different circulation models are employed (Wu, 1993).

River in Sarnia (see Fig. 9.10) are contaminated by fecal pollution originating from upstream sources (i.e., outflow from Lake Huron), dry weather sources (effluents from two sewage treatment plants in Point Edward and Sarnia), and wet-weather sources, including combined sewer overflows and storm sewer discharges (Marsalek

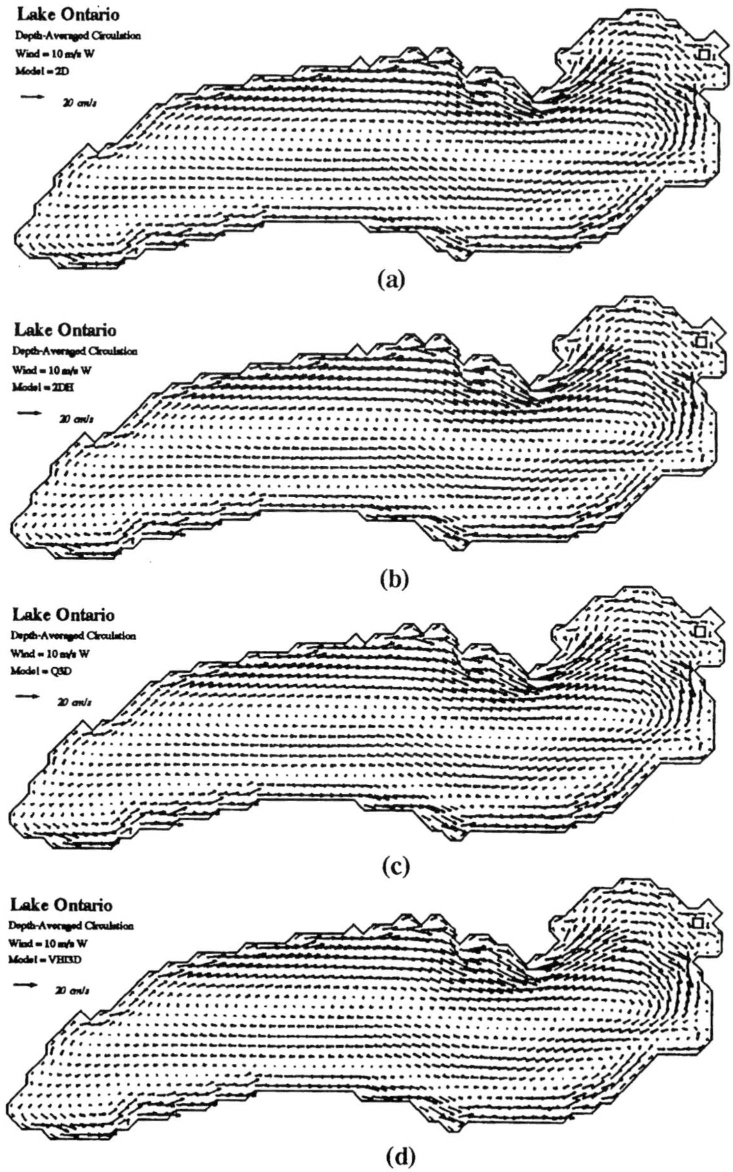

Fig. 9.7. Simulated multi-layer circulation pattern under a 10 m/s west wind in Lake Ontario. (a) Surface layer, (b) 5th layer—1/5 depth below the surface, (c) 4th layer—2/5 depth, (d) 2nd layer—4/5 depth (Wu, 1993).

et al., 1992). The results of such surveys indicate that these sources of fecal bacteria impact adversely on recreational water use in this area and increase the incidence of closed swimming beaches. Such adverse impacts could be reduced by remedial measures. The feasibility and effectiveness of such measures could be established by

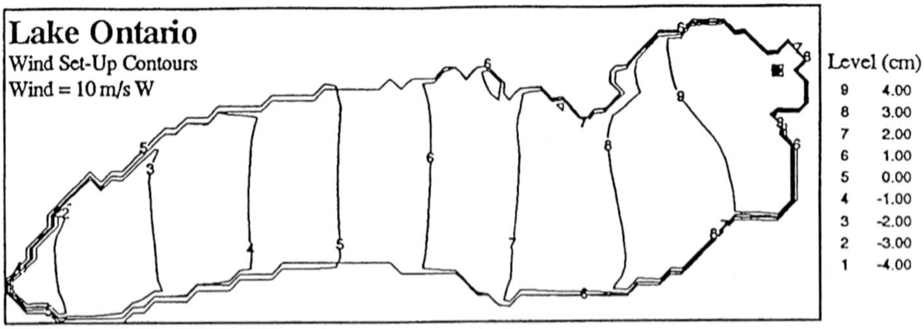

Fig. 9.8. Wind set-up under a 10 m/s west wind in Lake Ontario, simulated by the Q3D model (Wu, 1993).

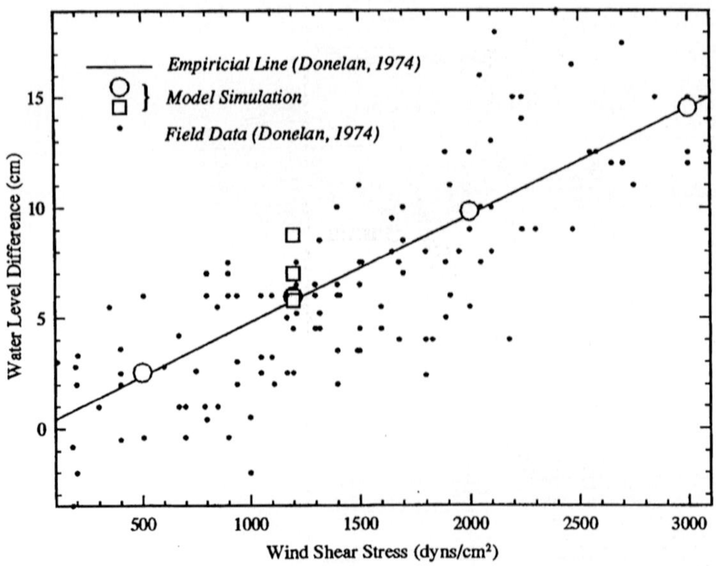

Fig. 9.9. Comparison of the simulated wind set-up along Burlington and Oswego with the empirical data (Donelan et al., 1974) in Lake Ontario (Wu, 1993).

Table 9.2. Wind-induced set-up along Burlington-Oswego for various winds.

Wind (m/s)	ζ_B (cm)	ζ_o (cm)	$\Delta\zeta$ (cm) from model	$\Delta\zeta$ (cm) from data*
15.81	−5.01	9.52	14.53	14.00
12.91	−3.29	6.53	9.82	9.00
10.00	−1.95	4.01	5.96	6.00
6.455	−0.87	1.67	2.54	1.50

Source: Wu, 1993
*The empirical value from the field data (Donelan et al., 1974).

Table 9.3. Wind-induced set-up along Burlington-Oswego for various wind drag coefficients.

C_D ($\times 10^{-3}$)	ζ_B (cm)	ζ_o (cm)	$\Delta\zeta$ (cm) from model	$\Delta\zeta$ (cm) from data*
2.00	−2.90	5.85	8.75	6.0
1.59	−2.30	4.69	6.99	6.2
1.35	−1.95	4.01	5.96	6.0
1.33	−1.88	3.89	5.77	5.5

Source: Wu, 1993
*See footnote in Table 9.2

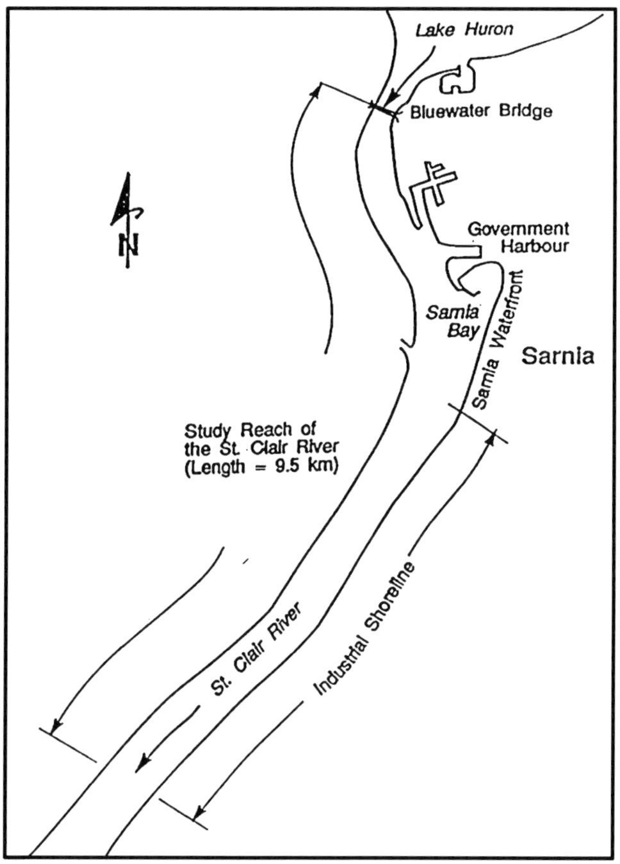

Fig. 9.10. Study area (Tsanis et al., 1995).

a modelling package comprising a loading model, to simulate wet-weather discharges of bacteria, and a receiving water model, to simulate bacterial densities in the river.

The modelling of bacteria in receiving waters is impeded by the complexity and dynamics of the processes occurring in nature. Even after simplification, the processes modelled may have to include bacterial effluent mixing and dilution, bacteria die-off or growth, and bacteria removal by sedimentation. In spite of such difficulties, the feasibility of modelling bacterial concentrations, for existing conditions, has been demonstrated. For example, Palmer and Dewey (1984, 1987), and Dewey and Palmer (1984), used the RAND hydrodynamic transport model to simulate currents and fecal coliform concentrations in St. John's Harbour, an urban beach on the Ottawa River, and the Eastern Beaches in Toronto. The simulated results were verified with limited field data. However, the RAND model, used to predict the currents, was not verified in terms of current pattern and magnitude, and problems with open boundary conditions and stability were reported (Palmer and Dewey, 1984). Therefore, a stable and verified circulation model is required to couple the transport model in these applications.

In the case of the nearshore zone of the St. Clair River in Sarnia, extensive simplifications in bacteria modelling are possible because of the special features of the receiving water body. In particular, the residence times in the main river channel are rather short, in the range of 45–90 min, and fast flows in the study reach preclude significant sedimentation. Consequently, it is safe to assume that the changes in bacterial densities caused by sedimentation can be neglected in the modelling analysis, and even though the die-off or growth can be included, using a first-order decay term, its effect should be negligible in view of short residence times.

The same assumptions, however, may not apply to the water bodies connected to the main river channel and characterized by limited circulation. Such bodies include the Government Harbour and Sarnia Bay, but only the latter is of interest in connection with water-based recreation. A detailed modelling of these water bodies would require extensive calibration data, which were not available at the planning stage of this study. Furthermore, to improve the bacteriological water quality in this area, substantial changes in the loading and transport regimes would have to be implemented, by means of physical changes of sewer outfalls and structures affecting circulation in the bay. Consequently, the models used could be calibrated and verified only for existing conditions, and the applicability of calibration parameters, derived from the measurements in Sarnia Bay (Marsalek et al., 1994), to future scenarios, was assumed. Thus, the main study was to examine the use of planning-level modelling, with limited calibration, for screening remedial measures. This is a common approach in the planning level of analysis of water management problems, and under such circumstances, the modeller's attention should focus on the sensitivity analysis of modelling results and the data needs for later, more advanced stages of analysis.

In simulations of remedial measures, fecal coliform was used as the primary indicator of bacteriological water quality. Following the completion of the study, the guidelines for recreational water quality of both the (Ontario) Ministry of Health (OMOE, 1984) and Health and Welfare Canada (Health and Welfare Canada, 1992) have been changed, and now recommend using *E. coli* as the primary indicator of fecal pollution. However, the Health and Welfare Canada Guidelines (1992) also note that fecal coliform may remain in use where "experience has shown that greater than 90% of the fecal coliforms are *E. coli*". This condition has been generally confirmed in Ontario waters, except where such waters receive effluents from pulp and paper mills or food processing plants (Health and Welfare Canada, 1992). Neither of these conditions apply to the receiving waters studied here, and, consequently, it is considered safe to assume that the presented methodology and its results would apply to *E. coli* as well, and would be useful in discussing the probability of compliance with the already mentioned recreational water quality guidelines. The main objective of the modelling work described here was to examine the feasibility of modelling indicator bacteria in the study area, with emphasis on enhancing the compliance with recreational water uses and the relevant water quality guidelines. In this process, the effectiveness of various remedial measures in mitigating fecal pollution problems in Sarnia Bay was simulated. These simulation results should be useful in planning controls of bacterial contamination in this area of concern (under the general water pollution control plan) and in planning future collection of bacteriological data.

9.2.1 Circulation and Pollutant Transport in Sarnia Bay

The St. Clair River section around Sarnia Bay was discretized in 142 × 71 square grids with a grid size $\Delta x = 25$ m; see Figure 9.11.

Figure 9.12 shows the current structure in the study area for an inflow/outflow river discharge of 6,000 m^3/s and a bottom friction coefficient $C_b = 0.0025$. The latter value of bottom coefficient corresponds to a range of Manning's coefficient between 0.016 and 0.028 for river depths of 1 and 25 m, respectively. This range of Manning's coefficient values is appropriate for the characteristics of the St. Clair River. The fecal coliform was selected as an indicator of fecal pollution and simulated for a number of storms. The decay rate $\lambda = 0.5$/day for the coliform bacteria is considered an appropriate value for a large river (Chamberlain and Mitchell, 1978; Swanson, 1986).

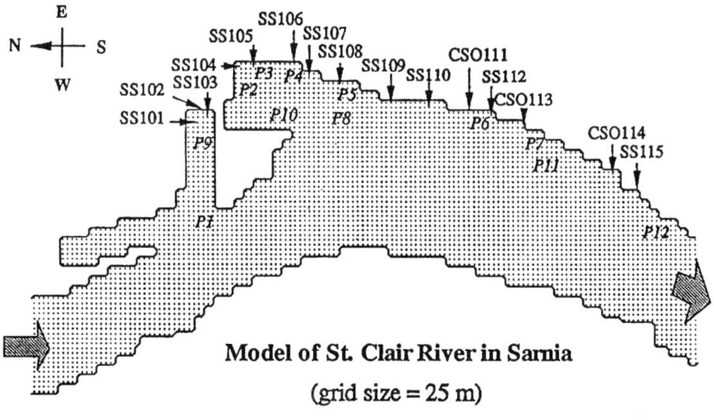

Fig. 9.11. The grid discretization of the St. Clair River reach with the locations of storm and combined storm sewers and the 12 points 1–12 where bacteria concentrations were simulated as a function of time (Tsanis et al., 1995).

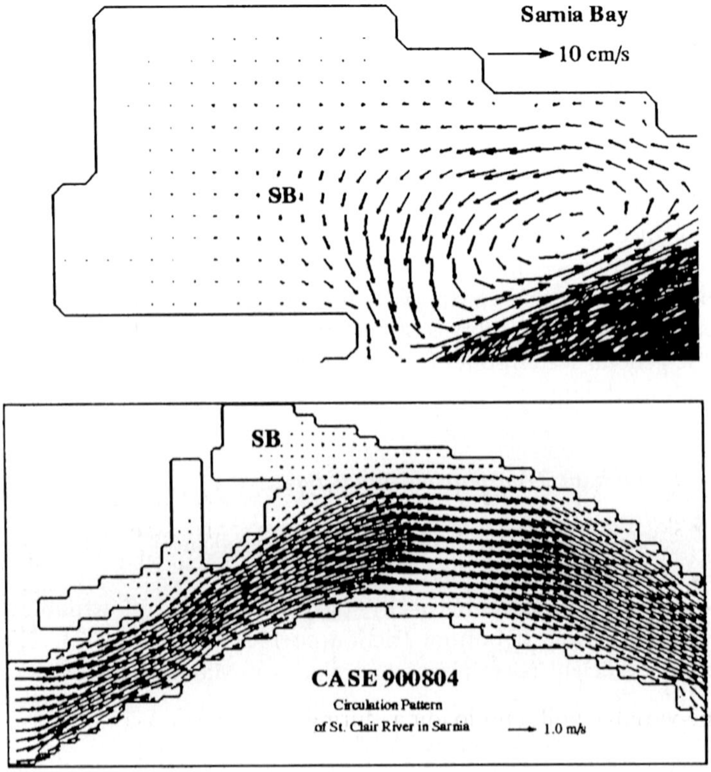

Fig. 9.12. Depth-averaged current distribution in the St. Clair River reach around Sarnia Bay for a 3 m/s west wind (Tsanis et al., 1995).

The time step used in this simulation was selected as 2 s in order to comply with the Courant criterion (Koutitas, 1988). The smoothing factor t_h was 0.995, and corresponded to a horizontal eddy viscosity of 0.75 m^2/s. A strong advective flow through the main river channel dominates the flow domain and contributes to the generation of counterclockwise eddies in two basins connected to the river channel, the Government Harbour, and Sarnia Bay. In the main river channel, the model produced current magnitudes similar to those measured by the U.S. Army Corps. of Engineers (1983).

The current structure obtained from the hydrodynamic model was used as input to the pollutant transport model. The horizontal dispersion coefficients of bacteria, D_x and D_y, in Sarnia Bay and Government Harbour, are taken as a constant 0.8 m^2/s (McCorquodale et al., 1986), which is similar to the diffusivity values for momentum. In the St. Clair River, however, they are calculated by (Vepsa et al., 1992)

$$D_x = D_L \cos^2 \phi + D_T \sin^2 \phi \tag{9.1}$$

$$D_y = D_L \sin^2 \phi + D_T \cos^2 \phi \tag{9.2}$$

where $\phi = \arctan(V/U)$ is the angle between the local flow direction and the x-axis. The D_L and D_T are the longitudinal and transverse dispersion coefficients in flow aligned local coordinate system. According to Fischer et al. (1979), they are evaluated by the following empirical formulas:

$$D_T = 0.9\ h\ u_* \tag{9.3}$$

$$D_L = 5.93\ h\ u_* \tag{9.4}$$

where h is the water depth; u_* the shear velocity $u_* = \sqrt{ghS}$; and S the slope of the river, calculated via the Manning's equation (Roberson and Croew, 1990):

$$S = \left(\frac{V\ n}{R_h^{2/3}}\right)^2 \tag{9.5}$$

where V is the flow velocity and R_h the hydraulic radius. For the St. Clair River at Sarnia, the average slope is 0.0000176 approximately, and the shear velocity is 0.051 m/s ($V = 1.0$ m/s, $h = 15$ m, $R_h = 14.40$, $n = 0.025$). From the foregoing formula, when the depths vary from 10 to 20 m, the transverse and longitudinal dispersion coefficients range from 0.45 to 0.90 and 2.9 to 5.8 m^2/s, respectively. The range of transverse dispersion coefficient values is close to that reported by McCorquodale et al. (1986).

A case with a time-variable pollutant loading, during wet weather, was simulated with the models discussed. The storm under study started at 15:00 on August 4, 1990 and lasted 7 h. Fifteen time–variable sources (storm sewers and combined sewer overflows) along the northern shoreline of Sarnia Bay, with flow rates, fecal coliform, and *E. coli* bacteria concentrations, as shown in Figure 9.13, were used as input to the pollutant dispersion model (Marsalek et al., 1992). A 48-h post-storm recovery period (from 7 to 55 h) was also simulated. The background fecal coliform bacteria concentrations for the dry and wet weather conditions were 50 and 200 counts/100 mL, respectively.

Figures 9.14(a)–(c) show the fecal coliform bacteria iso-concentration contours at the end of the storm, 12 h and 24 h after the storm. These results indicate important implications of the circulation patterns in the bay for the transport of contaminants discharged into the bay. Strong velocity gradients between the advective flow in the river channel and the eddies in the bay result in a limited interaction between the river and the bay. Thus, contaminants released in the bay will remain in the bay for a long time after their release. Figure 9.14(b) indicates the presence of a high level of bacteria even 12 h after the cessation of rainfall.

The prevailing circulation, in the counterclockwise direction, provides a mechanism for transport of pollutants discharged from the shore. In less than 5 h after the start of a continuous contaminant discharge, the contaminant traveled to the northwest corner of the bay and attained concentrations equal to about 30% of the concentration released from the outfall. Thus, stormwater discharges along the east bay shore, characterized by fecal coliform densities in the range of

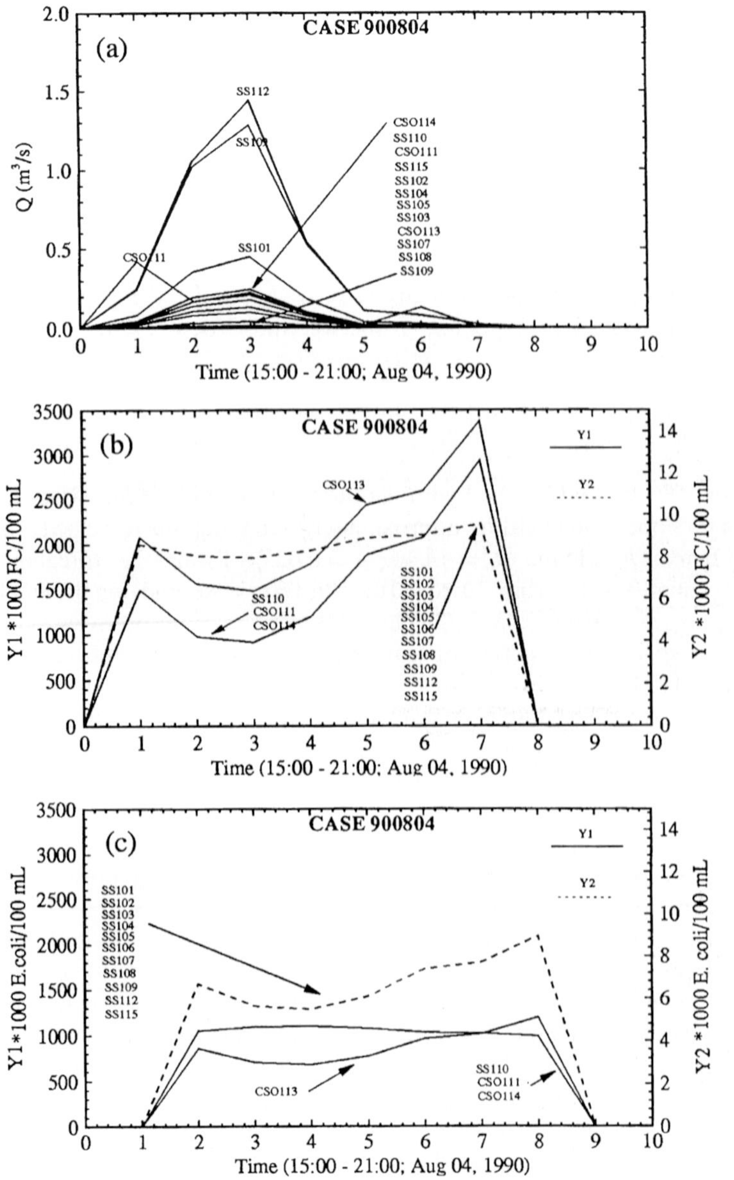

Fig. 9.13. Model inputs for case 900804: (a) measured flow rates; (b) fecal coliform concentrations; and (c) *E. coli* concentrations of storm sewer discharges and combined sewer overflows during the August 4, 1991 storm (Tsanis et al., 1995).

1200–5000/100 mL, would be transported by the described circulation throughout the bay. About 5 h after the start of continuous discharge, the bacterial densities in the bay would range from 400 to 1700 fecal coliforms/100 mL. For any contaminants discharged into the bay, the residence times would be quite long.

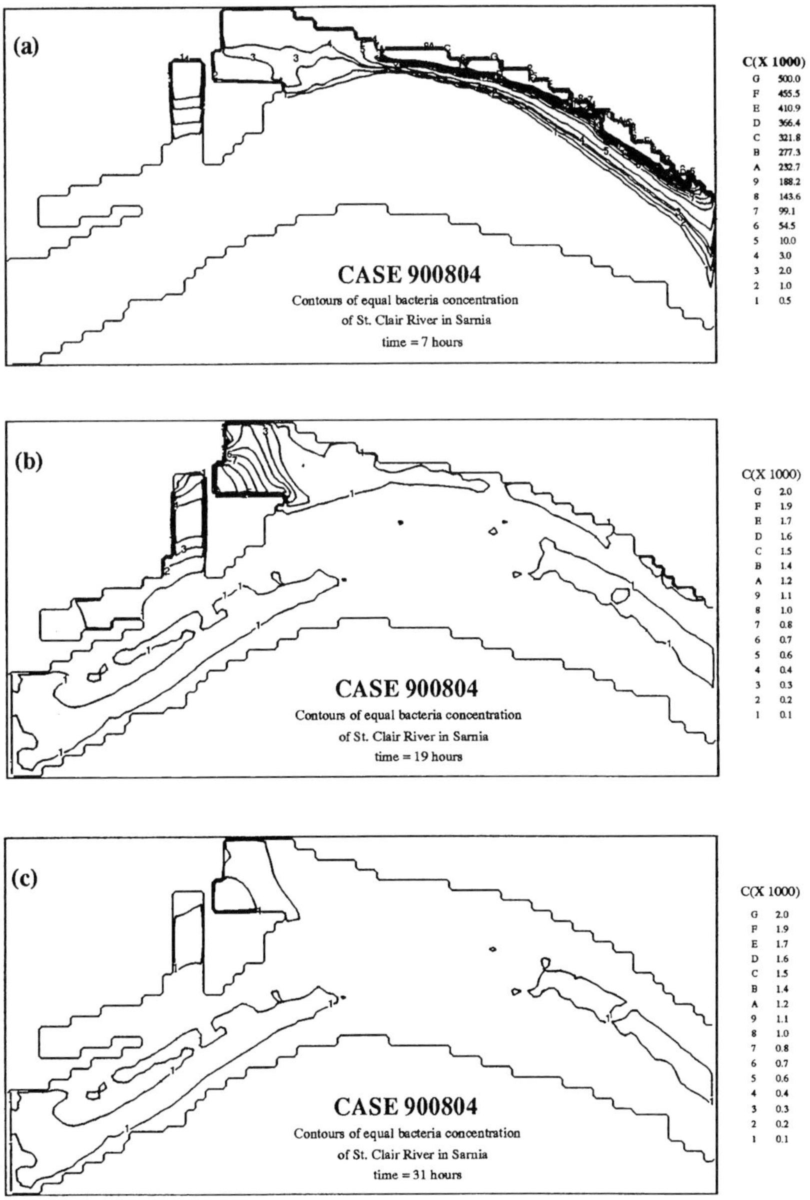

Fig. 9.14. (a–c) Modelled contours of equal bacteria concentrations at the end of the August storm, and 12 and 24 h after the storm (Tsanis et al., 1995).

9.2.2 Duration of the Aftereffects of Bacteria Releases

The field observations of bacterial densities showed great differences between dry and wet weather densities (Marsalek et al., 1992). The ability to distinguish between dry and wet weather data in field surveys is impaired by the fact that wet weather impacts may extend into dry periods long after the cessation of rain. Such aftereffects can be caused by two factors: slow flushing of the receiving waters (or some zones of receiving waters), or by malfunctions of sewerage systems, allowing sanitary discharges even during dry weather (Marsalek et al., 1992). Only the first issue can be effectively addressed here, because the second one would require detailed surveys of all sewer outfalls in both dry and wet weather.

The duration of wet weather aftereffects in the receiving water can be studied by the hydrodynamic model used in this example. In such numerical experiments, high bacterial density inputs (10^4–10^5 organisms/100 mL) were introduced into the model and, after the cessation of such inputs, the times required to reduce bacterial densities to the 100 counts/100 mL level (equal to the old water quality guideline for fecal coliform in swimmable waters) were determined for various locations in the study area.

Figure 9.15 shows the logarithm of bacteria concentrations (both for fecal coliform and *E. coli*) for the August 4, 1992 storm, at 10 nearshore locations close to pollutant sources, as a function of time during and after the storm.

Locations (2), (3), (4), and (10) are in Sarnia Bay; locations (5) and (8) are at the boundary between the river and Sarnia Bay, close to the shore, and locations (6), (7), (11), and (12) are in the advective path of the flow close to the northern shore. From this plot, the duration of aftereffects of bacteria releases, which give good indications of flushing times for various parts of the receiving water system, can be estimated. The duration of aftereffects of bacterial releases are summarized in Table 9.4.

In the main river channel, the decline of bacterial densities after the rain cessation is very fast and points to strong advection transport. Any aftereffects of wet weather bacterial pollution inputs would disappear within 10 h after the cessation of polluted discharges. In the basins with limited circulation, such as Sarnia Bay, the duration of aftereffects is up to 30 h. This slow flushing effectively extends the duration of wet weather impacts by the above long periods. Other case simulations also support the earlier conclusion.

When assessing the feasibility of modelling indicator bacteria, it appears that, for recreational water use, the ability to reproduce the magnitude of peak bacterial densities is not critical at this stage of analysis. Obviously, large exceedances of the swimming water quality guidelines will occur in wet weather and the degree of exceedance is not critical when assessing the probability of non-compliance with the guideline. The speed of decline of these high bacteria densities is, however, of greater interest, because it indicates how long the wet weather impacts persist in the receiving waters (Marsalek et al., 1992).

The proposed modelling approach, based on a planning-level modelling of bacterial loads and a detailed modelling of the receiving waters, is feasible for assessing the frequency and duration of non-compliance with swimming-water quality guidelines. The difficulties with assessing the dry weather conditions follow from the lack of knowledge of dry weather fecal bacterial discharges into the river.

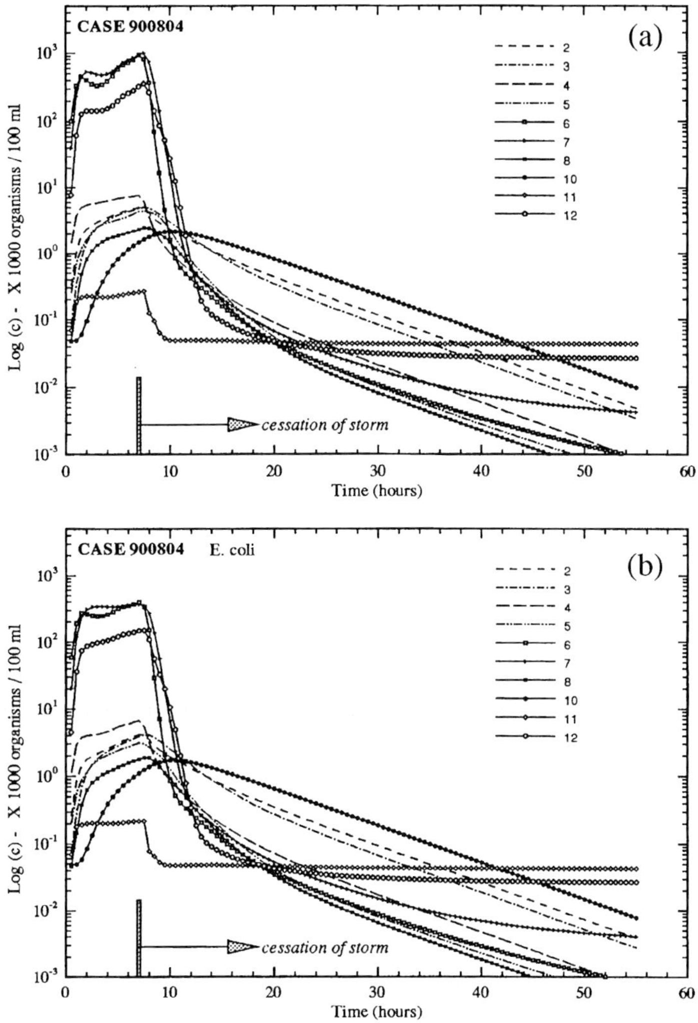

Fig. 9.15. (a, b) Modelled bacteria concentrations at eight locations in the St. Clair River in Sarnia Bay for the case 900804 (results in Table 9.5 refer to this case): (a) for fecal coliform, and (b) for *E. coli* (Tsanis et al., 1995).

9.2.3 Sensitivity Analysis

Different values of the model parameters, such as wind speed and bottom friction, can alter, to a certain extent, the simulated circulation features. Two values of bottom friction coefficient were chosen, 0.0025 and 0.01, that correspond to Manning coefficient values of 0.026 and 0.052, for the river depth of 20 m. One value of wind speed, 10 m/s, was chosen for two different directions, east and west. The circulation results for these values are given in Figure 9.16.

Table 9.4. Modelled durations (hours) of bacterial densities persisting above the 100 counts/100 mL limit after cessation of rainfall; Case 900804.

Point no.	Location	Fecal coliform	*E. coli*
2	GH, 250 m east of entrance, 50 m south of north pier	31.5	29.5
3	Sarnia Bay, 150 m south and 50 m east of NW corner	28.8	26.7
4	Sarnia Bay, 50 m south and 100 m east of NW corner	20.0	18.0
5	Sarnia Bay, 150 m south and 25 m east of NW corner	18.2	16.5
6	Sarnia Bay, 250 m south and 25 m east of NW corner	18.2	15.2
7	Sarnia Bay, sampling St. G, 25 m offshore	17.0	16.5
8	Sarnia Bay Inlet, opposite Derby Lane, 175 m offshore	17.0	15.4
10	River, 50 m south of Wellington St. CSO, 50 m offshore	37.2	34.0
11	River, 150 m south of Wellington St. CSO, 200 m offshore	8.8	8.0
12	River, 1 km downstream of site 11, 175 m offshore	14.4	13.9

Source: Tsanis et al. (1995).

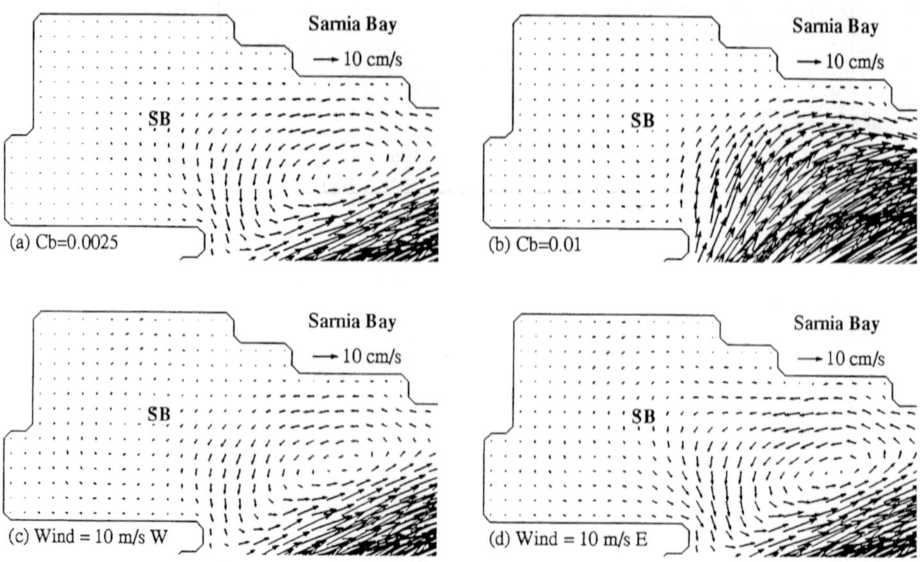

Fig. 9.16. Modelled depth-averaged circulation in Sarnia Bay for different bottom friction coefficients and wind speeds (Tsanis et al., 1995).

The depth–averaged circulation in Sarnia Bay is altered significantly by different bottom friction coefficients and, to a lesser extent, by different wind conditions. A large value of bottom friction coefficient does not produce a counterclockwise eddy at the interface between the St. Clair River and Sarnia Bay. This is caused by higher resistance imposed on the current by the bottom friction; the flow acts in this area like a plug flow without any recirculation zones.

On the other hand, different wind conditions, represented by 10 m/s east and west winds, do not have much effect on the circulation pattern, because of the very strong river current. The current magnitudes in the eddy are higher for the case of the east wind when compared to that of the west. The resulting circulation is enhanced by winds of favorable directions, that is, east winds result in a more intense flow exchange between the river and the bay. This will affect the dispersion of bacteria at location P2 in Sarnia Bay, where the bacterial concentration diminishes for east wind faster than for west wind (see Fig. 9.17). The same

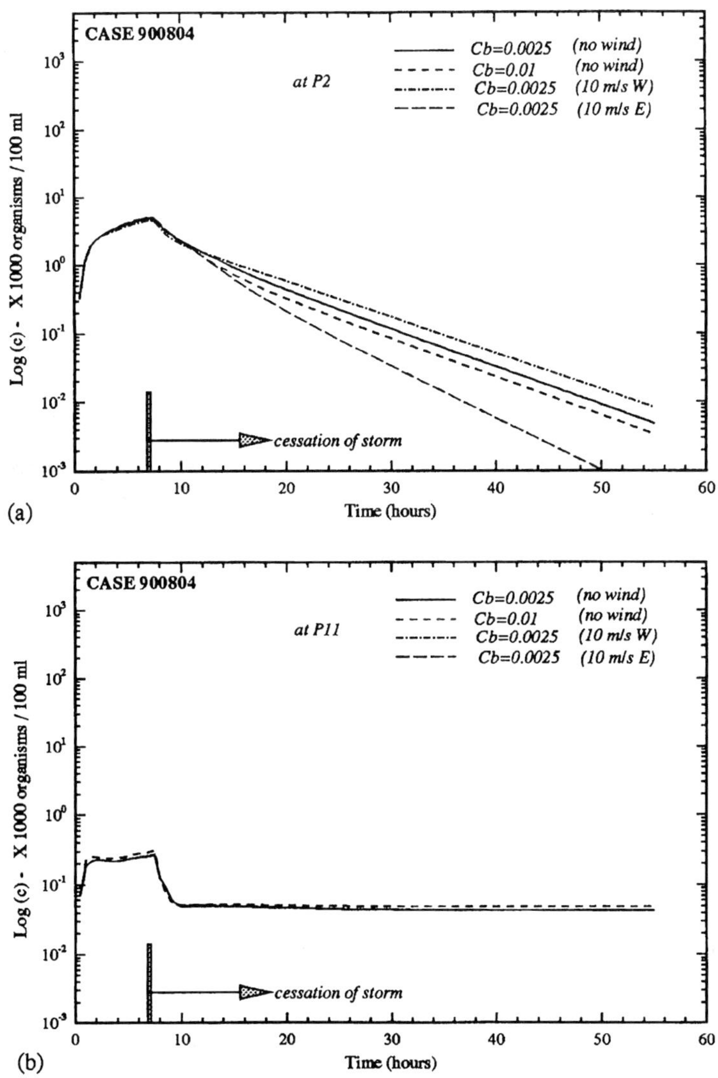

Fig. 9.17. (a, b) Modelled bacteria concentrations at locations P2 and P11, during and after the August storm, for different bottom friction coefficients and wind speeds (Tsanis et al., 1995).

behavior is observed at P2 for the higher value of bottom friction coefficient. At location P11, in the river, the effects of wind and bottom friction coefficient on the bacteria transport are insignificant.

Obviously, the flushing time (defined as the time when the concentration has dropped to 100 ppm level) depends on the loading concentration. The higher the loading concentration introduced, the longer the expected flushing time. In order to examine the effect of the loading conditions on the flushing times in Sarnia Bay, three numerical experiments were conducted for various loading conditions, that is, the existing load magnitude, one and half times the existing load and two times the existing load. It was found that the change in the flushing time was proportional to the difference in the load concentration logarithms. The characteristic flushing constant was equal to 7.9. For example, if the load concentration increases from 5000 to 10000 FC/100 mL, the flushing time is extended by 5.4 h.

Given a certain circulation pattern in the St. Clair River and Sarnia Bay, model parameters, such as bacteria decay and horizontal diffusivity coefficients, will alter the spatial and temporal distributions of the bacteria concentrations introduced into this area from a number of point sources.

Figures 9.18 and 9.19 show the bacteria concentrations as a function of time (during the first 48 h after the storm) for locations P2 (in Sarnia Bay) and P11 (in the River), respectively, for decay coefficients λ of 0.1, 0.5, 1.0 and 0.0/day, and for horizontal diffusivities D of 0.5, 0.8 and 1.0 m^2/s for Sarnia Bay and Government Harbour. Increases in the decay coefficient or the horizontal diffusivity result in a faster decrease of bacteria concentrations, with time, at location P2. The modelled duration of bacterial densities persisting above 100 counts/100 mL decreases from 36 to 31.5 h, for an increase of λ from 0.0 to 0.5, with $D = 0.8\,m^2/s$. A similar decrease in duration is observed for an increase in D from 0.5 to 0.8 m^2/s. Because of high advective flows in location P11, there are no significant differences caused by changes of the decay or horizontal diffusivity coefficients.

9.2.4 Model Verification

Since the assessment of microbiological pollution is based on probabilistic distributions of bacteria, model verification should be based on comparisons of probabilistic distributions derived from field measurements and model simulations. For such comparisons (Marsalek et al., 1994), five storms (Nos. 900804, 900812, 900815, 900819, and 900828) at location P2, inside Sarnia Bay (see Fig. 9.11), were considered. Observations of fecal coliform concentrations at location P2 and their probabilistic distribution were reported earlier by Marsalek et al. (1992). To derive the corresponding distribution from simulations, a sampling scheme similar to that used in field work was employed. For each of the simulated events, a random sampling starting time was selected, in half-hour increments between 0 and 2 h after the storm start. For the above storm series, these starting times were 1, 0.5, 1, 0, and 0 h. The subsequent samples were drawn at 2 h intervals, as done in the field, and this procedure yielded the following numbers of samples for the discussed events:

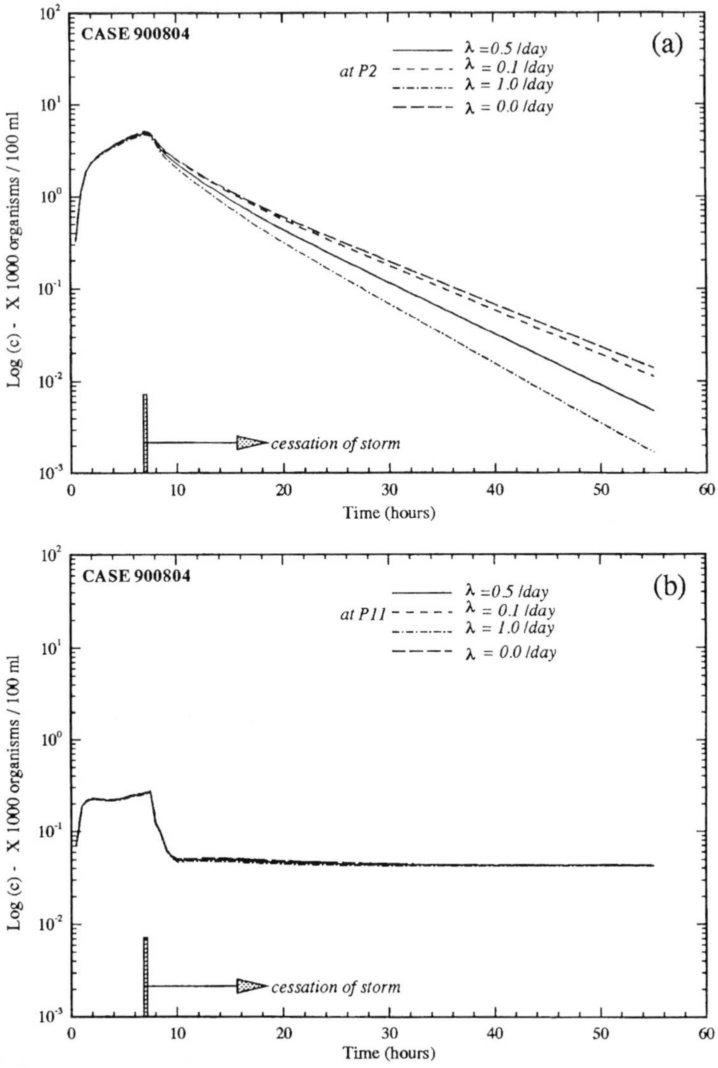

Fig. 9.18. (a, b) Modelled bacteria concentrations at locations P2 and P11, for different values of bacteria decay (Tsanis et al., 1995).

15, 16, 8, 18, and 15. For the fecal coliform concentrations determined this way at location P2, a lognormal distribution was derived and was characterized by the log-transformed mean of 2.8129, and the corresponding standard deviation of 0.6850. From these parameters, the lognormal distribution was plotted in Figure 9.20. It was noted that the mean of simulated concentrations was 650 FC/100 mL, slightly higher than the observed mean of 568 FC/100 mL. However, the overall agreement between the observed and simulated distributions is very good and justifies the use of the proposed modelling package in screening remedial measures.

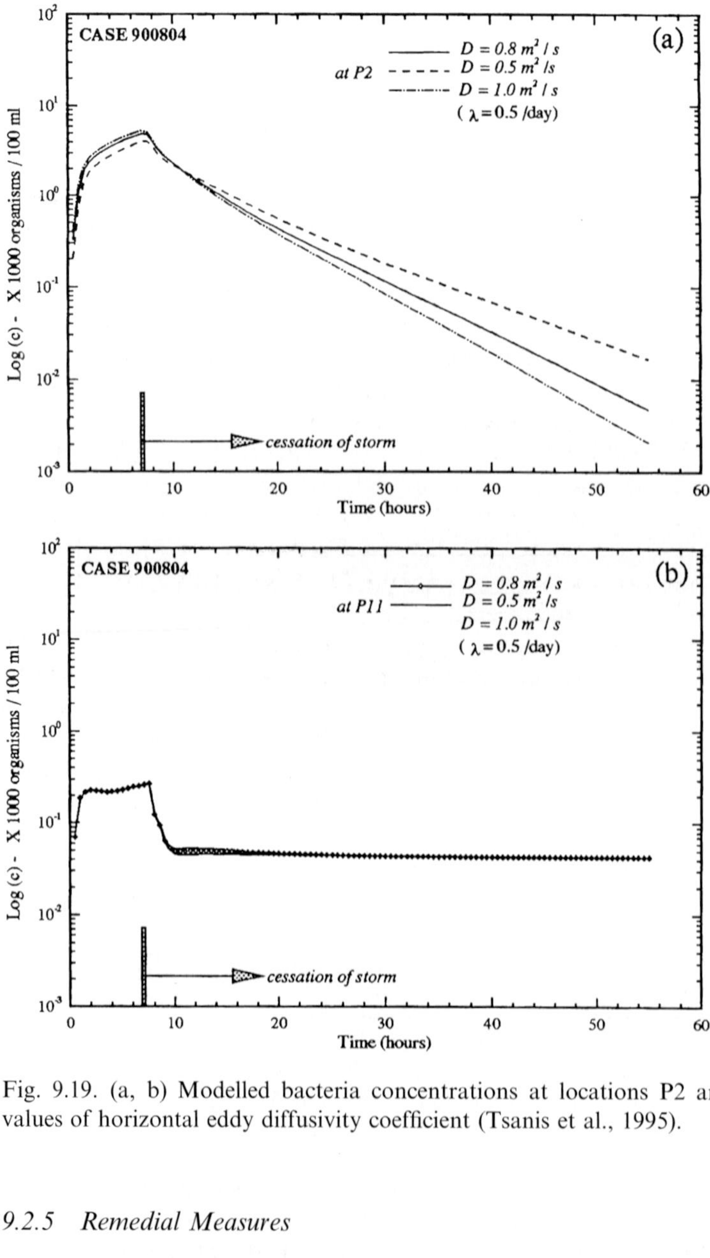

Fig. 9.19. (a, b) Modelled bacteria concentrations at locations P2 and P11 for different values of horizontal eddy diffusivity coefficient (Tsanis et al., 1995).

9.2.5 Remedial Measures

A number of remedial measures could be proposed to control bacteriological contamination and impairment of recreational water use in the study area, and particularly in Sarnia Bay, which is widely used for recreation. Field observations in the bay indicated frequent bacteriological contamination well in excess of the provincial and federal recreational water guidelines of 100 and 200 *E. coli*/100 mL, respectively. The sources of such contamination include one storm sewer

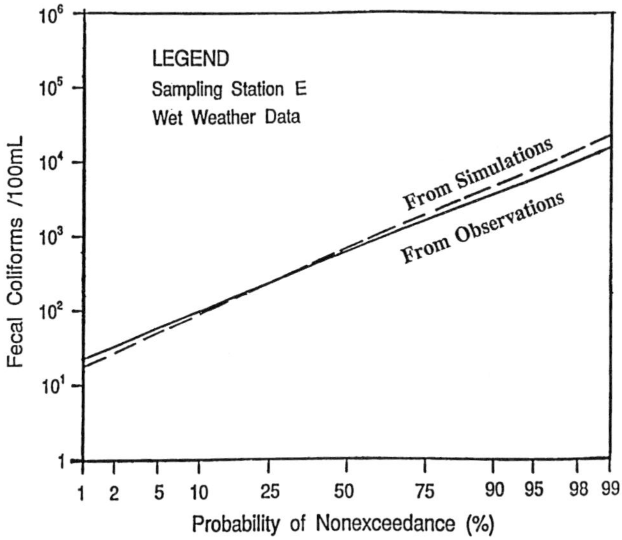

Fig. 9.20. Fecal coliform density distributions at Sarnia Bay (Tsanis et al., 1995).

Table 9.5. Hypothetical remedial measures for Sarnia Bay (St. Clair River).

Case	Remedial measure (RM)
	Reference case (900804)
RM1	1 m^3/s flushing flow in the northwest corner of Sarnia Bay
RM2	Disconnect SS 104, relocate SS105-SS107 out of Sarnia Bay
RM3	Build a straight barrier
RM4	RM2 + RM3

discharging along the north shore and four storm sewers discharging along the east shore of the bay.

The earlier proposed five remedial measures (Marsalek et al., 1992) were tested for effectiveness in reducing bacterial concentrations. The measures studied are listed in Table 9.5 and include (a) enhancing the flushing of the Bay by pumping in riverine water (1 m^3/s) in the northwest corner of the Bay (RM1); (b) disconnecting and/or relocating storm sewer outfalls (RM2); (c) building a deflector barrier diverting sewer discharges from the Bay and preventing counterclockwise circulation in the Bay (RM3); and (d) combination of (b) and (c) options (RM4). The assessment of individual remedial measures was demonstrated for storm 900804, which was used as the reference case.

The depth-averaged circulation patterns for the existing situation (storm 900804) and the four RM cases are shown in Figure 9.21. Figure 9.21(a) presents results for 900804 (RM0—no action case) and RM2, Figure 9.21(b) for RM1, and Figure 9.21(c) for RM3 and RM4. It is obvious from these results that (a) increased flushing by pumping water into the bay does not significantly change

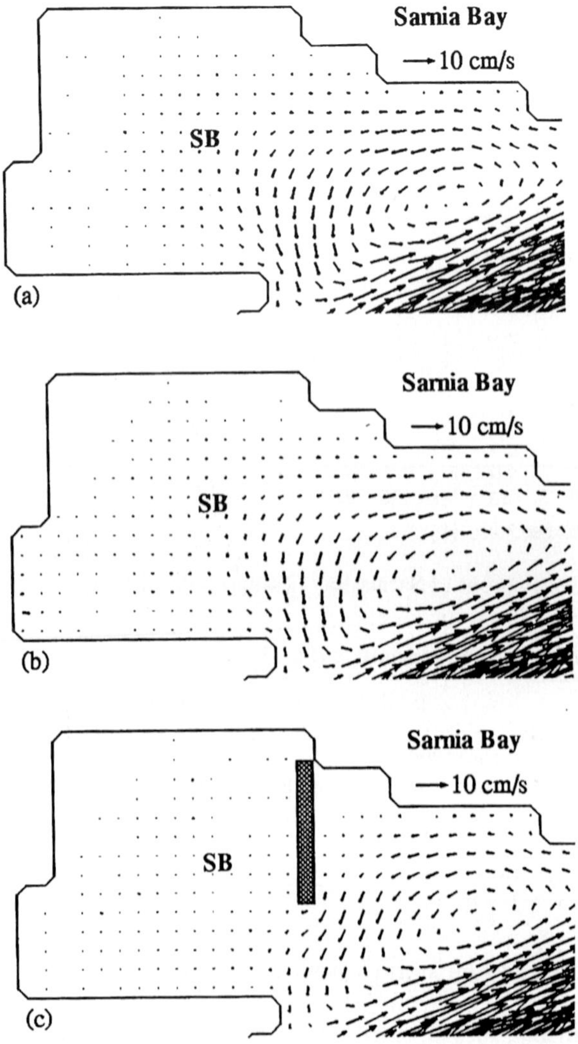

Fig. 9.21. (a–c) Modelled depth-averaged circulation in Sarnia bay for cases 900804, RM1 and RM3 (Tsanis et al., 1995).

circulation patterns; and (b) placement of a straight barrier at the bay entrance reduces the size of the eddy formed at the interface between the main river channel and Sarnia Bay.

In Figure 9.22, the logarithms of fecal coliform concentrations at locations (2), (3), (4), and (10) are shown, for all the remedial measures studied as functions of the time measured from the beginning of the storm 900804 until about 48 h after the storm end. In the first remedial case (RM1), the effect of pumping riverine water on the time series of bacteria concentrations was negligible at all locations in the bay.

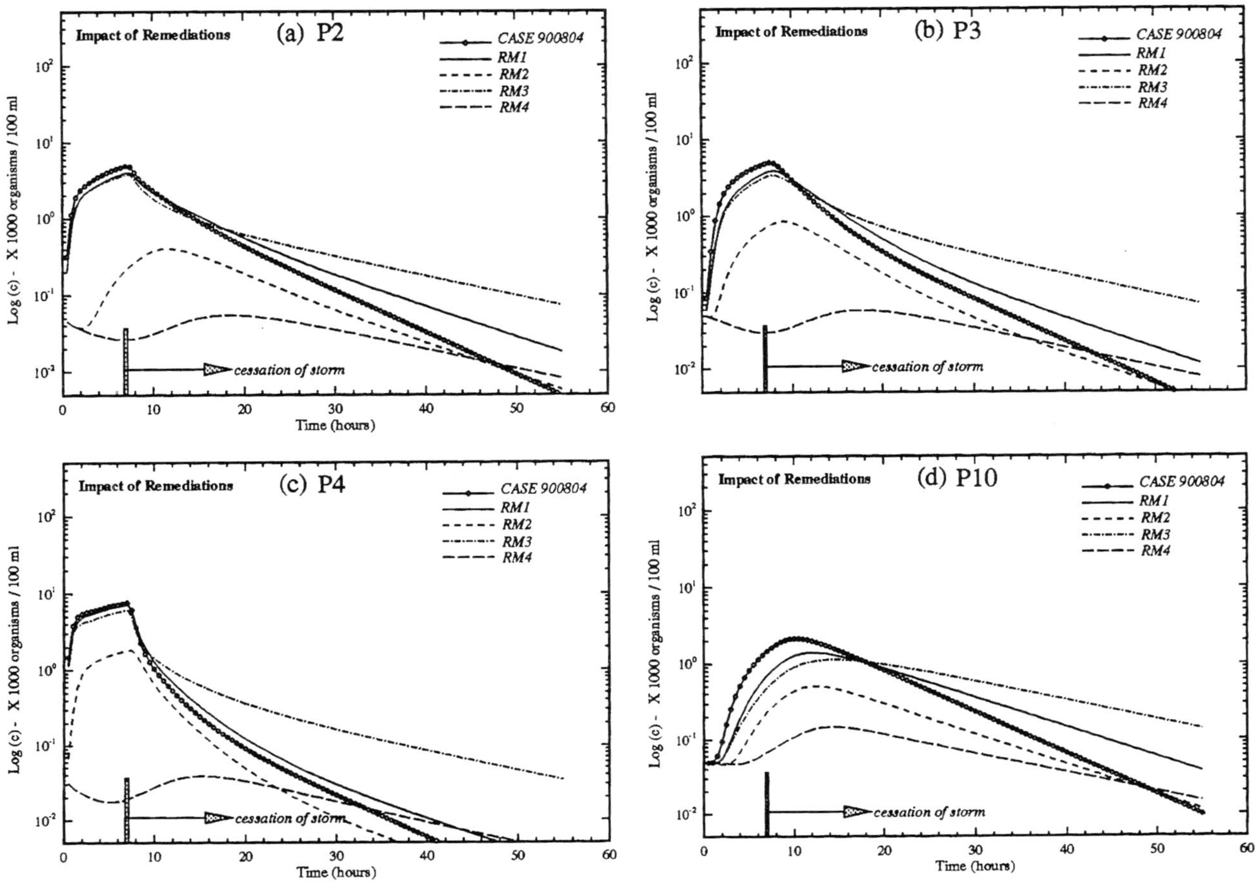

Fig. 9.22. Modelled bacteria concentrations at locations P2, P3, P4 and P10, for the reference and four RM cases (Tsanis et al., 1995).

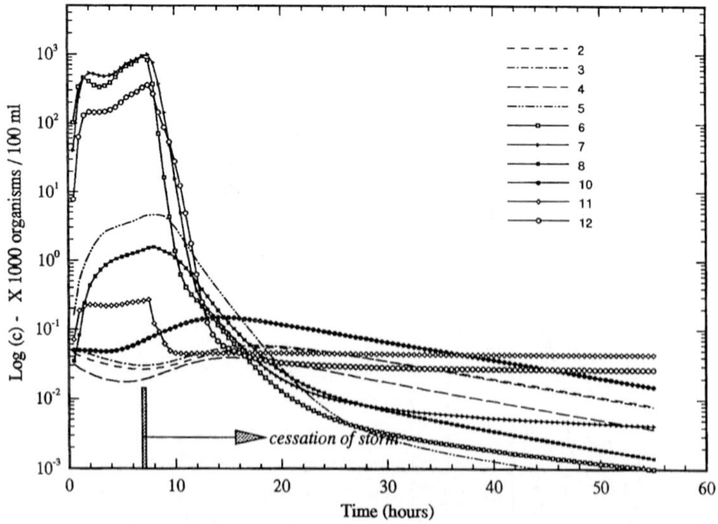

Fig. 9.23. Modelled bacteria concentrations at eight locations in St. Clair River and Sarnia Bay for the RM4 case (Tsanis et al., 1995).

In the second remedial case, RM2, the effect of disconnecting storm sewer SS104 is evident only at location P2, where the bacteria concentrations were lowered during the storm and dropped off faster after the storm end.

In the third remedial case, RM3, a barrier was placed at the mouth of the Bay to divert sewer discharges from the bay and prevent counterclockwise circulation in the bay. At the same time, however, this barrier retains the pollutants discharged into the bay for a longer time. These effects counteract each other and result in a negligible reduction in the bacteria concentrations when compared to the reference case.

In the fourth remedial case, RM4, the bacteria concentrations in Sarnia Bay were reduced to less than 100 counts/100 mL during and after wet weather, as shown in Figure 9.23. The storm sewer SS104 was disconnected and the barrier at the bay mouth prevented diffused pollutants, discharged close to the bay's mouth, from entering the bay.

9.3 TORONTO WATERFRONT RECEIVING WATER STUDY[10]

The Metropolitan Toronto waterfront is the most densely populated stretch of shoreline on Lake Ontario and has been designated as one of the 17 areas of concern in Ontario by the International Joint Commission. Discharges of contaminants from storm sewers, combined sewer overflows, four water pollution control plants, and six area tributaries have been identified as contributing to the degradation of the nearshore aquatic environment along the Metropolitan Toronto

[10]Material in Section 9.3 is reproduced from Shen et al. (1995), copyright with permission from the International Association of Great Lakes Research.

waterfront (Environment Canada et al., 1988). In order to evaluate the relative impact of these sources on the receiving water and the effectiveness of various abatement options, a 3D receiving water model is applied to provide water circulation and pollutant transport predictions for the nearshore area of the waterfront.

A comprehensive data collection program across the entire waterfront was completed in 1993 in support of the development of a model extending across the entire Metropolitan Toronto waterfront. In advance, a pilot study directed at an area near the mouth of a tributary was used to provide a preliminary evaluation of data which could then be used in the development of the model and in a comparison of plume dynamics under isothermal and stratified conditions. This application presents a summary of results in the current phase of model development, with an emphasis on the Lake Ontario nearshore area near Mimico Creek.

A 3D nested model was used to predict circulation patterns across the Metropolitan Toronto waterfront. A coarse-grid model ($2\,km^2$) was used for the whole of Lake Ontario, and a fine-grid model ($500\,m^2$) was used for the Metropolitan Toronto waterfront. The grid used for discretizing Lake Ontario was rotated by 45° clockwise, to improve the grid alignment with the Metropolitan Toronto shoreline. The grid extended west of Etobicoke Creek and east of the Rouge River; a 60 s time step was used for both models.

9.3.1 Field Study

In May 1992, the Ontario Ministry of Environment and Energy (MOEE) deployed six Aanderaa Instruments RCM4S current meters in Lake Ontario, along a 750 m transect. The transect consisted of four stations, spaced at 250 m and extended easterly from the mouth of Mimico Creek (see Fig. 9.24). The current meters were

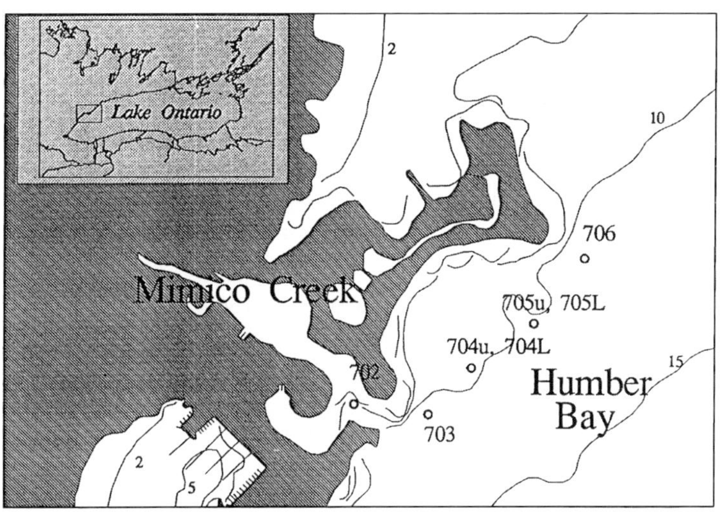

Fig. 9.24. Meter deployment locations (Shen et al., 1995).

used to collect time series data consisting of current speed, current direction, temperature, and conductivity. The measurements were time averaged at 10 min intervals and the data were collected for the period May to October 1992. Current meters were placed as close to the surface as possible (at a depth of approximately 2 m) so as not to pose a navigational hazard at each of the four stations (Station Numbers: 703, 704, 705 and 706). Current meters were also placed 2 m above the lake bottom at stations 704 and 705. A continuous recording Hydrolab Datasonde III water quality meter was placed at a 2 m depth near the mouth of Mimico Creek (Station 702). The meter was used to record time-series data consisting of temperature, conductivity, pH, and dissolved oxygen. Thirty-three rainfall events were recorded during the monitoring period.

A detailed analysis and discussion of field data is presented in Tsanis and Shen (1994c). Two data sets collected during lake stratified and isothermal conditions, respectively, were used for model calibration.

9.3.2 Model Results

The 3D nested hydrodynamic/pollutant transport simulation model was applied to the Lake Ontario nearshore, near the mouth of Mimico Creek. This involved applying a model for the whole of Lake Ontario, using a coarse grid (2 km^2) consisting of 12 layers, whereby model predictions were used to provide the necessary boundary conditions for a finer grid (500 m^2), applied to the Metropolitan Toronto waterfront. The model boundaries extended approximately 46 km along shore and 18 km perpendicular to shore. Predictions from this model were used to provide the necessary boundary conditions for a high resolution (100 m^2) grid applied to the nearshore area around the mouth of Mimico Creek. The model boundaries of this nearshore area extended approximately 5500 m along shore and 3100 m perpendicular to shore.

Two case studies were selected to provide hydrodynamic and pollutant transport predictions for representative storm event and receiving water conditions. The simulation conditions consisted of the following, as shown in Table 9.6.

Wind data used as model input should be collected as close to the study area as possible. Wind data which may be applicable for a circulation model of the

Table 9.6. Summary of simulation conditions.

Case	Period of simulation	Total rainfall (mm)	General current direction	Lake conditions	Wind data	Model configuration
1	July 17–20, 1992	14.6	Easterly	Stratified	Headland	(i) isothermal (ii) Stratified
2	September 18–22, 1992	27.4	Easterly	Isothermal	Headland	Isothermal

Source: Shen et al., 1995

Metropolitan Toronto waterfront is recorded at climatological stations located on the Leslie Street Spit (Toronto Headlands), Toronto Island Airport, and Pearson International Airport. Complete daily records were not available from the Toronto Island Airport station; data records from Pearson International Airport were not always complete and, spatially, this station is quite remote from the study area. Data collected at the Toronto Headlands station was considered more appropriate because of the station's proximity to the study area and the availability of complete data records. Discharge flow data collected at Mimico Creek through the Water Survey of Canada Station (02HC033), and conductivity data collected by MOEE at Station 702, were used as boundary conditions for the hydrodynamic and transport equations respectively. Background conductivity in Lake Ontario was taken, based on field measurements, to be about 320 and 300 umho/cm for Cases 1 and 2 respectively.

Case 1 (July 17–20)

Source discharge and wind time-series data is presented in Figure 9.25. Mild, southwesterly winds prevailed for most of the simulation. Mimico Creek discharge flows and conductivity peaked at about 40 m^3/s and 600 umho/cm, respectively, toward the end of July 17, in response to a 14.6 mm rainfall event.

The 3D model was applied in two modes: isothermal and stratified. In the 3D isothermal mode, the Mimico Creek plume was assumed to discharge into a vertically well-mixed (i.e., uniform temperature distribution of 20°C) receiving water body. In the 3D stratified condition, the receiving water was modelled assuming an initial vertical temperature profile as close to that measured in the field as possible. This profile consisted of: 17°C from the surface to a 1 m depth; 16.9°C from 1 to 3 m depths; 16.5°C from 3 to 7 m depths; and 12°C from 7 to 11 m depths. In all cases, the temperature of the Mimico Creek discharge was assumed to be 20°C. Conductivity contours superimposed on circulation patterns at 8:00 p.m. on July 17, for predictions using the 3D isothermal and stratified models, are presented in Figures 9.26 and 9.27, respectively. The very weak currents observed in these figures were attributed to the lack of wind at this moment in the simulation. The shape of the conductivity plume was affected by the Mimico Creek discharge and the west wind during the simulation. The surface currents were stronger under stratified conditions and the plume's areal coverage was larger. Conversely, the plume depth was shallower under stratification than under isothermal conditions, because buoyancy forces tended to spread and thin the warmer Mimico Creek plume along the lake surface.

Figures 9.28 and 9.29 provide a comparison of model-predicted velocity and conductivity to field measurements at Stations 703 to 705u, respectively. The 3D stratified model predictions are in better agreement with the field measurements. In comparison to the 3D isothermal model, the velocity time series predicted by the 3D stratified model appeared dampened. As expected, estimates of conductivity are generally underpredicted by the 3D isothermal model. This demonstrates that the vertical temperature distribution has a strong influence on the pollutant transport, and a 3D stratified model is necessary for the simulation of current and pollutant transport in this nearshore area during the summer.

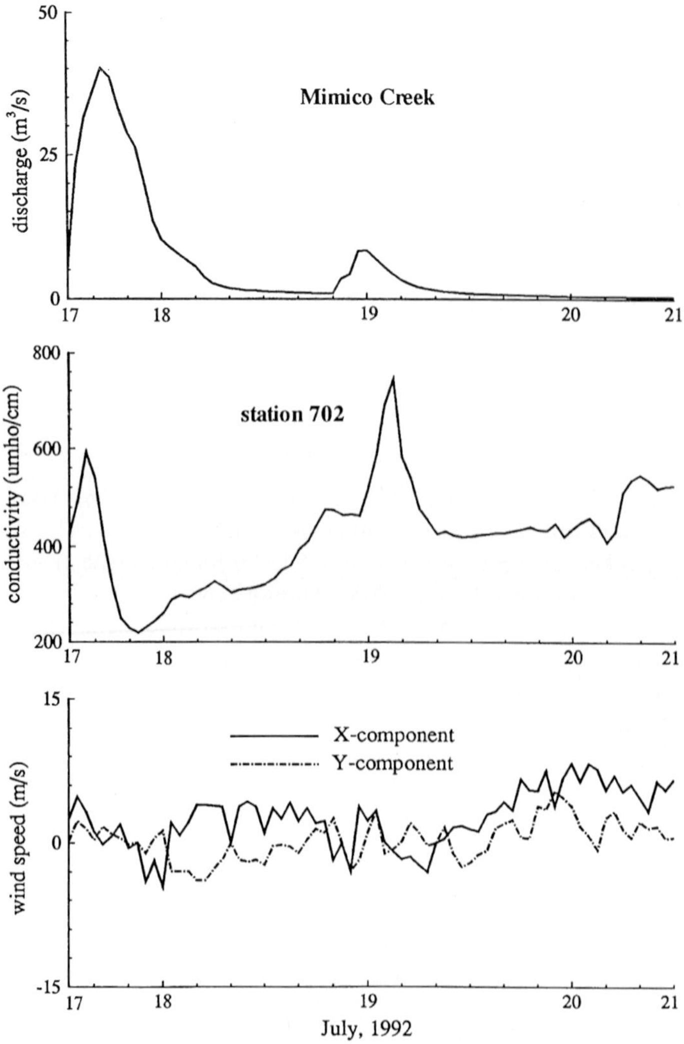

Fig. 9.25. Source discharge and wind time series for Case 1 (Shen et al., 1995).

Case 2 (September 19–22)

Source and wind time series data are presented in Figure 9.30, and the model-predicted conductivity contours, superimposed on the circulation patterns, at 10:00 a.m. on September 22 and 0:00 a.m. on September 23, are presented in Figures 9.31(a) and (b), respectively. Figure 9.31(a) shows the plume extending east to the Toronto Western Beaches, due to a strong east-tending current. Toward the end of the simulation the source discharge is weak, however, the plume is now confined to an embayment west of the Humber River in Humber Bay Park East. This demonstrates that embayments along the waterfront can be impacted by other less direct discharges.

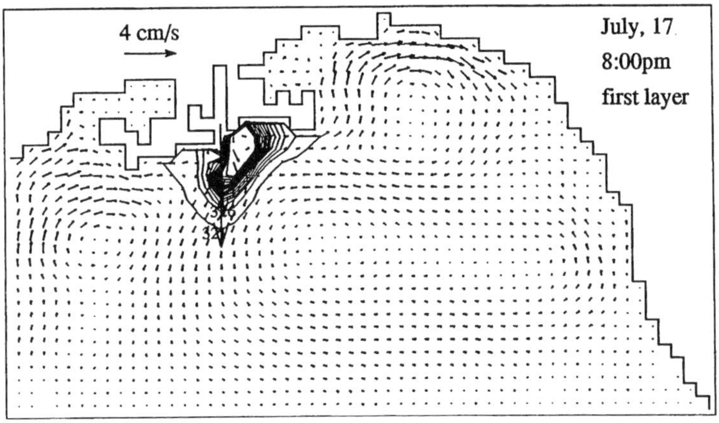

Fig. 9.26. Conductivity contours superimposed on circulation patterns for isothermal conditions, 8 h into the simulation (Shen et al., 1995).

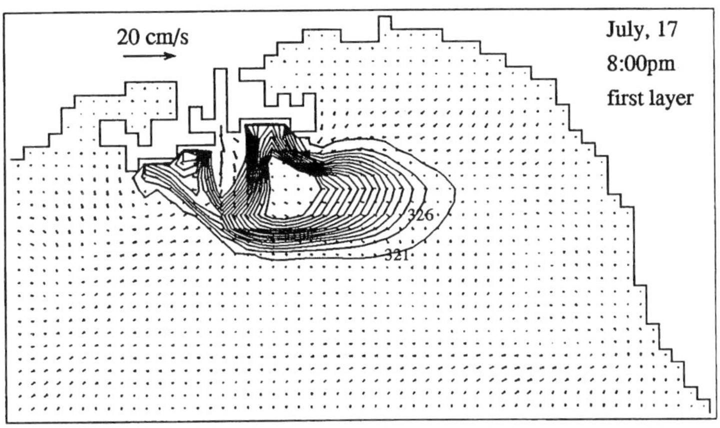

Fig. 9.27. Conductivity contours superimposed on circulation patterns for stratified conditions, 8 h into the simulation (Shen et al., 1995).

Time series of conductivity and velocity for field and model simulations, at Station 703, are presented in Figure 9.32 where the predictions are in good agreement with field measurements. The conductivity time series are in exceptional agreement where peaks in conductivity occurring early on September 21 and mid-September 22 match the field measurements in time and magnitude. It is interesting to note that a decrease in field measurements of conductivity toward the end of September 22 is coincident with strong northerly winds which generated an upwelling of lower conductivity, hypolimnetic water to the surface. This is substantiated by current meter time series data which show a rapid drop in temperature and conductivity. The reference conductivity for the nearshore area of the Metropolitan Toronto waterfront was taken to be 300 umho/cm, based on field

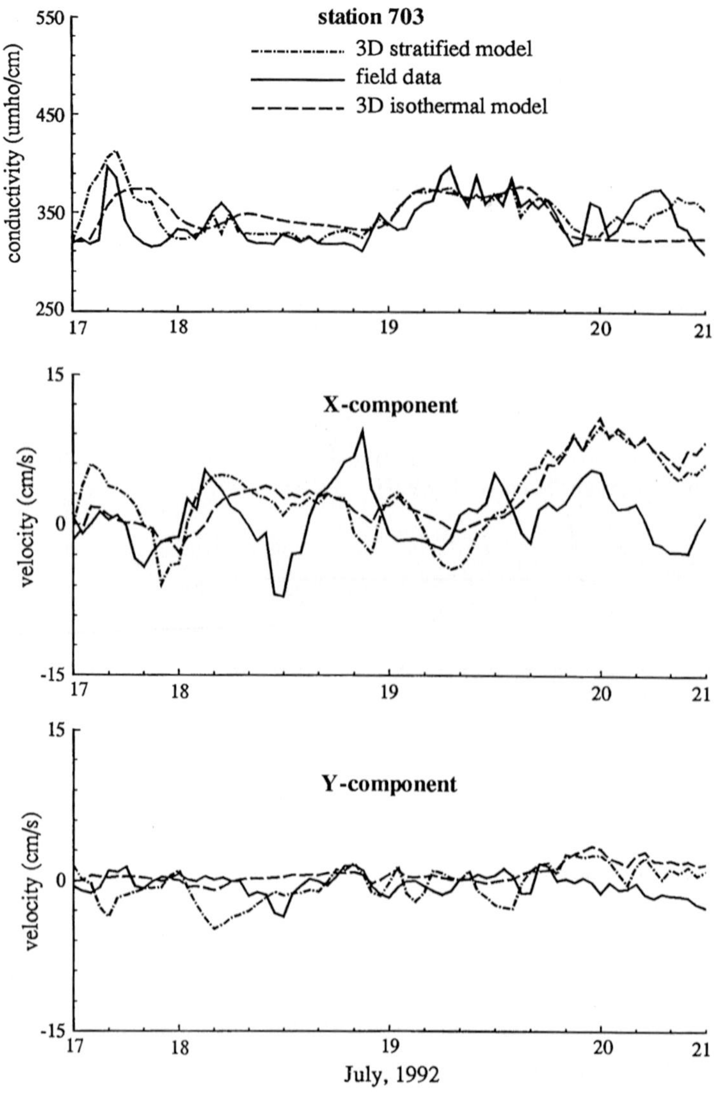

Fig. 9.28. Comparison between field measured and predicted conductivity and velocity time series at Station 703 (Shen et al., 1995).

measurements of the open water. The model predicted conductivity dropped to this level, which was the lowest level possible, based on this boundary condition.

The nested model used in this example provided predictions of circulation and pollutant transport, at high spatial resolution, which were in good agreement with field measurements. In comparison to the isothermal model, the 3D stratified model provided improved estimates for simulations during the summer period when the lake was stratified. Lake stratification prevails during the summer months and provides a worst case condition which reduces vertical mixing and increases

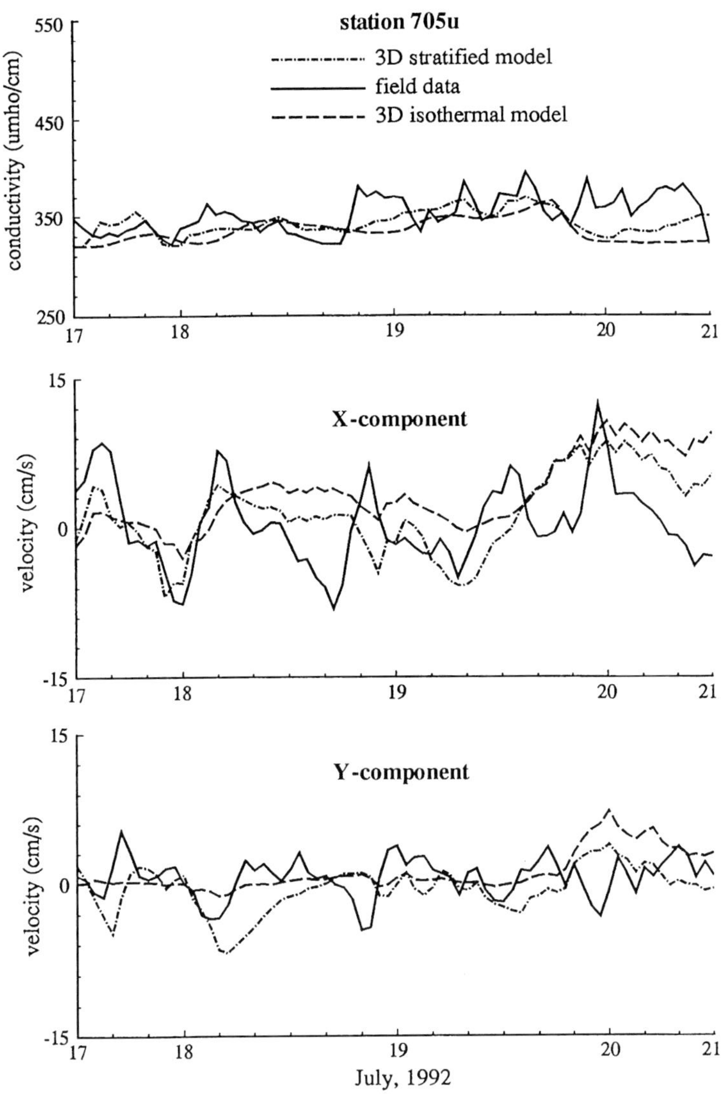

Fig. 9.29. Comparison between field measured and predicted conductivity and velocity time series at Station 705u (Shen et al., 1995).

horizontal dispersion of pollutants. This was demonstrated in Case 1 where improvements in model predictions were realized when the fully 3D model was compared to predictions from the isothermal model, which assumed vertically well-mixed conditions.

The wind field is the most important data input, and every effort should be made to use data as close to the study area as possible. In addition, whole lake circulation patterns, necessary for the nested model, should use a wind field generated by interpolating wind data over a number of stations across the whole lake.

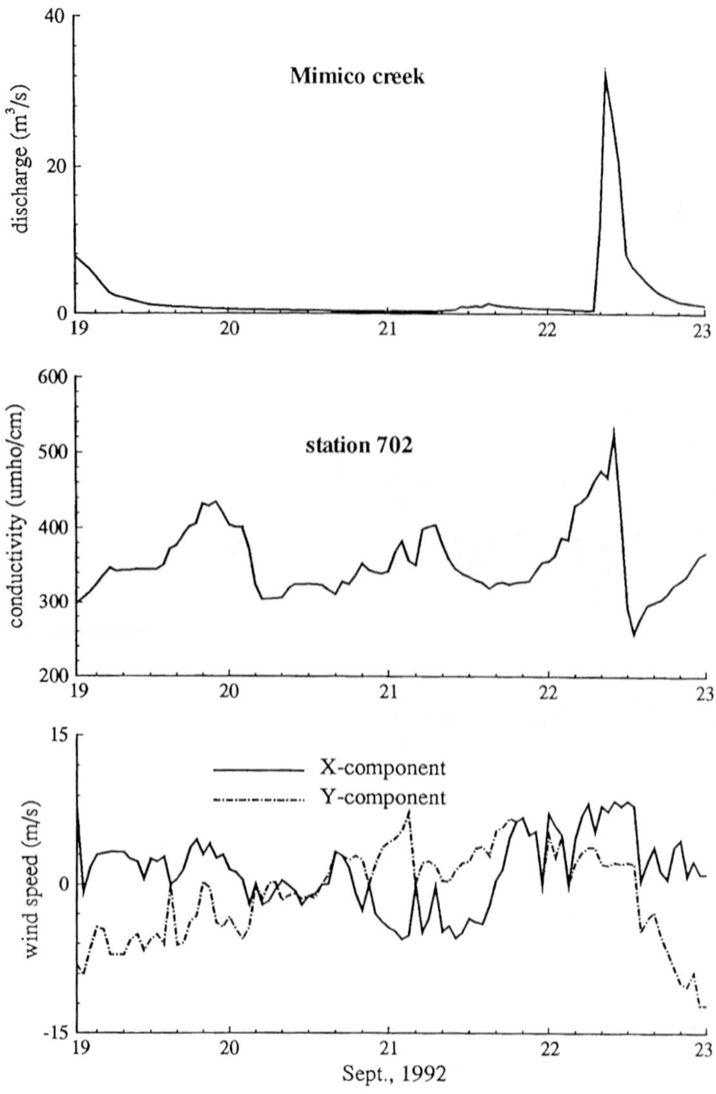

Fig. 9.30. Source discharge and wind time series for Case 2 (Shen et al., 1995).

The study has demonstrated that conductivity time series, recorded by Aanderaa Instruments RCM4S current meters and continuous recording water quality meters, such as the Hydrolab Datasonde III, can be used effectively for the calibration of the pollutant transport model. They offer several advantages over the chemical analysis of conventional grab samples. The data is continuous, and important peaks (or troughs), which are often missed through grab sampling, are recorded.

As discussed previously, modelling the lake stratification period should be a priority and under these conditions most surface discharges across the waterfront will not impact the deeper hypolimnetic waters. Conductivity measurements, collected

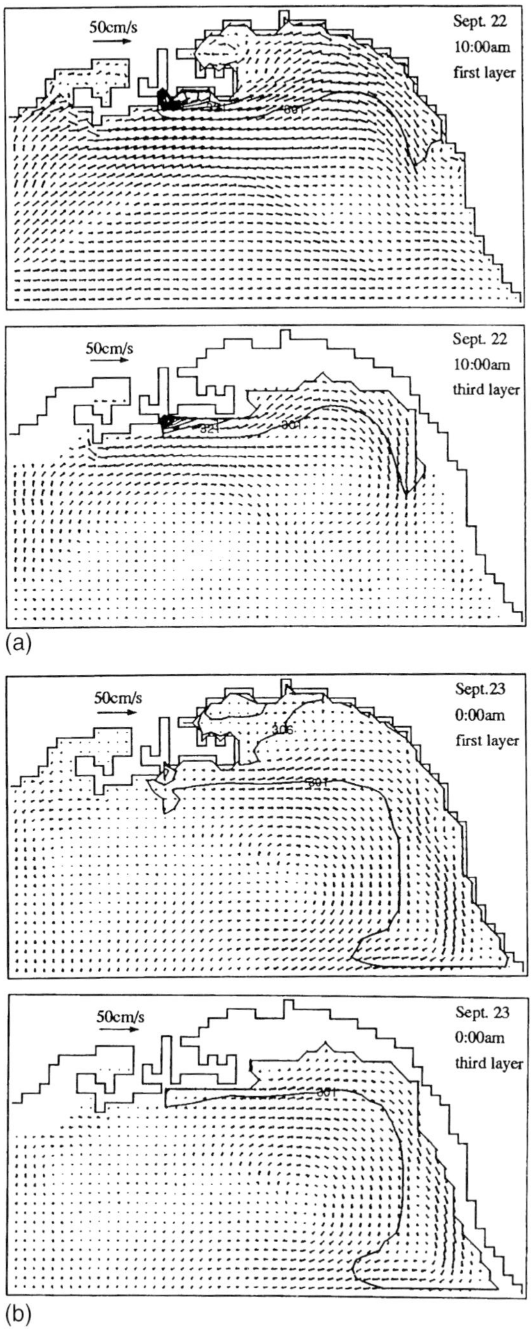

Fig. 9.31. Conductivity contours superimposed on circulation patterns for stratified conditions. (a) 82 h into the simulation, and (b) 96 h into the simulation (Shen et al., 1995).

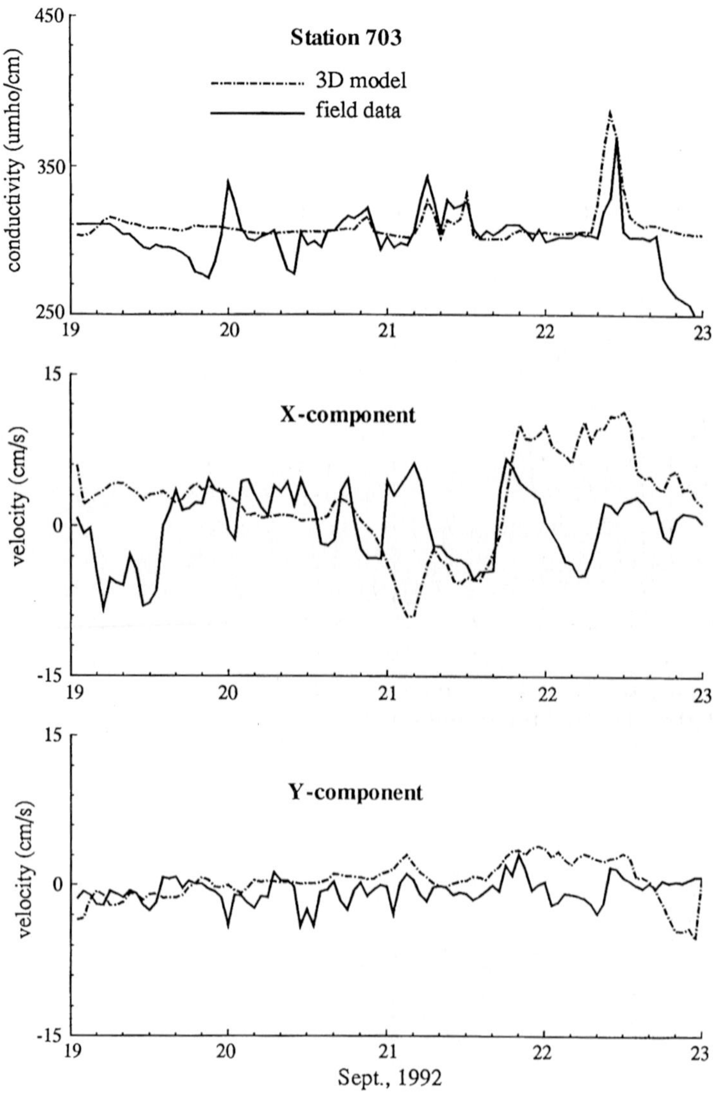

Fig. 9.32. Comparison between field measured and predicted conductivity and velocity time series at Station 703 (Shen et al., 1995).

from near bottom meters positioned close to the source, showed no response during lake stratification. In addition, because the meters were positioned along an east tending transect from the mouth of Mimico Creek, calibration was only possible for those cases where currents were east tending. Data collected during some large events where the currents were west tending were not suitable for model calibration purposes. It is therefore recommended that, for model calibration purposes, measurements of conductivity, temperature and current velocity should be taken using instruments positioned near the surface. In lake applications, the meters should be positioned in the nearshore environment, on both sides of source monitoring stations, to maximize the data capture potential for calibration purposes.

9.4 HAMILTON HARBOUR STUDY

Hamilton Harbour, located at the western end of Lake Ontario (see Fig. 9.33), is one of 40 areas of environmental concern in the Great Lakes. In recent years, many projects in the Harbour required a fine-resolution hydrodynamic model so that the current structure could be simulated and used as an input to water quality and sediment transport models (MOE, 1992).

There were a limited number of studies conducted on modelling the current movements in the Harbour (MOE, 1974; James and Eid, 1978; Rasmussen and Badr, 1979). The 2D models used by MOE (1974) and Rasmussen and Badr (1979) were unable to simulate the current structure in the vertical direction. A 3D model is necessary to simulate this phenomena. In the 3D model by James and Eid (1978), the non-linear horizontal advection terms were not included and the vertical eddy viscosity coefficient was taken as a constant, which is not realistic for the wind-induced flow (Tsanis, 1989). The coarse-grids used in all previous models were unable to describe in detail the shoreline and the Harbour's bathymetry. Furthermore, the model verification was insufficient. As a result, a high-resolution model must be employed and detailed model verification is required. A high-resolution grid of 100 m was used for Hamilton Harbour in this study, as shown in Figure 9.34. The system is rotated 23.8° clockwise from the north, to better

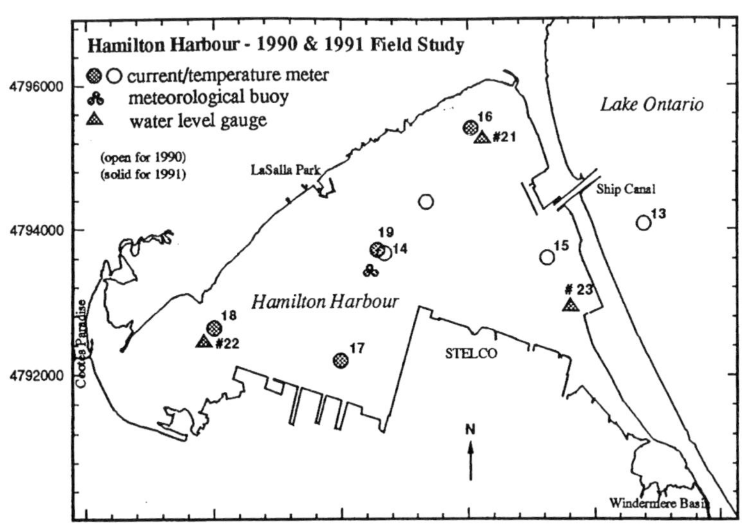

Fig. 9.33. Locations of meteorological and limnological observation stations in Hamilton Harbour during the 1990 and 1991 field study seasons (Wu, 1993).

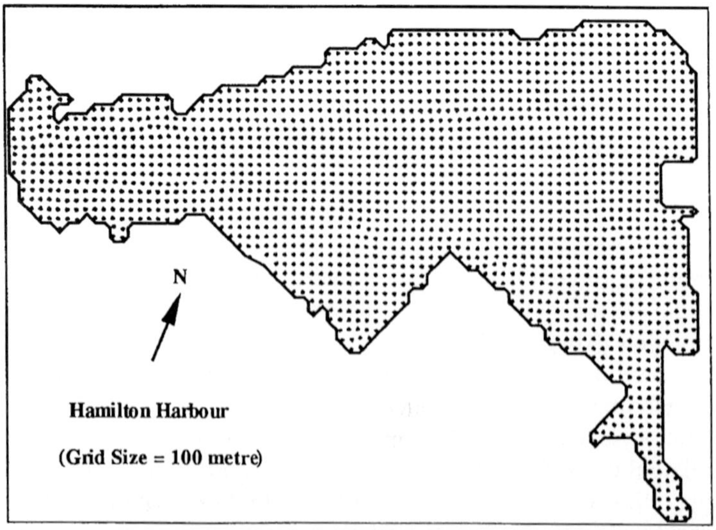

Fig. 9.34. A high-resolution (100 m grid) finite difference model for Hamilton Harbour. The grid rotated 23.8° anticlockwise from the north to better represent the shorelines (Wu, 1993).

represent the eastern and northeastern shorelines. This grid system is sufficient to represent the shorelines and verify the models.

The 3D hydrodynamic model used for Hamilton Harbour is a multi-layered stratified 3D model. A sigma coordinate transformation is used in the vertical plane within the 3D model for a smooth representation of the topography. A fractional step method is used to split the vertical diffusion terms from the rest of the momentum equations so that a semi-implicit scheme can be applied to offset the limit on the time step. A particle trajectory model that uses the random-walk approach is employed to simulate the trajectories of particles released in the Harbour.

Model verification using field data is a very important aspect in model development and application. Although much data has been obtained in Hamilton Harbour (MOE, 1992b), there is no existing model that utilizes these data for verification purposes. An extensive field study was undertaken by the scientists at the National Water Research Institute (NWRI) of the Canadian Centre for Inland Waters (CCIW) during 1990 and 1991 (Boyce and Chiocchio, 1991). The field data included currents from 17 current meters, trajectories from over 10 drogue experiments, and water levels from three water-level gauges. The locations of these instruments are presented in Figure 9.33. The features of the current and water-level data are presented in Boyce and Chiocchio (1991) and Wu (1993).

In this study, two methods will be used for the model verifications, using measured current, drogue trajectory, and water-level data. The first method is to compare model results with field data in the time domain (point-by-point comparison). A circulation model is judged to be accurate if it yields velocities closely resembling the observed velocities. However, a quantitative verification of the model with current data is particularly difficult because a meaningful comparison of observed and computed vector quantities must involve both direction and magnitude. The comparison depends on the location of a station and its meter depth because the current meter is sometimes located in a region of flow reversal, whereas, in other cases, the meter is located in a region of unidirectional current. The second method is to compare simulated trajectories with the field measurements (spatial comparison). This method is easily visualized and can be statistically evaluated.

9.4.1 Verification with Current Meter Data

The latest data from current meters were collected during the 1990 and 1991 field seasons in Hamilton Harbour. The simulated currents at five mooring meter locations, #16d5 m, #17d5 m, #18d5 m, #19d6 m, and #19d10 m (see Fig. 9.33), are compared with the measured data d5m, d6m, d10m are the depth of current meters. The comparisons were made by using the point-by-point method, and three statistical criteria were used to evaluate the degree of agreement between the simulated and measured data.

The scenario HH910717 is a three-day west-wind process, from 16:00 July 16 to 14:00 July 19, 1991 (see Fig. 9.35). The wind directions were between 210 to 310°, and the maximum wind speed reached 9.3 m/s on July 17. A comprehensive water quality measurement effort was undertaken at 25 stations, amongst which, three were outside the Ship Canal in Lake Ontario (MOE, 1992b). Temperature profiles were obtained at these stations from which an initial temperature field for the 3D stratified model was interpolated. During this period, the temperatures were quite uniform within the top 5 m of the water column and below the 20 m depth. A large

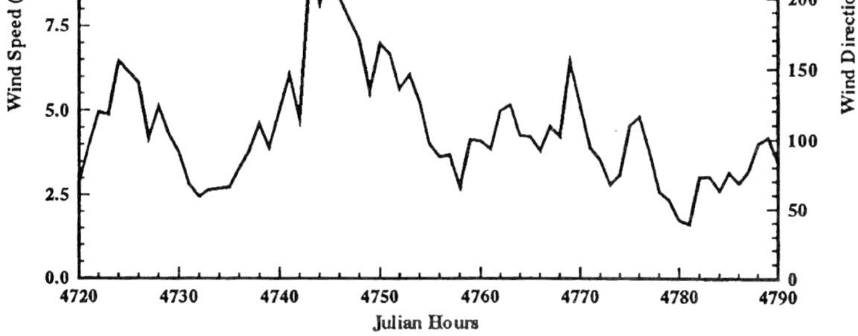

Fig. 9.35. Wind speed and direction of scenario HH910717 (16:00 16–16:00 19 July, 1991) in Hamilton Harbour (Wu, 1993).

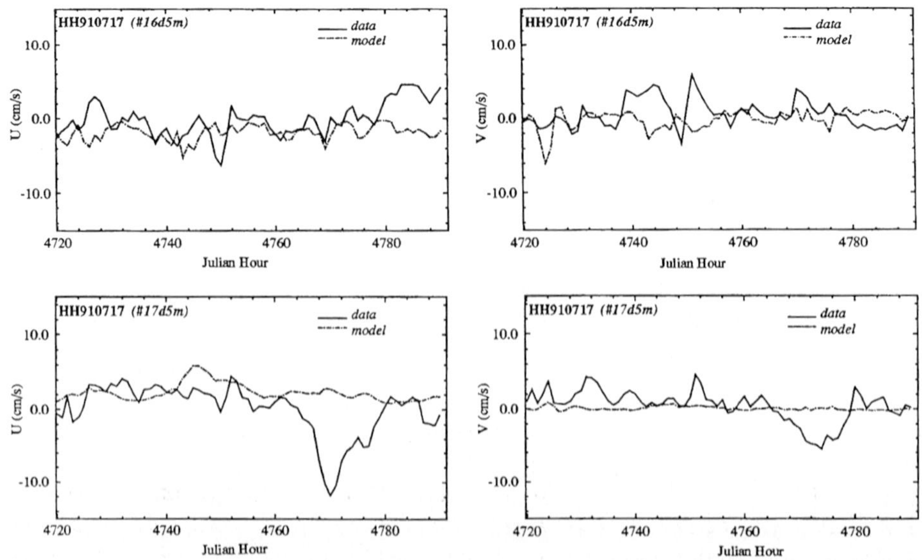

Fig. 9.36. Measured (solid line) and simulated (dashed-line) current components (U, in the west–east direction; V, in the Harbour (Wu, 1993), south–north direction), at the mooring stations #16d5 m, #17d5 m, #19d6 m and #19d10 m, scenario HH 910717, Hamilton.

gradient existed between the 5–15 m depth, where the temperature difference was about 12°C (1.2°C/m). This is the typical temperature profile in the Harbour during the summer season.

Figure 9.36 presents the simulated current velocity components ($U =$ west-east component and $V =$ south–north component), along with measured data at #16d5 m, #17d5 m, #19d6 m and #19d10 m. The simulations are in good agreement with the field data at most stations. However, many large peaks in the measured data are not found in the model simulations. Such peaks do not correspond with the wind peaks, such as the large peak around Julian hour 4770 at #17d5 m.

The following statistical criteria (Schwab et al., 1989) were used to evaluate the above comparisons: (1) α_1 represents the ratio of the energy in the time-variable part of the current to the total energy in the observed current:

$$\alpha_1 = \frac{\sum\left[(u_0 - \bar{u}_c)^2 + (v_0 - \bar{v}_c)^2\right]}{\sum(u_0^2 + v_0^2)} \tag{9.6}$$

where (u_0, v_0) are the eastward and northward components of the observed currents; (2) α_2 compares the differences between observed and computed values for the total flows;

$$\alpha_2 = \frac{\sum\left[(u_0 - u_c)^2 + (v_0 - v_c)^2\right]}{\sum(u_0^2 + v_0^2)} \tag{9.7}$$

Table 9.7. Comparison of observed and calculated currents in Hamilton Harbour (HH910717-16:00 16–16:00 19 July, 1991).

Ratio	#16d5m	#17d5m	#19d6m	#19d10m
α_1	0.96	0.99	0.39	0.96
α_2	2.22	1.21	2.81	1.65
α_3	1.07	0.54	1.53	0.64

and (3) α_3 is the ratio of the root mean square value of the computed currents to the root mean square value of the observed currents:

$$\alpha_3 = \frac{\sqrt{\sum(u_c u_c + v_c v_c)}}{\sqrt{\sum(u_0 u_0 + v_0 v_0)}} \tag{9.8}$$

The statistical comparison between computed and measured currents are listed in Table 9.7. It can be seen that almost all the energy is in the time-variable current because α_1 exceeds 0.95 at all stations. The typical value of α_2 is about 1.2, and α_3 ranges from 0.3 to 1.2. Overall, the model verification with the current meter data is satisfactory. A large variation in the field data that did not follow the wind is not well simulated in the numerical model. Good simulations are only made in some mooring locations. The wind field over Hamilton Harbour is taken as a uniform field in the model but this may not be justified and may contribute to the variation in the current meter data. Another reason may be that the horizontal and vertical distributions of water temperature, which may contribute to the current variation, were insufficiently included in the numerical model.

9.4.2 *Verification with Drogue Data*

There were over 10 drogue experiments conducted during the 1990 field season. The drogue experimental data provides an excellent opportunity to verify the 3D circulation model because the comparison is easily visualized and statistically analyzed.

The locations for each drogue in the Harbour's grid of the northing and easting in metres, are first converted into the model grid system (rotated 23.8° clockwise from the north). For the convenience of comparison, the data are interpolated so as to have the same start and end times and the same time interval (a half hour in this study) for all the drogues in each experiment. The 100 m grid of Hamilton Harbour is used in the coupled circulation and trajectory prediction model. Hourly wind speed and direction are input to the model. Because the drogues were deployed at either 2 or 5 m below the water surface (the water depth is about 10–20 m), the drogue movements represented the currents in the upper layer. Each experiment lasted for less than 5 h. As a result, the change in thermal structure could be ignored, and the thermal stratification only had an effect on the vertical viscosity and baroclinic terms in the fully 3D model. The current fields at 2 and 5 m below the water surface, which were depths for the drogues, were interpolated from multi-layered current fields. The trajectories for each drogue are predicted using the trajectory model without the random-walk term in each time step (a time step of 4 s was used in the numerical model), while only recorded at half-hour intervals.

The drift velocity v_d is assumed to be the sum of the current velocity v_c and an unknown windage, that is,

$$v_d = v_c + a\, v_w \tag{9.9}$$

where v_w is the wind velocity, and a the windage factor. A least-squares procedure is used to determine the optimal value of the windage factor for each drogue. For statistical comparison purposes, the following ratios were designed (Schwab et al., 1989): γ_1—the ratio of the variance of the difference between the observed drogue velocities and the computed currents to the variance of the observed drogue velocities; γ_2—the ratio of the variance of the difference between the observed drogue velocities and the wind velocities multiplied by windage factor to the variance of the observed drogue velocities; (γ_3—the ratio of the variance of the difference between the observed drogue velocities and the combined computed velocity (current plus windage) to the variance of the observed drogue velocities; that is,

$$\gamma_1 = \frac{\sum\left[(u_d - u_c)^2 + (v_d - v_c)^2\right]}{\sum(u_d^2 + v_d^2)} \tag{9.10}$$

$$\gamma_2 = \frac{\sum\left[(u_d - a\, u_w)^2 + (v_d - a\, v_w)^2\right]}{\sum(u_d^2 + v_d^2)} \tag{9.11}$$

$$\gamma_3 = \frac{\sum\left[(u_d - u_c - a\, u_w)^2 + (v_d - v_c - a\, v_w)^2\right]}{\sum(u_d^2 + v_d^2)} \tag{9.12}$$

Six scenarios were selected from the 1990 field experiments. Figures 9.37, 9.38, and 9.39 show the observed drogue trajectories (solid line, solid circle) and the predicted trajectories (dashed line, open circle), along with the duration, mean wind velocity, and direction. The statistical comparisons are shown in Tables 9.8 and 9.9, in which the averaged velocities derived from trajectories for both field and model data, windage a, and ratios γ_1, γ_2, and γ_3 are listed for each drogue in each scenario. No attempt is made to analyze all scenarios. Only two of them are presented in detail. In the scenario TR900620 (on June 20, 1990), nine drogues were released in the southeastern basin of Hamilton Harbour. Among them, seven drogues were deployed at 2 m below the surface, and two drogues at 6.5 m.

Winds were about 1.2–1.9 m/s from 213° to 216° clockwise from the north (it was a very weak wind condition). In general, all drogues followed the wind direction, with a slight deflection for the drogues released at the 6.5 m depth. The velocity ranged from 2 to 4 cm/s (see Tables 9.8 and 9.9). The computed trajectories are

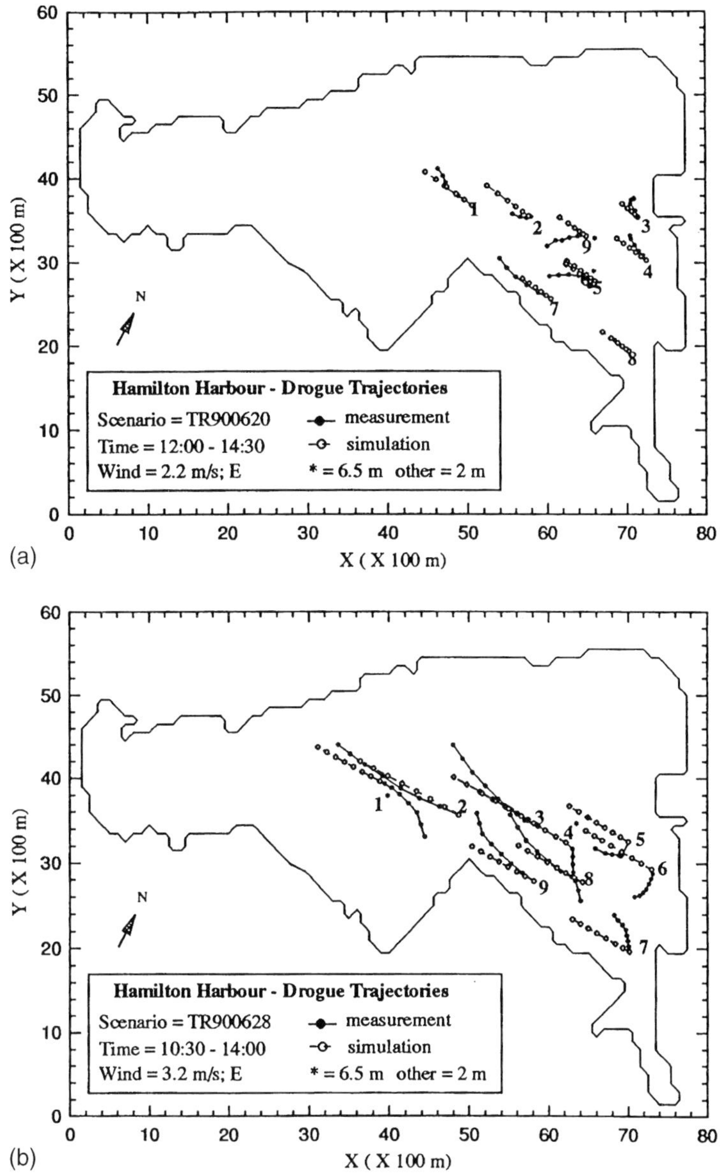

Fig. 9.37. The measured (solid line with solid circle) and the simulated (dashed line with open circle) drogue trajectories for: (a) TR900620, (b) TR900628, in Hamilton Harbour (Wu, 1993).

quite close to the observed trajectories for drogues #1, #4, and #5, and the ratio of variance of difference (γ_1) is less than 0.5. Extremely small currents were measured for drogue #6. Except for #2 and #6, the ratios of the variance of the difference between the observed drogue velocities and the computed currents to the variance

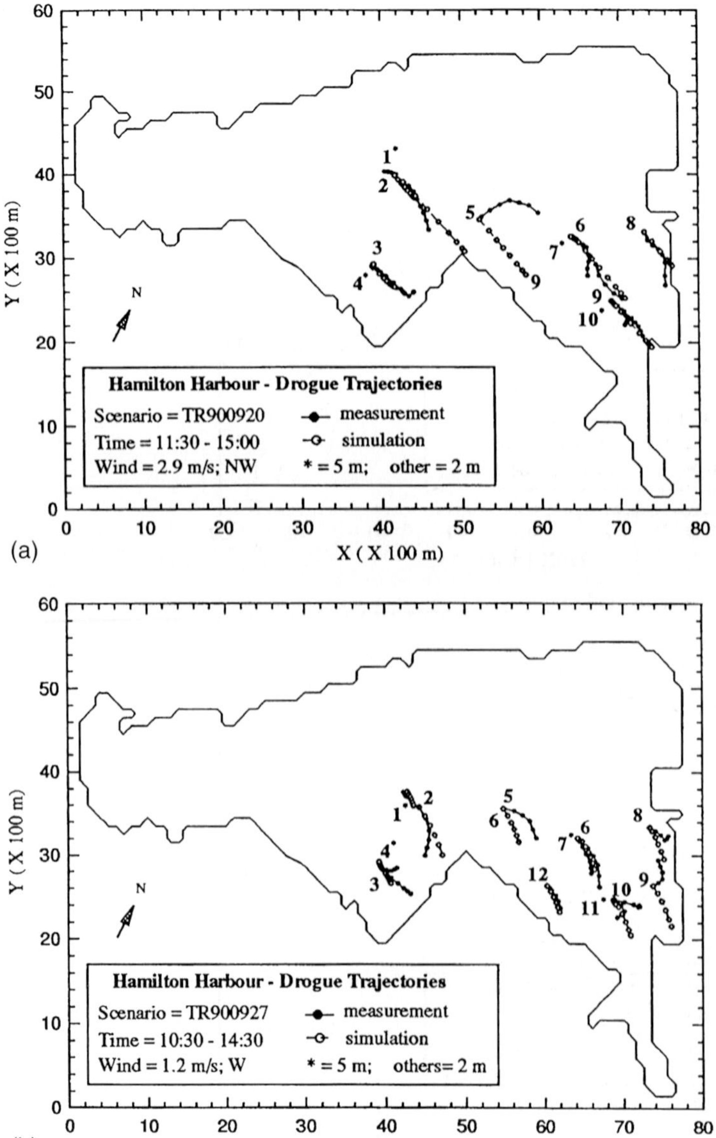

Fig. 9.38. The measured (solid line with solid circle) and the simulated (dashed line with open circle) drogue trajectories for: (a) TR900920, (b) TR900927, in Hamilton Harbour (Wu, 1993).

of the observed drogue velocities γ_1 are below 1.3 (below 0.7 for six drogues). In scenario TR901102 (on November 2, 1990), a 2.4 m/s southwestern wind prevailed. Six drogues were released at 2 m and four at 5 m; the observed and computed drogue trajectories are presented in Figure 9.39(b). Excellent simulations are obtained and all γ_1 values are below 0.9 except for one drogue.

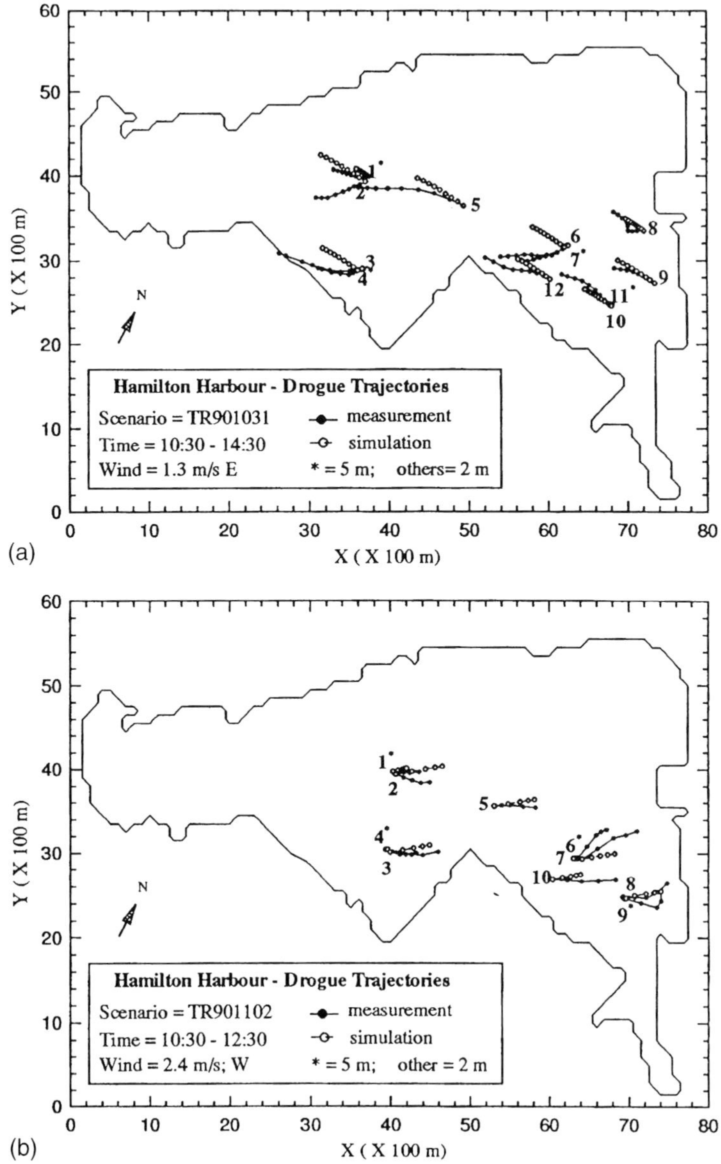

Fig. 9.39. The measured (solid line with solid circle) and the simulated (dashed line with open circle) drogue trajectories for: (a) TR901031, (b) TR901102, in Hamilton Harbour (Wu, 1993).

It is observed from these figures that, for many drogues, the model simulated trajectories were very close to those from the field observation, while, for some drogues, there was an angle within them. As we pointed out in the description of the trajectory model, two particles released in the same location may not follow the same path, because of the turbulence in the current field. Therefore, the comparison

Table 9.8. Comparison of drogue measurements and model simulations (Group 1).

Scenario	#	D(m)	u_d (cm/s)	v_d (cm/s)	u_c (cm/s)	v_c (cm/s)	α	γ_1	γ_2	γ_3
TR900620	1	2.0	−3.94	4.03	−5.41	3.69	−0.0106	0.4792	1.3062	0.4155
	2	2.0	−1.92	0.25	−4.84	3.33	−0.0207	2.1014	34.225	1.1467
	3	2.0	−0.42	1.94	−1.81	1.41	−0.0087	1.3755	6	1.1467
	4	2.0	−1.93	2.74	−3.49	2.42	−0.0080	0.3303	1.3773	0.1826
	5	2.0	−5.31	0.56	−3.25	2.24	−0.0002	0.3092	1.5513	0.3099
	6	6.5	−0.33	−0.16	−3.25	2.24	−0.0227	14.8716	1.0079	4.6807
	7	2.0	−6.03	4.53	−3.35	2.28	0.0173	0.2541	10.836	0.1395
	8	2.0	−1.68	1.22	−3.06	2.07	−0.0100	0.6110	9	0.1516
	9	6.5	−4.53	−1.11	−3.07	2.07	−0.0093	0.7468	0.5902	0.7161
									2.4599	
									1.3579	
TR900628	1	6.5	3.83	−4.52	−5.39	2.84	−0.0399	3.5757	1.0655	1.7029
	2	2.0	−10.39	5.82	−8.50	4.43	0.0075	0.0579	0.8224	0.0389
	3	2.0	−7.38	6.66	−7.34	3.98	0.0060	0.1100	0.8392	0.1035
	4	6.5	1.28	−4.75	−7.34	3.98	−0.0428	5.8159	2.2116	2.7097
	5	2.0	−2.87	−0.56	−5.12	2.88	−0.0122	1.3191	1.8446	1.0366
	6	2.0	−1.49	−2.26	−5.75	3.14	−0.0232	5.5217	3.8904	3.0700
	7	2.0	−1.32	2.93	−4.91	2.58	−0.0080	1.2718	1.8341	0.8495
	8	2.0	−6.32	5.50	−5.61	3.00	0.0077	0.1367	0.7645	0.1117
	9	2.0	−4.97	5.55	−5.30	2.84	0.0052	0.2090	0.8244	0.2046
TR900920	1	5.0	0.34	0.34	1.68	−1.87	−0.0111	6.7133	2.6417	3.4104
	2	2.0	−3.94	−3.94	5.29	−5.74	−0.0117	1.1069	2.0453	0.8052
	3	2.0	−2.03	−2.03	1.84	−1.60	0.0018	0.3576	0.9199	0.3511
	4	5.0	−1.97	−1.93	0.04	0.25	0.0111	1.0699	0.5357	0.5258
	5	2.0	0.57	0.57	4.07	−4.52	−0.0102	1.2501	1.3967	1.1013
	6	2.0	−4.61	−4.61	4.03	−4.55	−0.0013	0.1034	1.0624	0.1014
	7	5.0	−2.93	−2.93	−0.47	0.24	0.0107	1.2100	0.6896	0.7638
	8	2.0	−4.32	−4.32	2.45	−2.75	0.0002	0.6309	0.9888	0.6307
	9	2.0	−2.89	−2.89	3.50	−3.82	−0.0035	0.3836	1.2977	0.3501
	10	5.0	−1.97	−1.97	0.30	−0.42	0.0059	0.6748	0.6400	0.4554

with some angles may also be considered as a good comparison. A quantitative criterion, which is not available in any previous literature, may be established with the help of the trajectory model, using the random-walk approach. It may also be clearly explained by using the conventional advection–diffusion model, that is, for the movement of the cloud of particle

$$\frac{\partial P}{\partial t} = -u_P \frac{\partial P}{\partial x} - v_P \frac{\partial P}{\partial y} + D_x \frac{\partial^2 P}{\partial x^2} + D_y \frac{\partial^2 P}{\partial y^2} \tag{9.13}$$

where P is the "concentration" particles, and D_x and D_y are the dispersion coefficients. Supposing that a 100 unit particle is continuously released at one

Table 9.9. Comparison of drogue measurements and model simulations (Group 2).

Scenario	#	D (m)	u_d (cm/s)	v_d (cm/s)	u_c (cm/s)	v_c (cm/s)	α	γ_1	γ_2	γ_3
TR900927	1	5.0	−0.45	−0.02	0.72	−1.51	0.0102	2.5806	1.0010	2.2026
	2	2.0	0.66	−5.39	2.65	−5.34	0.0141	0.2258	1.0750	0.1803
	3	2.0	2.91	−2.19	0.64	−1.04	−0.0108	0.4987	0.8992	0.4347
	4	5.0	1.72	0.12	−0.36	0.84	−0.0093	1.6564	0.8738	1.4583
	5	2.0	3.87	−3.26	1.81	−3.72	−0.0047	0.2717	0.9665	0.2661
	6	2.0	2.02	−5.11	1.47	−3.17	−0.0001	0.1795	0.9994	0.1795
	7	5.0	1.36	−3.70	−0.21	0.25	−0.0092	1.1565	0.9702	1.1157
	8	2.0	2.16	−0.96	1.66	−3.55	0.0074	1.4304	1.1418	1.3853
	9	2.0	0.58	2.81	2.66	−4.41	0.0264	4.7111	11.046	4.2659
	10	2.0	2.92	−0.59	1.93	−3.86	0.0104	1.0452	6	0.9902
	11	5.0	0.46	−1.93	0.24	−0.49	0.0087	0.7923	1.0723	0.7281
	12	2.0	1.54	−2.46	1.44	−2.87	0.0012	0.2233	0.9621	0.2220
									0.9807	
TR901031	1	5.0	−2.75	0.52	−0.99	0.57	−0.0067	0.4299	0.7988	0.3453
	2	2.0	−3.84	−1.16	−3.47	1.99	0.0146	0.6792	1.3765	0.6336
	3	2.0	−6.03	1.22	−2.72	1.58	−0.0137	0.4486	0.8341	0.3905
	4	5.0	−3.96	0.31	−0.09	0.06	−0.0165	0.9717	0.7985	0.7804
	5	2.0	−8.10	1.34	−3.55	2.04	−0.0173	0.3519	0.8147	0.2856
	6	2.0	−5.02	−0.71	−2.49	1.44	0.0011	0.4711	1.0187	0.4752
	7	5.0	−3.56	−1.06	0.06	−0.01	−0.0091	1.0275	0.8864	0.9097
	8	2.0	−2.34	1.35	−1.43	0.88	−0.0059	0.3694	0.8409	0.3309
	9	2.0	−3.13	1.08	−2.85	1.64	0.0018	0.0790	1.0617	0.0794
	10	2.0	−3.66	2.18	−1.88	1.11	−0.0023	0.2993	0.7412	0.2173
	11	5.0	−0.99	0.55	0.01	0.01	−0.0046	0.9957	0.9080	0.9035
	12	2.0	−5.07	1.54	−2.51	1.43	−0.0110	0.2708	0.8106	0.2117
TR901102	1	5.0	3.59	−0.05	1.86	0.41	−0.0007	6.7133	1.0328	0.3955
	2	2.0	4.70	−1.12	6.46	1.05	−0.0033	1.1069	1.1520	0.4681
	3	2.0	6.36	0.06	5.46	0.91	−0.0033	0.3576	1.0967	0.1755
	4	5.0	4.19	−0.48	0.20	0.04	0.0022	1.0699	0.9300	0.8570
	5	2.0	5.82	−0.19	5.67	0.87	−0.0016	1.2501	1.0599	0.2852
	6	5.0	3.96	3.77	−0.61	−0.03	0.0010	0.1034	0.9736	1.1270
	7	2.0	7.56	3.66	4.53	0.71	0.0002	1.2100	0.9947	0.2818
	8	2.0	6.19	1.87	5.26	0.77	−0.0017	0.6309	1.0589	0.1775
	9	5.0	5.15	1.08	0.22	0.03	0.0074	0.3836	0.8435	0.7997
	10	2.0	8.82	−0.02	3.95	0.62	0.0033	0.6748	0.9226	0.3262

location, the contours will present the most likely distribution for the movement of particles released at this location.

Figure 9.40(a) shows the trajectories (open circle) of 1000 drogues released at the same point, that is, the first point in scenario TR900620 using the random-walk approach. The concentration at each mesh could be calculated by adding the number of drogues located in the mesh. The solid circles are presented for comparison purposes. Contours from the advection–diffusion modelling are shown in Figure 9.40(b). It is observed that both methods provide similar results.

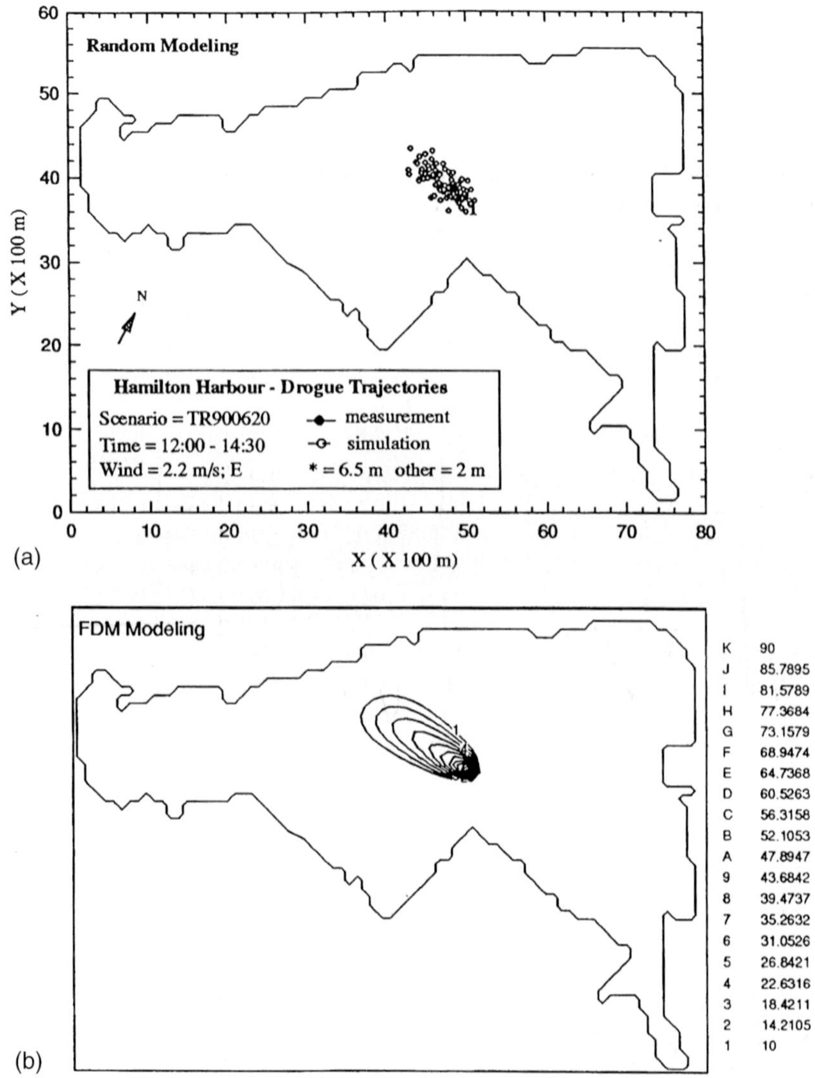

Fig. 9.40. Experiments on the trajectory predictions: (a) movement of 1000 particles released once the first drogue location in Scenario TR900620 in Hamilton Harbour, using the trajectory prediction model coupled with a Monte-Carlo randomwalk approach; (b) contours of equal concentration for an emission of the pollutant (equivalent to a cloud of particles) from the first drogue location in Scenario TR900620, using the finite difference advection-diffusion model (Wu, 1993).

9.4.3 Verification with Water-Level Data

The water-level change in a lake may be induced by the wind (wind-induced set-up), in/outflows, precipitation, evaporation, etc. For most lakes, except some small lakes with high volumes of in/outflows, wind-induced change to the water level is dominant for short periods. The magnitude of wind-induced water-level change depends on the wind speed and fetch, or size of the lake; for example, in Lake Ontario (300 km × 80 km), there is a 12 cm difference in water level between the two ends of the basin with a 10 m/s westerly wind over several hours. It was found from field data in Hamilton Harbour that water-level change did not follow wind change, and the magnitude of change was much higher than expected (for such a small-sized basin). The only possible reason is the influence of Lake Ontario (Wu, 1993).

To confirm the above observation and to verify the model, the scenario HH910814 (August 14–16, 1991) was simulated. This was a three-day west wind storm, with the wind speed varying between 2 and 9 m/s, as shown in Figure 9.41(a).

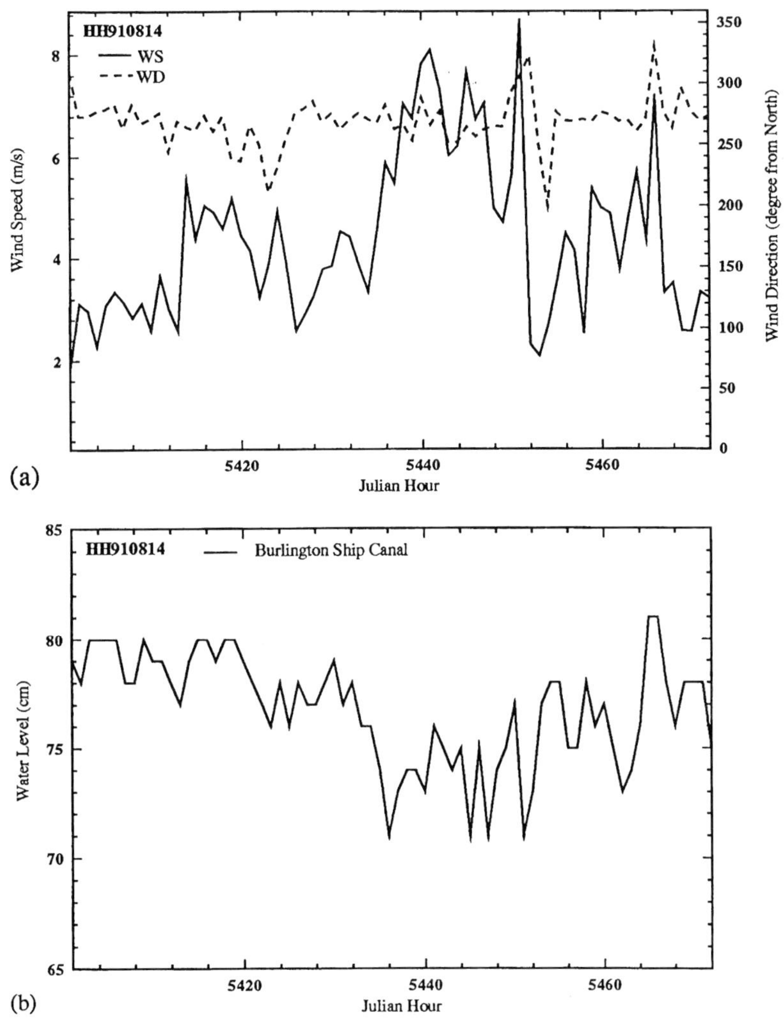

Fig. 9.41. Time series of (a) wind speed and direction, (b) water level at Burlington Ship Canal during August 14–16, 1991, Hamilton Harbour (Wu, 1993).

The water-level time history at the Burlington Ship Canal during this period is presented in Figure 9.41(b). The water-level change was between 1 and 8 cm over three days. About an 8 cm decrease in water level occurred in a couple of hours during the second day, when the wind increased from 2 to 9 m/s.

At first, the water-level change in Burlington Ship Canal was not included in the simulation. As expected, the water level changes by about 0.1–0.5 cm, mainly following the wind change. Then, the water-level change in the canal was included as a boundary condition in the simulation. Figure 9.42 presents the simulations (dashed-line) and the field measurements (solid line) at three stations (station #21, #22, #23, see Fig. 9.33). The simulations are in good agreement with the measurements and the water-level changes are as high as 10 cm. In conclusion, the influence of Lake Ontario must be included in predicting the water-level change in Hamilton Harbour.

In summary, two 3D hydrodynamic models were developed and used to simulate the currents and water levels in Hamilton Harbour. The models were verified using the field data collected during the 1990 and 1991 field seasons. Two scenarios, HH810717 and HH911002, were selected to verify the models by comparing the simulated current results point-by-point with the measured current meter data. Overall, the comparisons are satisfactory. Results from the trajectory model are in good agreement with the drogue experimental data. A quantitative criterion to evaluate the trajectory comparison was established with the help of the trajectory model, using the random-walk approach. We found that, under an averaged eddy viscosity of 5.0 m^2/s, the ratio γ_1 of 1.651 for simulation is acceptable. By using the water-level changes in the Burlington Ship Canal, the model predictions were validated with measurements at three water-level stations in the Harbour. The simulations are in good agreement with the field data (see Table 9.10).

9.5 COOTES PARADISE STUDY

Cootes Paradise is a small marsh at the western tip of Hamilton Harbour. The open water area measures 2.5 km^2, and, prior to the 1940s, it was abundant with emergent aquatic plants and home to a diverse animal population. The steady degradation of the marsh is due to many factors, and, today, the water quality is extremely poor, the emergent vegetation is virtually non-existent and the diversity of the flora and fauna greatly reduced.

The Department of Fisheries and Oceans Canada has taken the leading role in co-ordinating the restoration of Cootes Paradise. In order to improve conditions for planting and germinating the natural seed bank, a number of ideas were proposed, including the large-scale installation of an "Aqua Dam" for dewatering sections of the marsh. Knowledge of the current patterns in Cootes Paradise is necessary for any future decisions regarding flow diversions, building dikes, etc. Therefore, modelling of the currents in Cootes Paradise should precede any of the above plans.

A 2D depth-averaged hydrodynamic/pollutant transport model will be used to predict the combined hydraulic and wind-induced circulations and

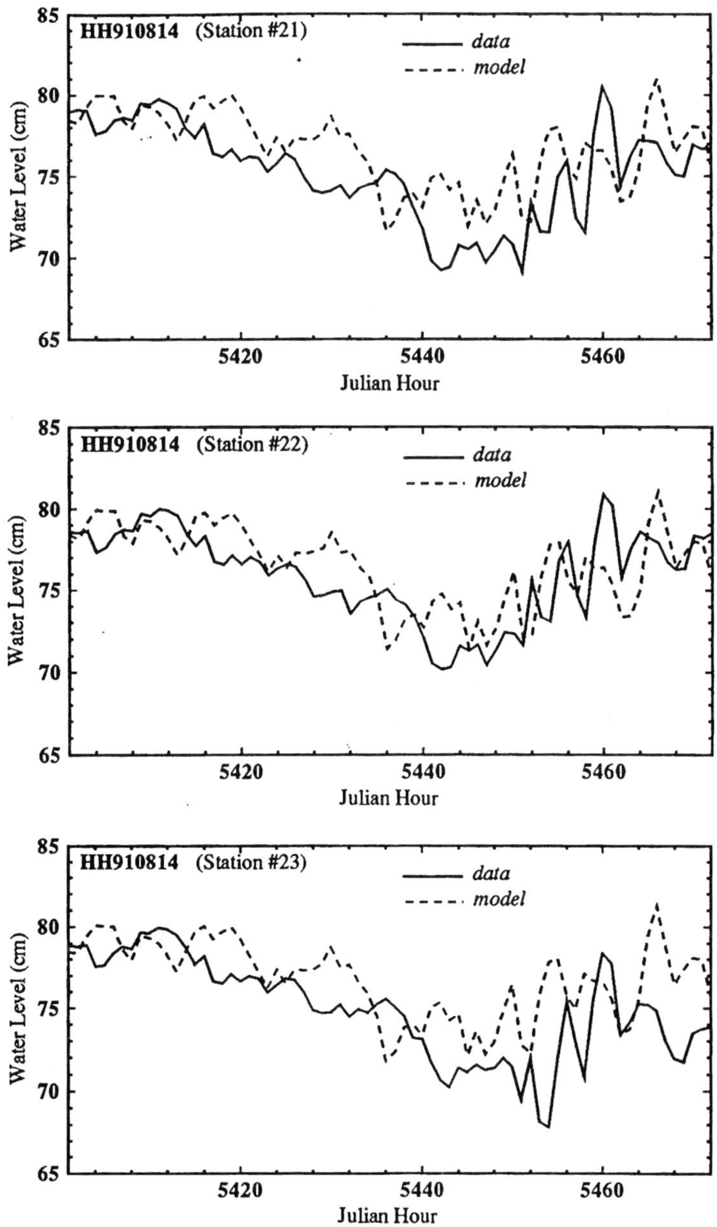

Fig. 9.42. Simulated (dashed line) and observed (solid line) water level at three stations #21, #22, #23 in Hamilton Harbour during August 14–16, 1991 (Wu, 1993).

pollutant transport in Cootes Paradise. The model will allow time-variable inputs from sewage treatment plants, storm sewers, and combined sewer overflows, in terms of discharges. The specific tasks of this project are discussed in the next paragraph.

Table 9.10. Theoretical value of γ_1 from random-walk modelling.

Ah (m^2/s)	Mean	Minimum	Maximum
0.1	0.099	0.002	0.579
1.0	0.555	0.016	5.429
5.0	1.651	0.073	17.29

Source: Wu (1993).

Set up a hydrodynamic/pollutant transport model for Cootes Paradise using both 10 and 20 m grids under different wind and water-level conditions. Model runs are to be carried out for different wind directions (e.g., the west and north) at a wind speed of 5 m/s. In addition, model runs for the dominant west wind direction will be carried out for three different wind speeds (5 and 10 m/s) in order to test the sensitivity of the model on the wind speed. Variable wind will also be used as an input to the model, in order to determine the sensitivity of currents to changes in the direction of winds. The current results will be used in a 2D advection–diffusion model to predict the pollutant transport from two point sources (Spencer Creek and Borer's Creek). The trajectory of pollutant particles from Spencer Creek under different wind conditions will be determined, and the residual times of pollutants will be calculated. Plots of the circulation and pollutant transport results for Cootes Paradise, for all the above cases, will be provided.

9.5.1 Current Patterns and Pollutant Transport

Simulations were performed for constant winds (5 and 10 m/s) blowing from four directions (north, east, south, and west) under high and low water-level conditions. The domain was discretized in 10 m × 10 m (381 × 136 grids) square grids, respectively. Pollutant particles were released from Spencer Creek continuously, with the concentration 10 ppm, and the discharge of Spencer Creek was assumed to be 2.8 m^3/s.

Figure 9.43 shows the depth-averaged velocity under 5 m/s constant west wind. Several gyres occurred in the central area, due to the bottom topography, and large currents occurred in nearshore areas, with the maximum velocity of 30 cm/s. A particle trajectory method was used to predict the path of the suspended sediments introduced at Spencer Creek. The time for the particles to leave the marsh was about 22.2 h. Figure 9.44 shows the time series of total kinetic energy. The time for the water in Cootes Paradise to reach steady state, under 5 m/s west wind, was approximately 6 h. Figures 9.45(a) and (b) show the pollutant plumes at different times (after 6 and 10 h, respectively). The pollutants released from Spencer Creek are transported northeast by the wind-induced current. After moving around Rat Island, the pollutants were divided into two parts; the first part kept moving along the north shoreline, due to strong currents in the nearshore area, and the other part moved toward the west, due to the gyre in the central area.

Figure 9.46 shows the depth-averaged velocity under 5 m/s constant north wind; the time for the particles to leave the marsh was about 31.1 h. Figure 9.47 shows the

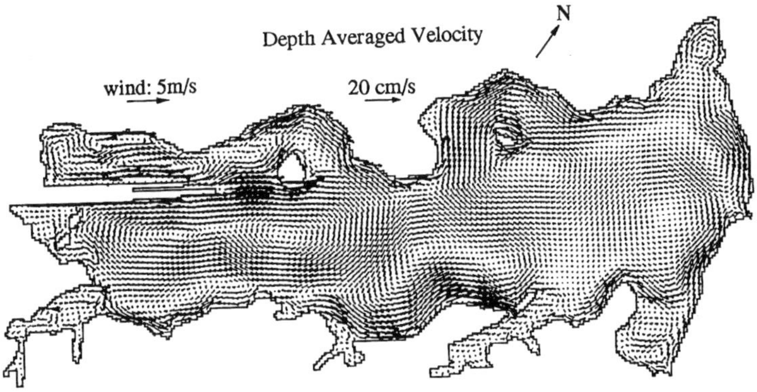

Fig. 9.43. Depth-averaged velocity in Cootes Paradise (10 m grid) under a 5 m/s constant west wind (Tsanis and Shen, 1994b).

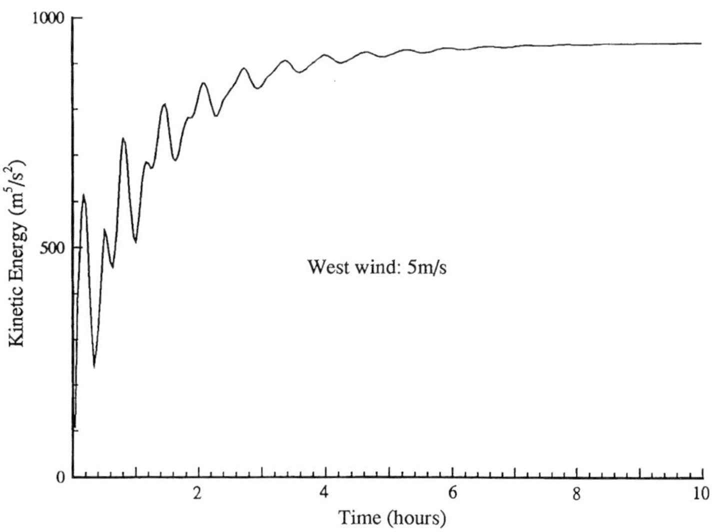

Fig. 9.44. Time series under total kinetic energy under a 5 m/s west wind (Tsanis and Shen, 1994b).

time series of total kinetic energy; the time to reach steady state under 5 m/s north wind was more than 10 h. Figures 9.48(a) and (b) show the pollutant plumes at different times. The pollutants from Spencer Creek were transported to the south first and then diffused to the east and west.

Figure 9.49 shows the depth-averaged velocity under 10 m/s constant west wind. The current structure was similar to that under 5 m/s west wind, with maximum current velocity of 60 cm/s. The time for the particles from Spencer Creek to leave the marsh was about 10.6 h. Figure 9.50 shows the pollutant plume after 10 h under the wind. Compared to Figure 9.45(b), the pollutant plume obviously moved toward the east faster than that under 5 m/s west wind.

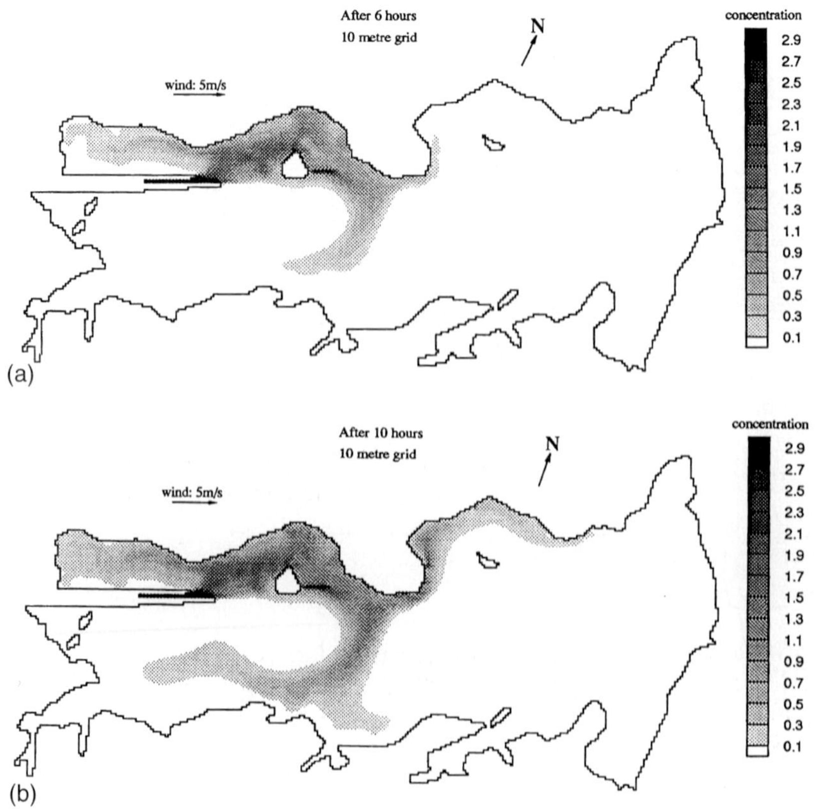

Fig. 9.45. Pollutant plume (10 m grid) under: (a) a 5 m/s constant west wind after 6 h, (b) after 10 h (Tsanis and Shen, 1994b).

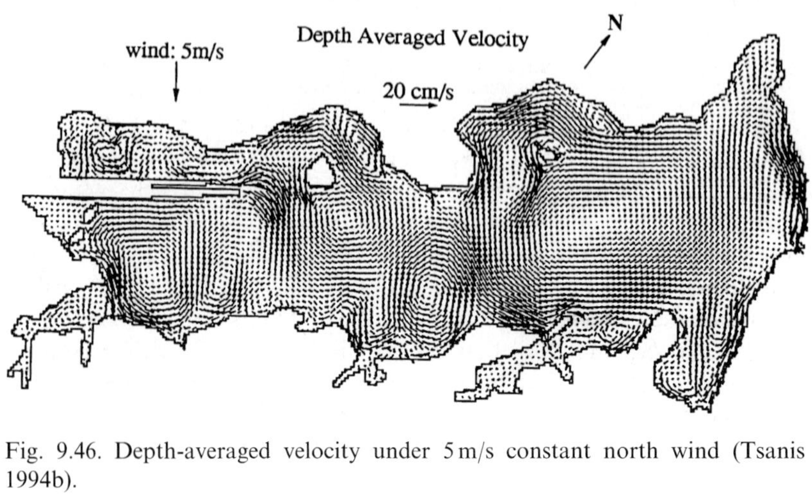

Fig. 9.46. Depth-averaged velocity under 5 m/s constant north wind (Tsanis and Shen, 1994b).

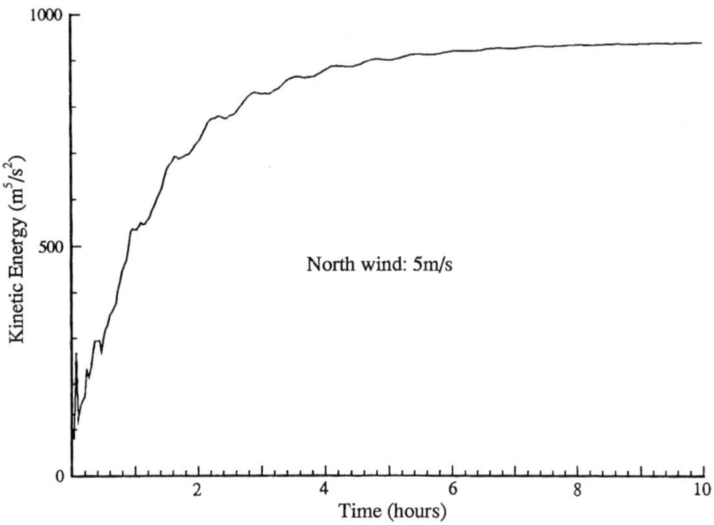

Fig. 9.47. Time series of total kinetic energy under a 5 m/s constant north wind (Tsanis and Shen, 1994b).

Figure 9.51 shows the depth-averaged velocity under 5 m/s constant west wind for a low water-level condition. The path of the pollutants introduced at Spencer Creek was different from that under a high water-level condition (see Fig. 9.43). The reason for this is the difference in water depth and the fact that some areas in Cootes Paradise were totally exposed under a low water-level condition. Figure 9.52 shows the pollutant plume after 10 h under 5 m/s west wind and low water-level condition. Compared to Figure 9.45(b), the pollutant plume under low water-level condition moved slower and was confined to a smaller area.

Long-term simulations

In order to run the hydrodynamic/pollutant transport model for long-term simulations, the Cootes Paradise basin was discretized in a 20 m × 20 m square grid (192 × 70 grids). The model was run for a 20 day simulation period under 5 and 10 m/s constant west winds and variable wind.

Figure 9.53 shows the pollutant plumes after 10 h under 5 m/s constant west wind, based on the 20 m grid. The pollutant plume in Figure 9.53 had the same pattern, but with lower resolution than that in Figure 9.45(c), which was based on 10 m grid. Figures 9.54(a) and (b) show the pollutant concentration distributions after 24 and 168 h under 10 m/s constant west wind. The pollutants entered Cootes Paradise from Spencer Creek and left the basin from the Desjardins Canal. The spatial concentration distribution of Cootes Paradise would never be uniform, even after reaching steady state, therefore Cootes Paradise is not a fully mixed basin.

Figure 9.55 shows the temporal variations of pollutant concentrations at different locations under 5 m/s constant west wind. The initial concentration was assumed to be uniform (the normalized concentration was equal to one) over the whole basin and no pollutants entered the basin during the simulation period. Since

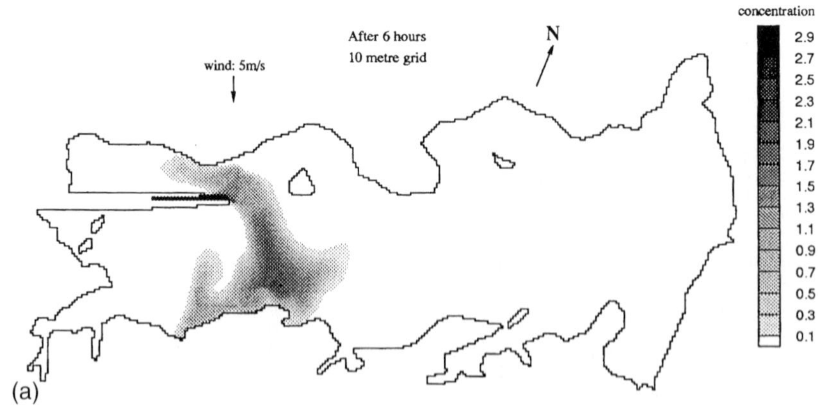

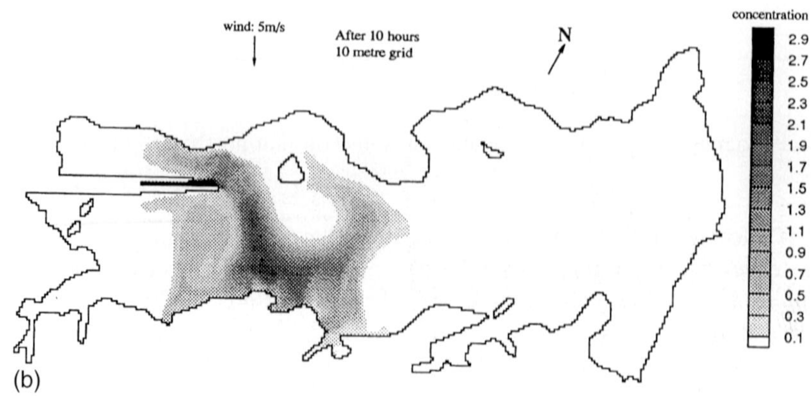

Fig. 9.48. Pollutant plume under a 5 m/s constant north wind: (a) after 6 h, (b) after 10 h (Tsanis and Shen, 1994b).

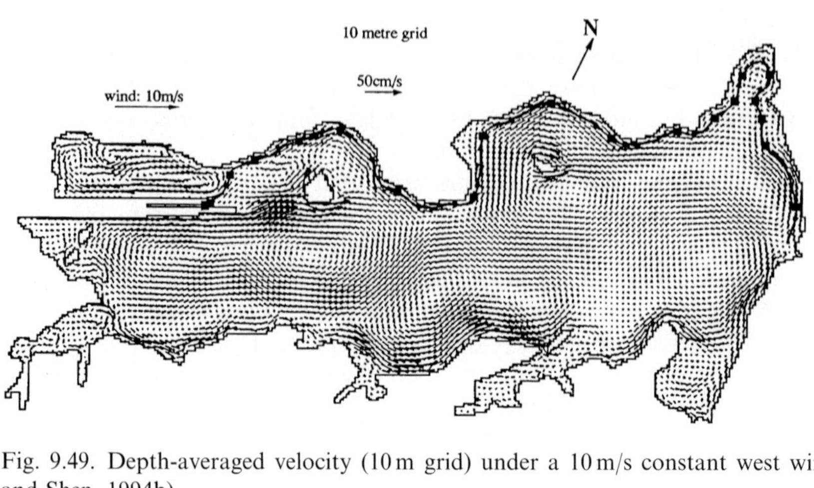

Fig. 9.49. Depth-averaged velocity (10 m grid) under a 10 m/s constant west wind (Tsanis and Shen, 1994b).

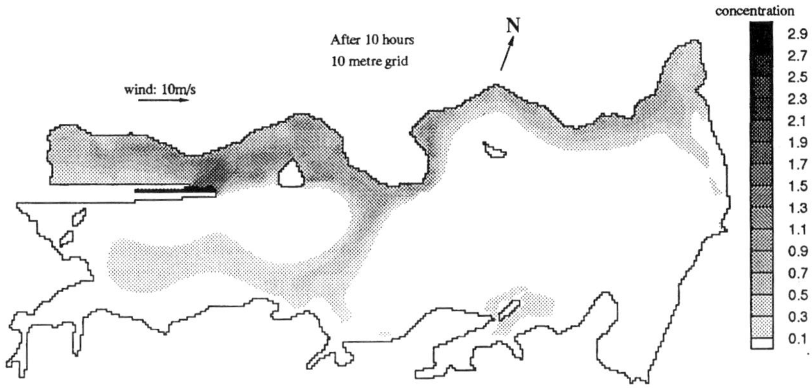

Fig. 9.50. Pollutant plume (10 m grid) under a 10 m/s constant west wind after 10 h (Tsanis and Shen, 1994b).

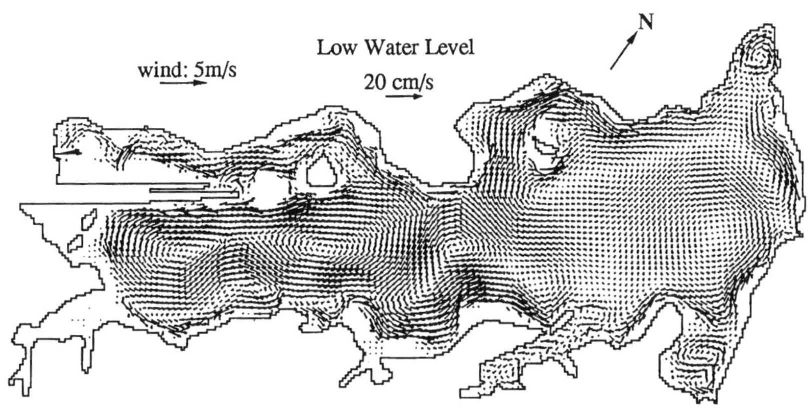

Fig. 9.51. Depth-averaged velocity (10 m grid) under a 5 m/s constant west wind at low water-level conditions (Tsanis and Shen, 1994b).

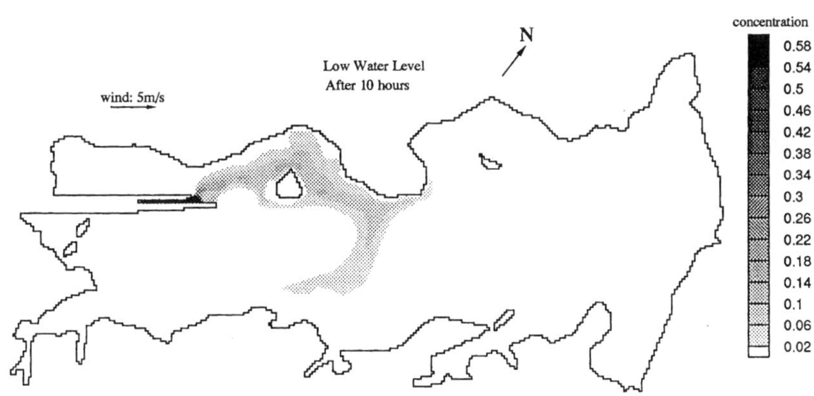

Fig. 9.52. Pollutant plume (10 m grid) under a 5 m/s constant west wind after 10 h at low water-level conditions (Tsanis and Shen, 1994b).

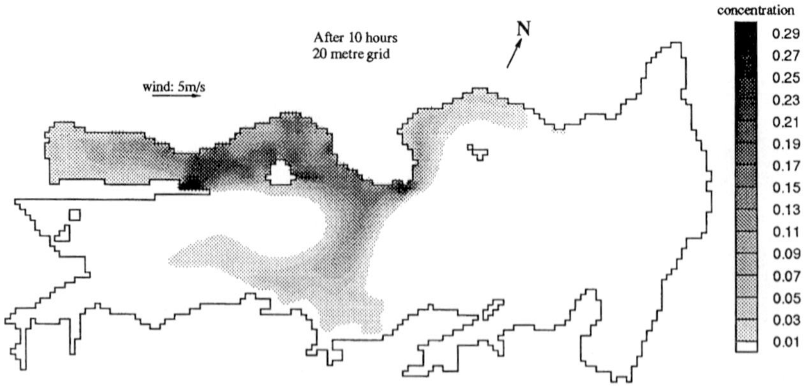

Fig. 9.53. Pollutant plume (20 m grid) under a 5 m/s constant west wind after 10 h (20 m grid) (Tsanis and Shen, 1994b).

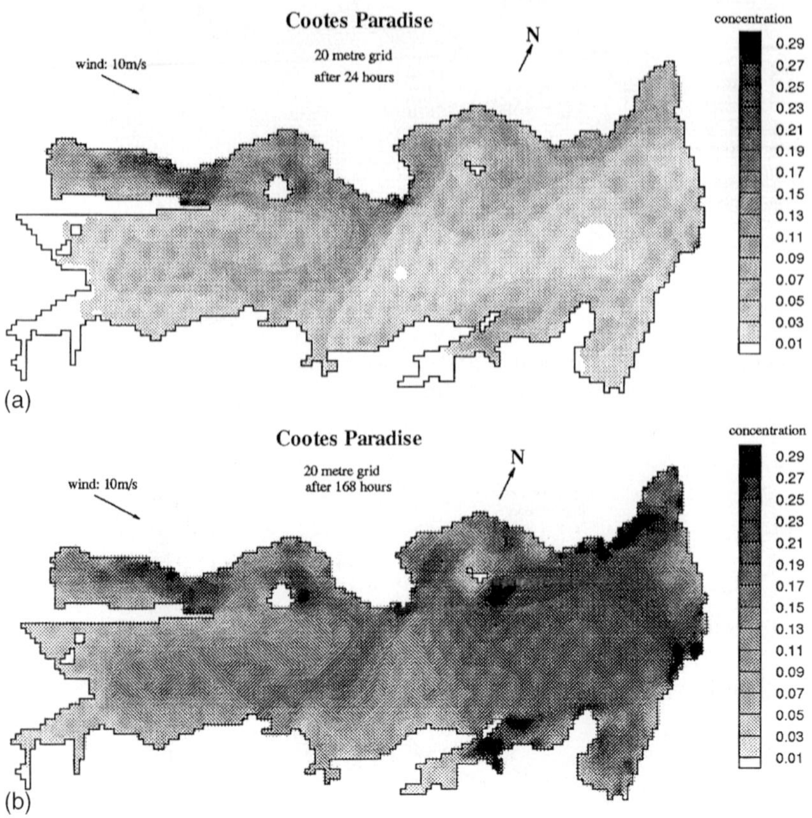

Fig. 9.54. Pollutant plume (20 m grid) under a 10 m/s constant west wind: (a) after 24 h; (b) after 168 h (Tsanis and Shen, 1994b).

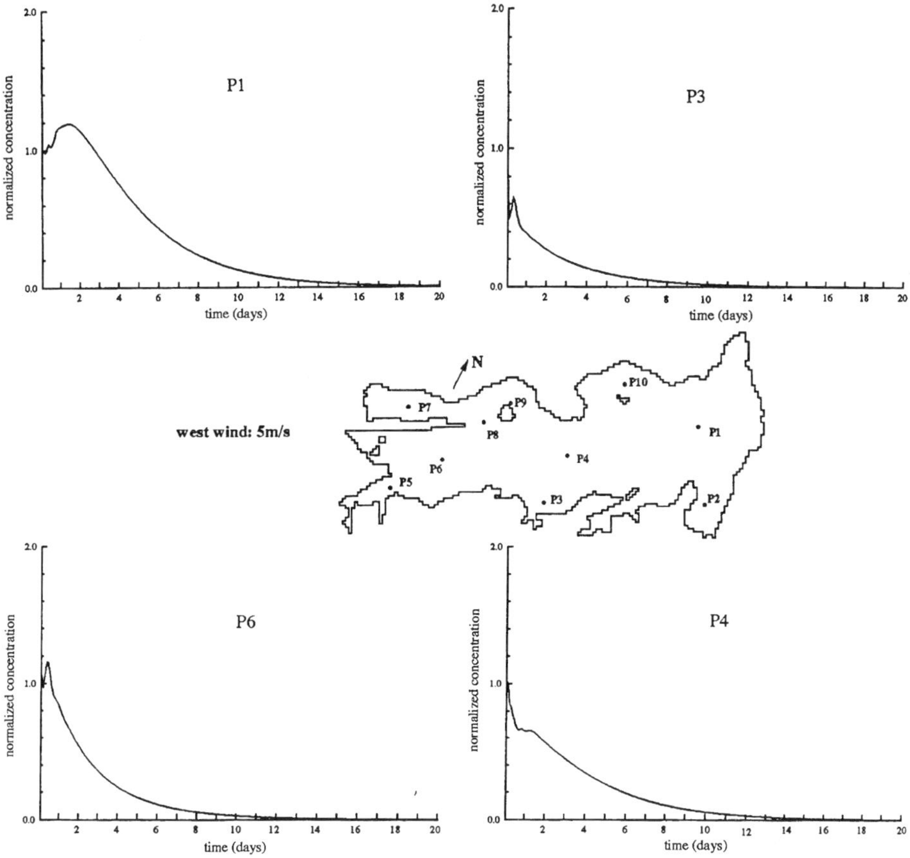

Fig. 9.55. Temporal variations of pollutant concentration at different locations under a 5 m/s constant west wind (initial concentration is one) (Tsanis and Shen, 1994b).

the pollutants left the basin from the Desjardins Canal, the concentrations at P1, P3, P8, and P4 decreased with time. It took approximately 18, 12, 14 and 16 days for the concentrations at P1, P3, P8, and P4 to reach zero, respectively. Figure 9.56 shows the time series of pollutant concentrations at different locations under 10 m/s constant west wind. The other computational conditions were the same as in Figure 9.54. The times for the concentrations at P1, P3, P8, and P4 to reach zero were 14, 10, 8, and 12 days, respectively, which were shorter times than those in Figure 9.55. Figure 9.57 showed a similar trend, but with high-frequency oscillation. This was because the variable wind pushed the pollutant particles back and forth.

Figure 9.58 shows the time series of pollutant concentrations at different locations under 10 m/s constant west wind. In this test case, the initial concentration was assumed to be zero over the whole basin, and pollutants with normalized concentration were introduced at Spencer Creek continuously. The times for the concentrations at P1, P3, P8, and P4 to reach steady state were approximately 12, 8,

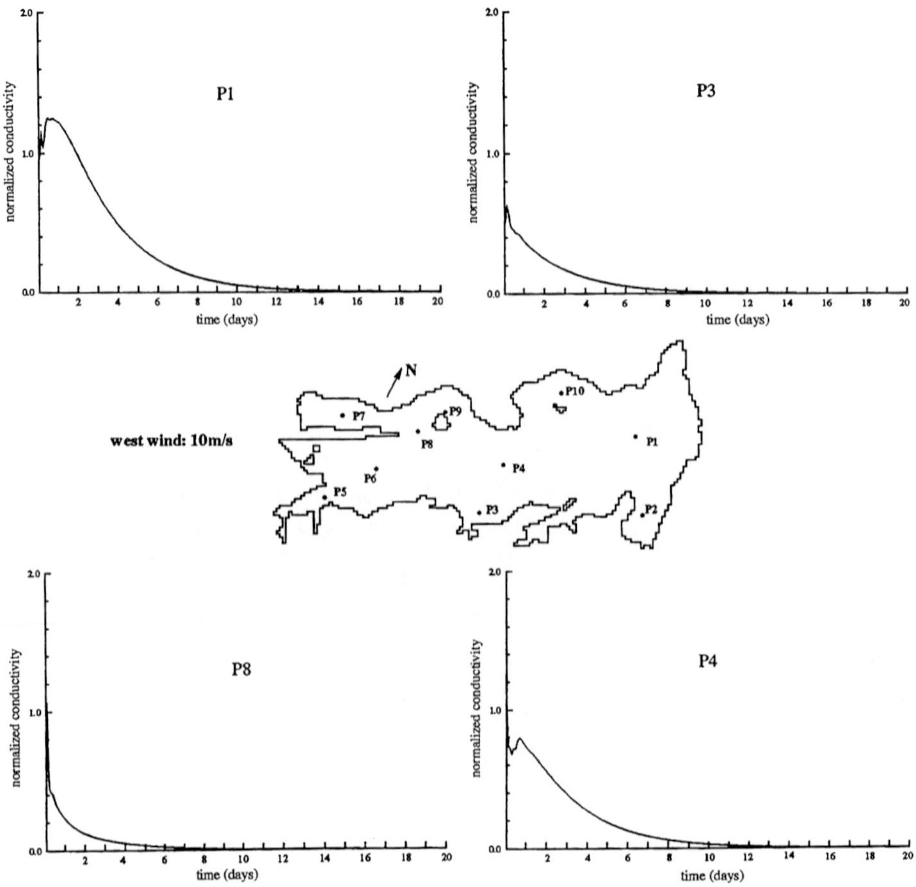

Fig. 9.56. Temporal variations of pollutant concentration at different locations under a 10 m/s constant west wind (initial concentration is one) (Tsanis and Shen, 1994b).

6, and 10 days, respectively. The concentrations at P1 and P8 were much higher than those at P3 and P4. This also proves that Cootes Paradise is not a fully mixed basin. Figure 9.59 shows the time series of pollutant concentrations at different locations. In this case, a 5×10^{-6} decay coefficient was given and the other computational conditions were the same as those of the last case. Compared to Figure 9.57, the pollutant concentrations at all points in this case were obviously lower, due to decay effect. Figure 9.60 shows the time series of pollutant concentrations at different locations under variable wind (January 1–20, 1990). The wind data at Hamilton Harbour were used in this simulation, and the other computational conditions were the same as those of the case in Figure 9.57.

Compared to Figure 9.57, the pollutant concentrations had the averaged concentration of the whole basin and the concentration near the Desjardins Canal. It took about 10–12 days to reach steady state, and the mean concentration was about 0.45. This implied that the dilution of Cootes Paradise was about 45%. The

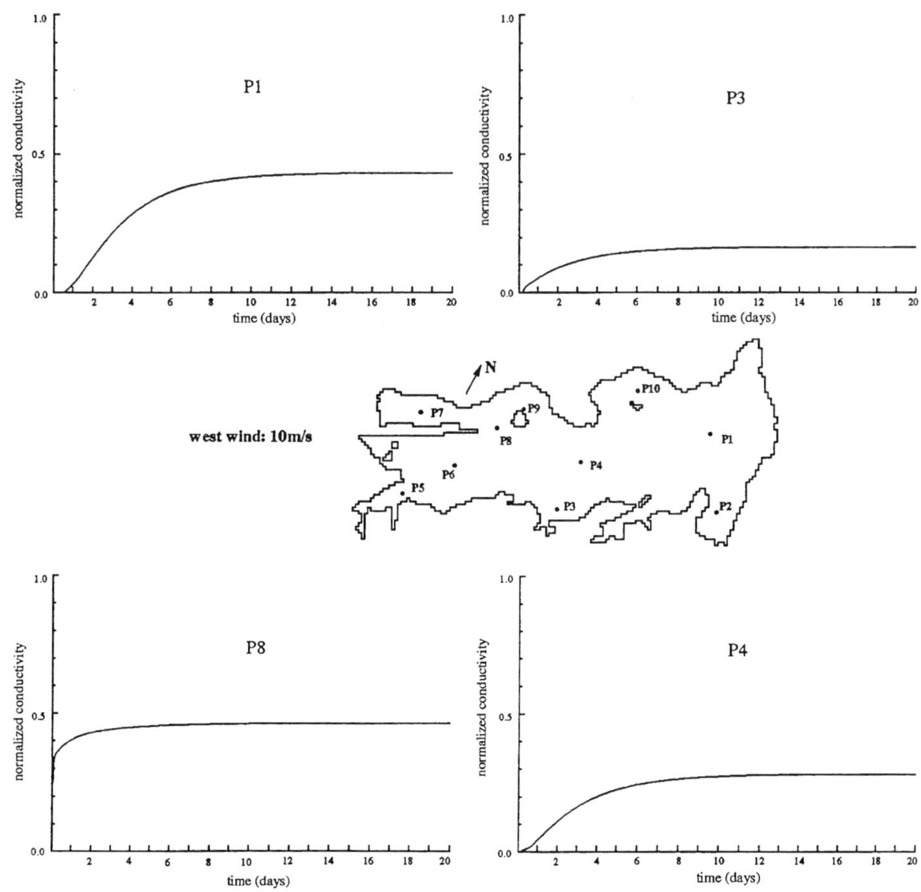

Fig. 9.57. Temporal variations of pollutant concentration at different locations under a 10 m/s constant west wind (initial concentration is one) (Tsanis and Shen, 1994b).

mean concentration of the whole basin was also calculated from field data analysis and was determined to be 0.466, with a standard deviation of 0.188 and a normalized deviation of 0.403. The concentration near the Desjardins Canal varied largely with the wind.

9.6 LAKE ST. CLAIR STUDY[11]

Lake St. Clair is located between Lake Huron and Lake Erie, and is connected with Lake Huron via the St. Clair River and with Lake Erie via the Detroit River. It is a small shallow lake with a mean depth of 4 m and a maximum depth of 6 m

[11]Material from Section 9.6 is reproduced from Brissette et al. (1993), copyright with permission from the International Association of Great Lakes Research.

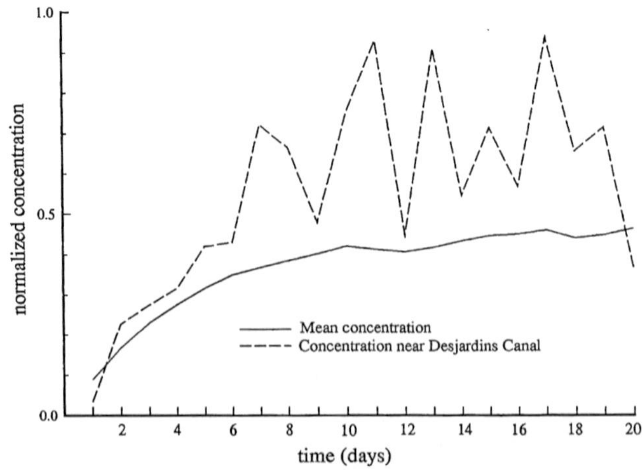

Fig. 9.58. Time series of the averaged concentration of the whole basin and the concentration near the Desjardins Canal (Tsanis and Shen, 1994b).

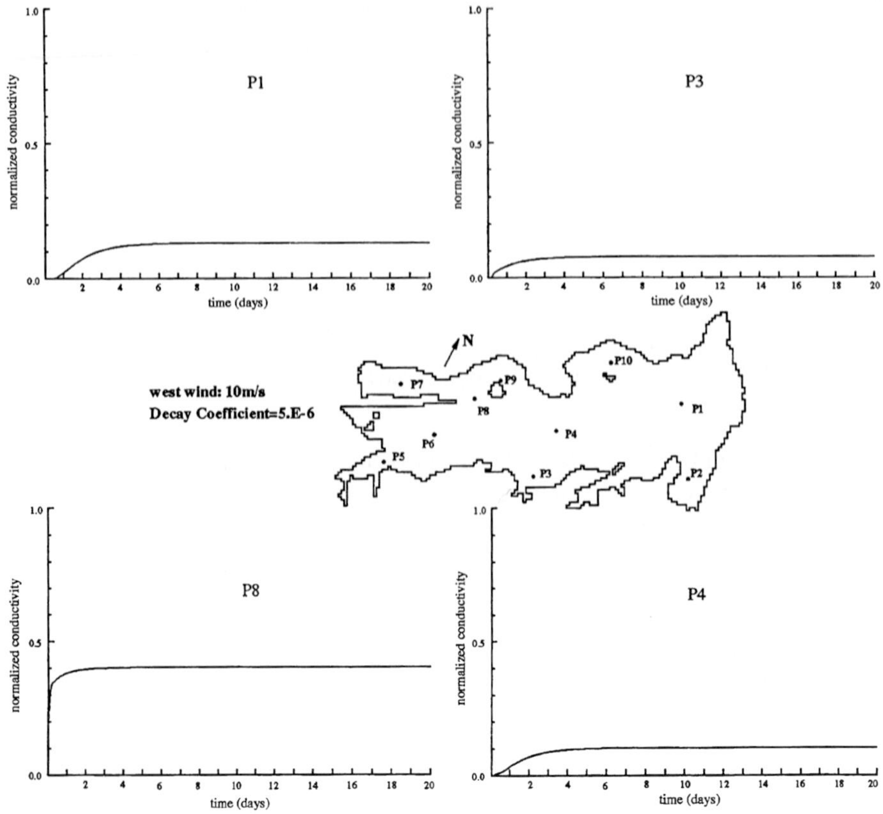

Fig. 9.59. Temporal variations of coefficient 5.0 × 10 pollutant concentration at different locations under a 10 m/s constant west wind with decay (initial concentration is zero) (Tsanis and Shen, 1994b).

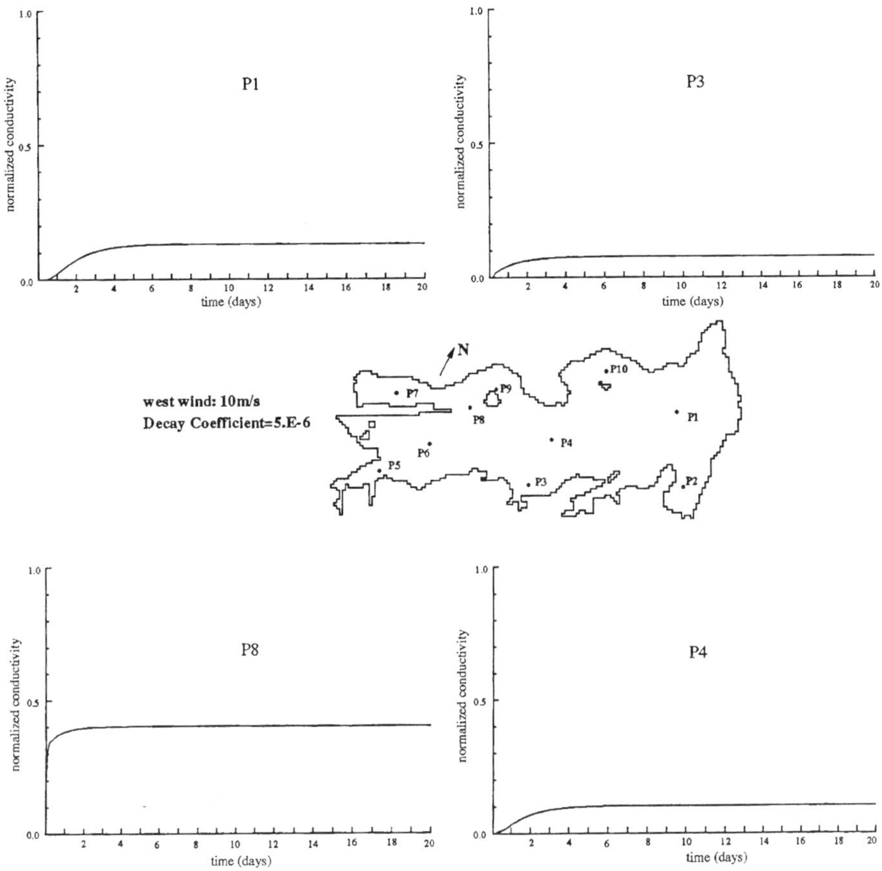

Fig. 9.60. Temporal variations of pollutant concentration at different locations under variable wind (January 1–20, 1990) (initial concentration is zero) (Tsanis and Shen, 1994b).

(Fig. 9.61a). It has the St. Clair River as an inflow and the Detroit River as an outflow, with an average flow rate of 5700 m^3/s. Its shallowness and high in/outflow result in a strong hydraulically induced current aligned roughly north to south, which is the direction of the St. Clair River to the Detroit River (Schwab et al., 1989). The hydraulic residence time is only about 9 days (100–500 days for the other Great Lakes).

Analysis of wave data collected during the 1985 field study on Lake St. Clair revealed significant differences in certain cases between the wind and wave directions (Donelan, 1980). These differences may be due to the interaction between the waves and the currents in the St. Clair-Detroit River system flowing through Lake St. Clair. Thus, there is a need to establish the water current structure in the lake under different wind conditions so that the problem of wave–current interaction can be studied (Brissette, 1992). The simulated current field is also useful in the water quality modelling in Lake St. Clair.

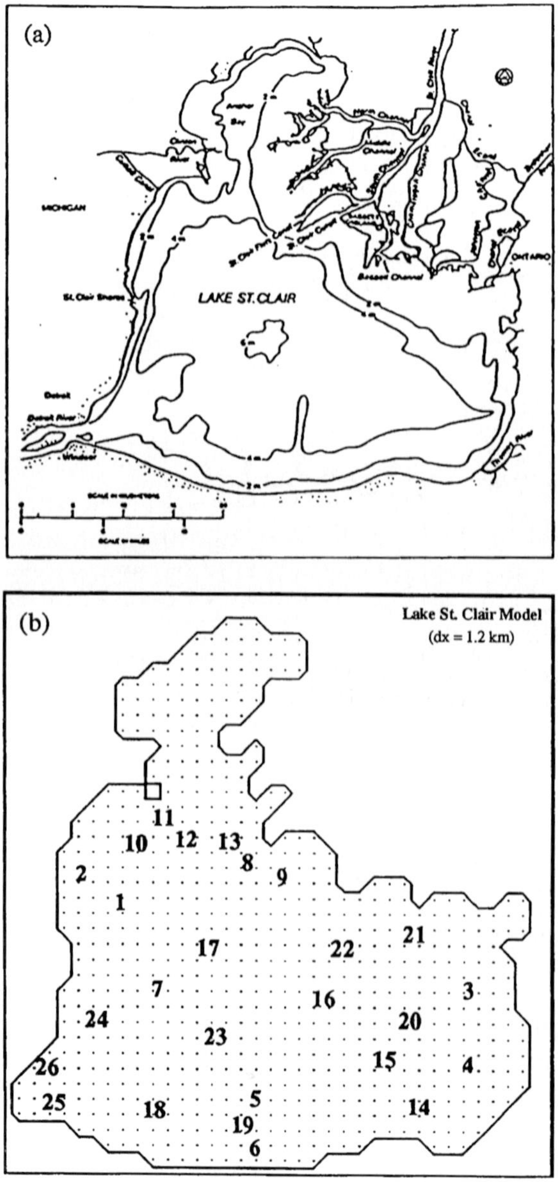

Fig. 9.61. (a) Bathymeric and location map of Lake St. Clair (Schwab et al., 1989); (b) 1.2 km grid discretization of Lake St. Clair. The numbers show location of lake current measurements (adapted from Ibrahim and McCorquodale, 1985).

Three types of models have been employed for the simulations of circulation and wind set-up in Lake St. Clair. These are: a time-dependent rigid-lid model (Schwab et al., 1989), a 2D depth-averaged hydrodynamic model (Simons and Schertzer, 1989a; Halfon et al., 1990); and a finite element circulation model (Ibrahim and McCorquodale, 1985). All the afore mentioned models ignore the non-linear advection and horizontal diffusivity terms. The first two models are unable to simulate the 3D current structure, while the third can only estimate the steady state 3D velocity field. Bottom friction was physically unrealistic in the conventional depth-averaged model for such a shallow lake, and no improvement can be made in such a 2DH model (Simons and Schertzer, 1989a). The coarse grid used in the third model makes it difficult to obtain a detailed current field. As discussed in Chapter 5, the Q3D model can simulate the 3D current field in an economical and efficient way. In this section, the typical circulation patterns will be simulated and, based on that, a wave ray analysis will be performed to examine the wave-current interaction in the lake.

9.6.1 Typical Circulation Patterns

The Q3D hydrodynamic model is used with a 1.2 km grid to simulate the combined hydraulically and wind-induced circulation in Lake St. Clair (Fig. 9.61b). In this simulation, the wind drag coefficient $C_D = 1.8 \times 10^{-3}$, and the mean hydraulic inflow/outflow = 5700 m^3/s (Schwab et al., 1989). The wind speeds and directions used in the simulation were selected to meet the requirement of the wave–current interaction study. The total kinetic energy evaluated in the simulations reaches the steady state condition in a very short period, within 2 h, since the lake is shallow and the bottom frictional damping is very large (Fig. 9.62). The 3D-mesh of a typical wind set-up (under 10 m/s SWW wind) in Lake St. Clair is shown in Figure 9.63 and it indicates a high water level in the north shore and a low level in the south.

The pure hydraulically induced depth-mean circulation simulated by the Q3D model is shown in Figure 9.64. An advective hydraulic pattern is evident between the inflow and outflow rivers, along with a large counterclockwise eddy covering the

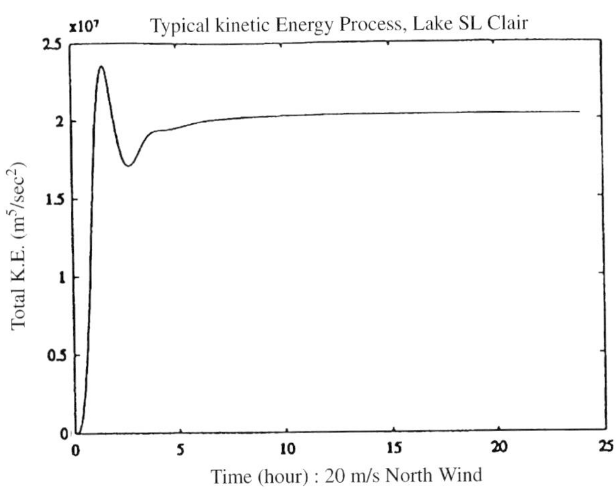

Fig. 9.62. Typical total kinetic energy process of simulation of Lake St. Clair (Wu and Tsanis, 1991).

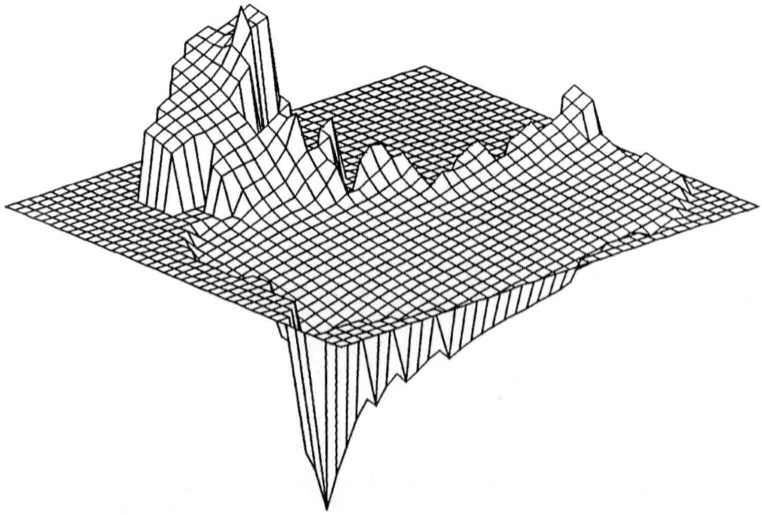

Fig. 9.63. Typical wind set-up in Lake St. Clair (wind: SWW 10 m/s) (Wu and Tsanis, 1991).

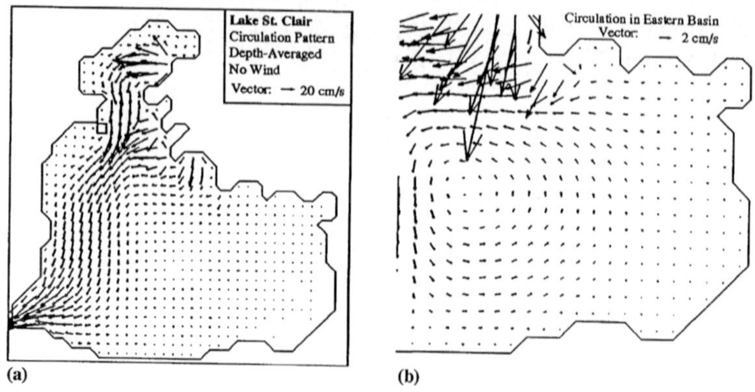

Fig. 9.64. Depth-averaged hydraulically induced circulation in Lake St. Clair. (a) In the whole lake (vector scale = 10 cm/s); (b) In the eastern basin (vector scale = 2 cm/s) (Wu and Tsanis, 1991).

eastern basin. The magnitude of the current is about 20–60 cm/s in the advective hydraulic flow, while below 2 cm/s in the eastern basin. This result is in agreement with the observation findings under no-wind conditions (Schwab et al., 1989). The combined (hydraulically induced and wind induced) circulation patterns for 10 m/s winds at four directions (east, west, north, south) are presented in Figure 9.65. The common feature in all of these patterns is that the hydraulic flow along the NE–SW direction is dominant in the central and western parts of the lake, while a wind-induced flow is dominant in the eastern part of the lake. In the eastern basin, under east and west wind conditions, the currents follow the wind in the north and south

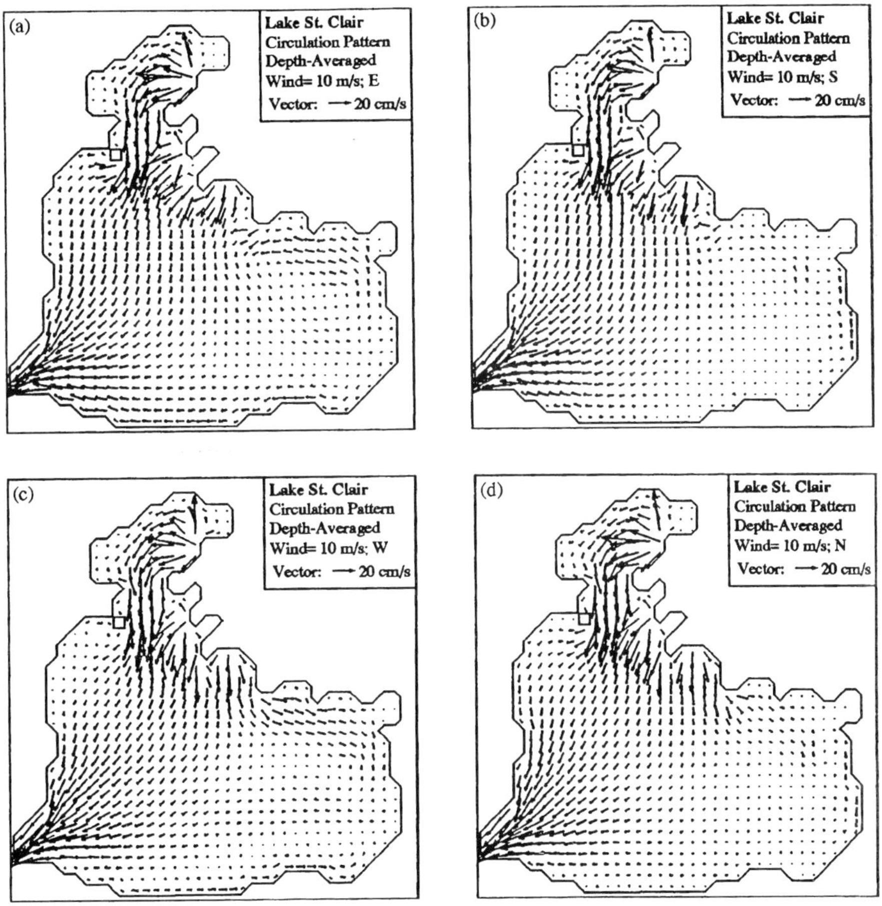

Fig. 9.65. Depth-mean circulation patterns in Lake St. Clair for a 10 m/s wind speed at four different directions: (a) east, (b) south, (c) west, (d) north (Wu and Tsanis, 1991).

shorelines and form a return flow in the central region; while, under north and south winds, the currents along the east shoreline follow the wind.

A typical multi-layer circulation pattern is shown, in Figure 9.66, for the case of a 12.3 m/s wind blowing from the SWW direction (87°). At the upper layer (5th layer), the current follows the wind, except in the advective hydraulic flow, where the flow is forced to turn in the wind direction. At the middle and bottom layers (3rd and 2nd layers), the currents in the eastern part of the lake are 180° out of phase with the wind direction, while the central and western parts are dominated by the hydraulically induced currents. It is interesting to examine how the wind affects the advective hydraulic flow in the upper layer. Figure 9.67 shows the surface currents for SWW winds at four speeds, that is, 3, 7, 10, and 20 m/s opposite to the hydraulic flow. In the eastern parts and nearshore areas of the lake, the surface currents are aligned with the wind, while in the hydraulically dominated flow area,

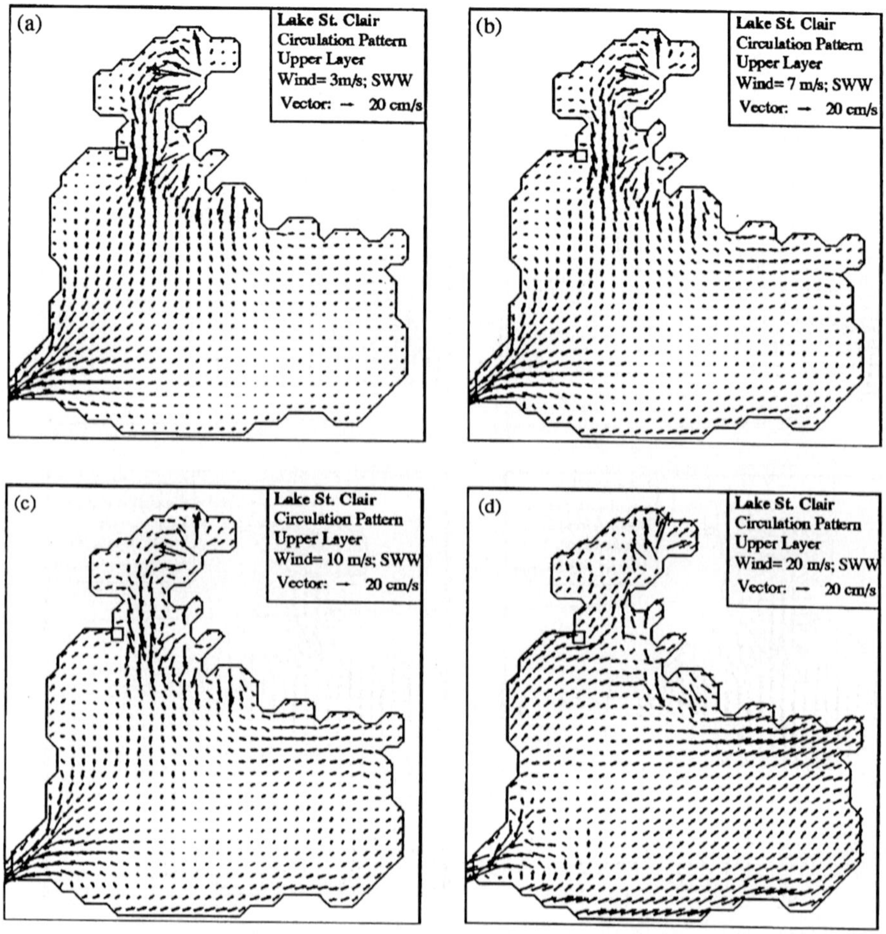

Fig. 9.66. Surface circulation patterns in Lake St. Clair for a SWW wind at four wind speeds: (a) 3 m/s, (b) 7 m/s, (c) 10 m/s, (d) 20 m/s (Wu and Tsanis, 1991).

the lake currents are weak because the wind-induced current is acting against the hydraulic flow. With a 20 m/s wind, the hydraulic flow is totally destroyed except in river mouths (see Fig. 9.66(d)).

Table 9.11 shows water current data collected at two depths ($z1$ and $z2$) from 26 stations on 4 days and under various wind conditions (Ibrahim and McCorquodale, 1985), which are used to verify the currents obtained from the Q3D model: Figure 9.68 presents the observed and calculated current vectors for these stations. It clearly indicates a general agreement in terms of current magnitudes and directions (the ratio of the variance of difference between the observed velocities and the computed currents to the variance of the observed velocities is 0.4127).

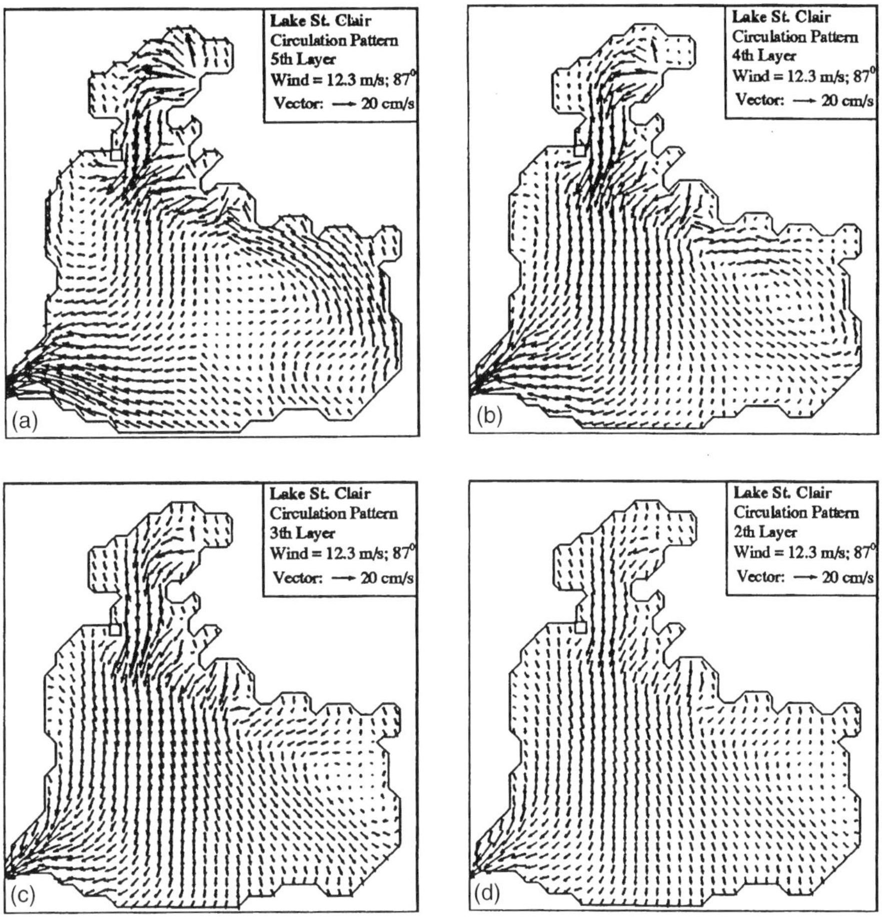

Fig. 9.67. Three-dimensional pattern in Lake St. Clair. Wind = 12.3 m/s, 87 degree: (a) 5th layer—1/5 depth below surface, (b) 4th layer—2/5 depth, (c) 3rd layer—3/5 depth, (d) 2nd layer—4/5 depth (Wu and Tsanis, 1991).

Table 9.11. Four episodes of current measurements in Lake St. Clair.

Data date	September 29, 1983	June 23, 1983	July 26, 1984	July 31, 1984
Wind (m/s)	−1.18, −1.18	1.49, 3.60	0.50,−0.98	2.35, 2.35
Station #	1 2 3 4 5 6 7	8 9 10 11 12 13	14 15 16 17	18 19 20 21 22 23 24
Depth z1 (ft)	4 3 2 2 2 2 1	1 0 x x x x	x x x x	3 1 3 3 2 4 1 3 3 3
Depth z2 (ft)	7 6 6 7 5 5 5	5 × 15 5 5 5	7 4 4 1	11 12 10 8 8 12 12 8 8

Note: x means that no measurement was made in the station.
Source: Ibrahim and McCorquodale (1985).

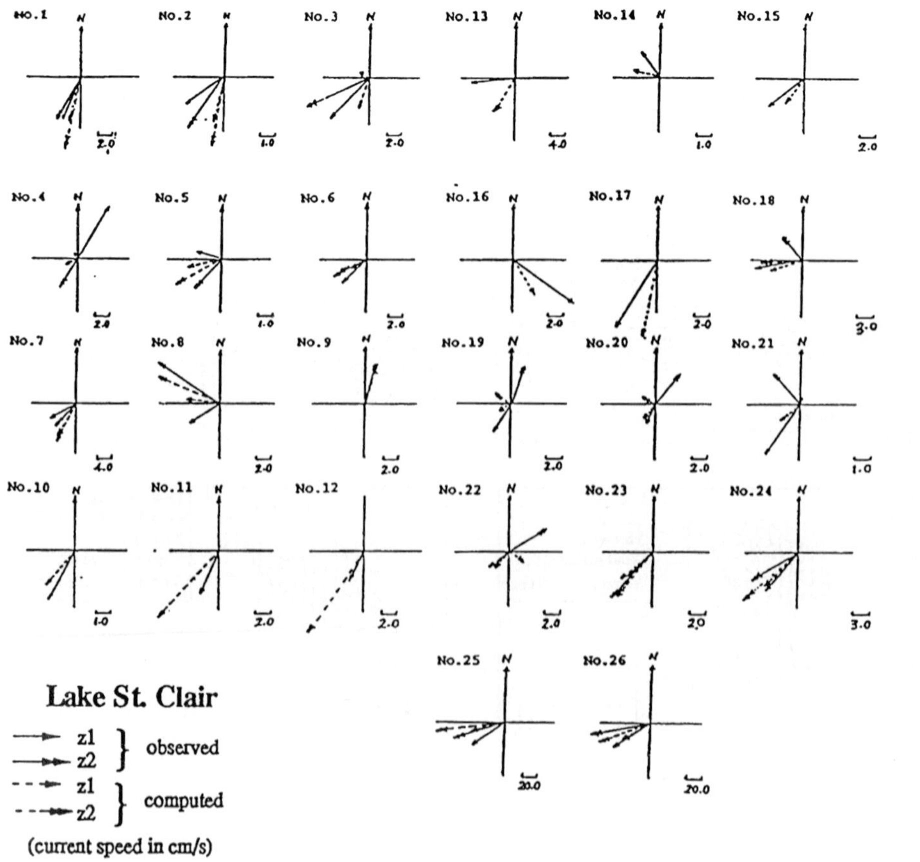

Fig. 9.68. Observed and calculated current vectors (Wu and Tsanis, 1991).

9.6.2 Current-Induced Wave Refraction

Analysis of wave data obtained from a three-wave staff array and collected during the 1985 field study on Lake St. Clair revealed, in some cases, significant differences between the wind directions found at the central location on the lake (C3, closed circle in Fig. 9.69), especially for south winds. The differences could be explained by the presence of a strong hydraulic current aligned, roughly, north to south, which would tend to deflect the waves toward the south if they travel with the current, or in the opposite direction if the waves travel against the current. The high-resolution directional spectrum estimations displayed significant directional effects; and the wave refraction model, which includes the various possible effects on wave refraction, revealed that the effect of fetch gradient and refraction, due to the bottom topography, were insignificant, while the current-induced refraction prevailed (Brissette et al., 1993).

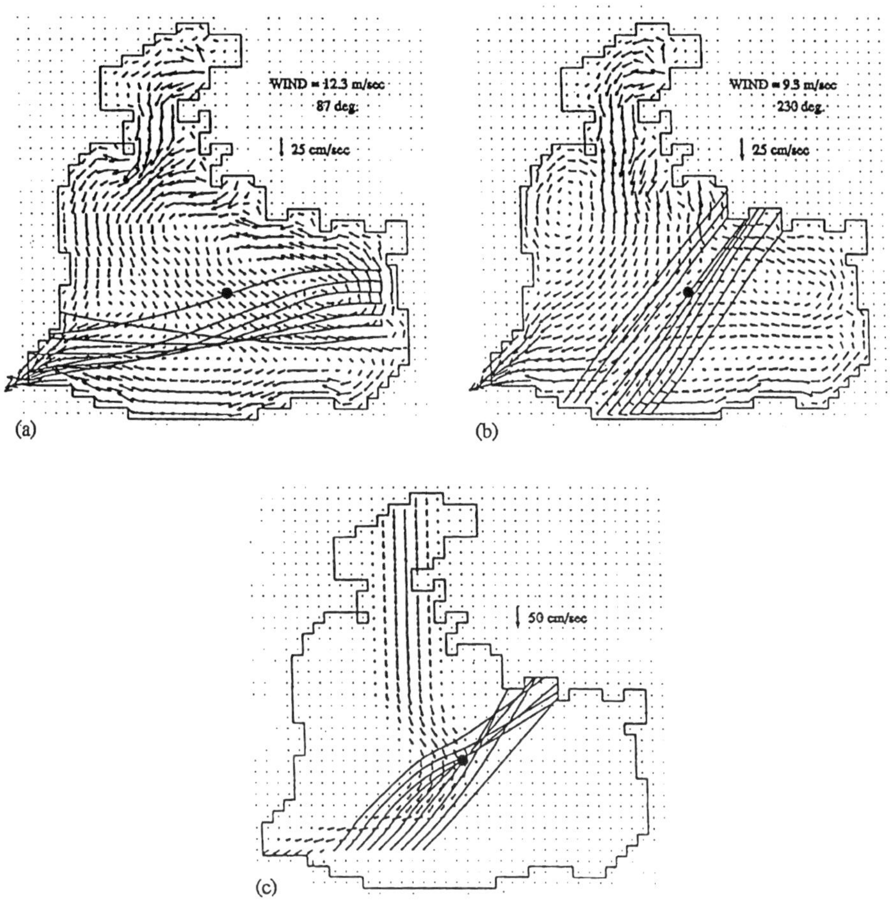

Fig. 9.69. Current-induced refraction as obtained from wave ray analysis and calculated circulation pattern in Lake St. Clair: (a) for a wind blowing from the east, refraction in excess of 20° is observed; (b) for a wind blowing from the southwest, little refraction is observed as the waves propagate directly against the shear current with a 0° angle of incidence; (c) for an idealized circulation pattern for a southwesterly wind, curvatures in the shear current can have a dramatic effect on the wave rays, refraction of 30° is observed (the closed circle represents the location of Tower C3 during 1985 wave observation in Lake St. Clair) (Brissette et al., 1993; Wu, 1993).

In this study, the wave-ray analysis method developed by Irvine (1987) is employed to examine the wave–current interaction in Lake St. Clair. The ray equations for waves propagating over a spatially varying current are as follows (Irvine, 1987):

$$\frac{dk_x}{dt} = -k_y \frac{\partial V}{\partial x}; \quad \frac{dk_y}{dt} = -k_x \frac{\partial V}{\partial y} \tag{9.14}$$

and

$$\frac{dR_x}{dt} = (C_g)_x + U; \quad \frac{dR_y}{dt} = (C_g)_y + V \tag{9.15}$$

where U and V are the depth-averaged velocity components in the x- and y-directions, respectively. $\vec{k}(k_x, k_y)$ is the wavenumber vector; C_g the group velocity; $\vec{R}(R_x, R_y)$ maps out the wave rays. Equation (9.15) indicates that the turning of waves caused by the current is due to both the advection and dispersive nature of the waves (Phillips, 1981). In the following analysis, wave rays were computed with Eqs. (9.14) and (9.15), using a calculated current structure in Lake St. Clair. The analysis was performed using 0.4 Hz waves, which are typical peak waves in Lake St. Clair for medium-to-strong wind conditions. This analysis is clearly an approximation as it assumes wave components with fixed frequencies traveling without dissipation or wind input of energy. However, the analysis should reveal some aspects of wave–current interaction in Lake St. Clair, and at least indicate if such interaction is compatible with the field observation.

Figure 9.69(a) presents the results of a wave-ray analysis for the case of an easterly wind. It can be seen that the wave rays are clearly refracted by the shear current at the east end of the lake, with the result that the main wave direction becomes about 50°, an observation consistent with the field data. Waves generated further away from the WNW flowing current will show a direction more consistent with the wind. It should be noted that this analysis neglects the tendency of the waves to realign with the wind once the action of the shear current decreases. In this case, the force would be counteracted by the presence of the SE current; as previously discussed, and the waves tend to be the direction normal to it.

Figure 9.69(b) presents results of the wave-ray analysis for the case of a southwesterly blowing wind, where the waves were observed to come from either side of the current but not directly against it. The ray analysis indicates that there is very little gained by considering the work of Irvine (1987) in which he showed that wave patterns on a current are strongly influenced by the curvature. He also demonstrated that very small changes in the current structure can have large effects on wave rays. The circulation patterns calculated in this work should adequately represent the general features of the circulation pattern in Lake St. Clair. However, they are computed for steady cases and are obviously subject to parameterization for various factors such as horizontal diffusivity, bottom friction, and wind stress. As such, the finer details of the circulation pattern, such as the curvature of the main current, cannot be considered exactly, especially when trying to apply a "steady state numerical circulation pattern" to wave data in unsteady wind conditions.

In order to look more closely at the effect of the curvature of the main current on the wave refraction pattern, the circulation pattern was idealized by slightly increasing the curvature of the main current. Wave rays for southwesterly winds were then traced, as shown in Figure 9.69(c). The results now indicate the significant refraction of the waves. The main effect of the refraction is the turning of the waves to the west, up to 30°, which is compatible with field data. This model is a simplification of the real phenomenon in Lake St. Clair, but it certainly outlines the fact that currents of magnitudes similar to those observed in Lake St. Clair are sufficient to create significant refraction, and are compatible with field observations. It should be noted that the choice of 0.4 Hz waves is probably a conservative one since it implicitly assumes that the shear currents had no effect on the growth of the waves (the wave's direction of propagation is not affected by the currents until they reach a period of 2.5 s). In most cases, the depth-averaged current velocities should be considered as minimum velocities since it is to be expected that the current close to the surface is the one that affects the waves. Considering simply a logarithmic velocity profile, the average velocity of the upper 2 m could be 20% higher than the velocity averaged over a 6 m depth.

To our knowledge, no similar wave-interaction cases have been observed elsewhere. This can be partly explained by the fact that Lake St. Clair is characterized by an unusually low residence time, causing strong currents and especially steep horizontal velocity gradients. The horizontal velocity gradients are fundamental in controlling wave refraction, and the gradients observed in Lake St. Clair (0–10 cm/s/km) are as high as the ones observed in major currents such as the Gulf Stream or Kuroshio, off Japan. In fact, the problem of swell refraction by such major currents (as discussed by Irvine, 1987) can be almost exactly scaled down to Lake St. Clair, when considering 0.4 Hz waves. Additionally, few studies of waves make use of high-resolution directional spectrum estimates which are needed to observe the complex energy structure of wave fields in strong shear currents.

9.7 LAKE SIMCOE STUDY

Previous studies of Lake Simcoe have documented the retention of phosphorus and metals in the sediments, along gradients from points of input to Cook Bay and Kempenfelt Bay (Johnson and Nicholls, 1988, 1989). Also, a feasibility study of the potential for artificially enhancing bottom water 29.88 dissolved oxygen concentrations in Kempenfelt Bay, with aeration/oxygen injection systems, has been carried out (Neil, 1990). Water quality monitoring and modelling of phosphorus loading and dissolved oxygen depletion in Kempenfelt Bay, and in the main basin of the Lake, are continuing (Draper et al., 1985; The Land Sub-Group, 1985; Snodgrass and Holubeshen, 1993). None of this work has had the benefit of information about current patterns in the lake, especially information relating to exchange of bottom water masses between Kempenfelt Bay and the main lake. At this stage of the Lake Simcoe limnology program, an understanding of water movement patterns in the lake and the flow exchange between Lake Simcoe and Kempenfelt Bay is essential. The mass balance of the water quality parameters in Kempenfelt Bay require knowledge of flow and pollutant exchange between Lake Simcoe and the bay. To achieve this goal, the model VHI3D was used to simulate the temporal and spatial current distribution in Kempenfelt Bay and Lake Simcoe. The study area is shown in Figure 9.70.

Figure 9.71 gives the wind speed and direction of a storm on December 3, 1990. The wind was blowing, for the first 24 h, from the east, with the wind speed gradually increasing from 4–14 m/s. During the last 14 h, the wind speed dropped to about 3–4 m/s and the wind shifted to the west. During the last 10 h, the wind speed was 5–6 m/s and the wind shifted to the north. Figure 9.72 shows the depth-

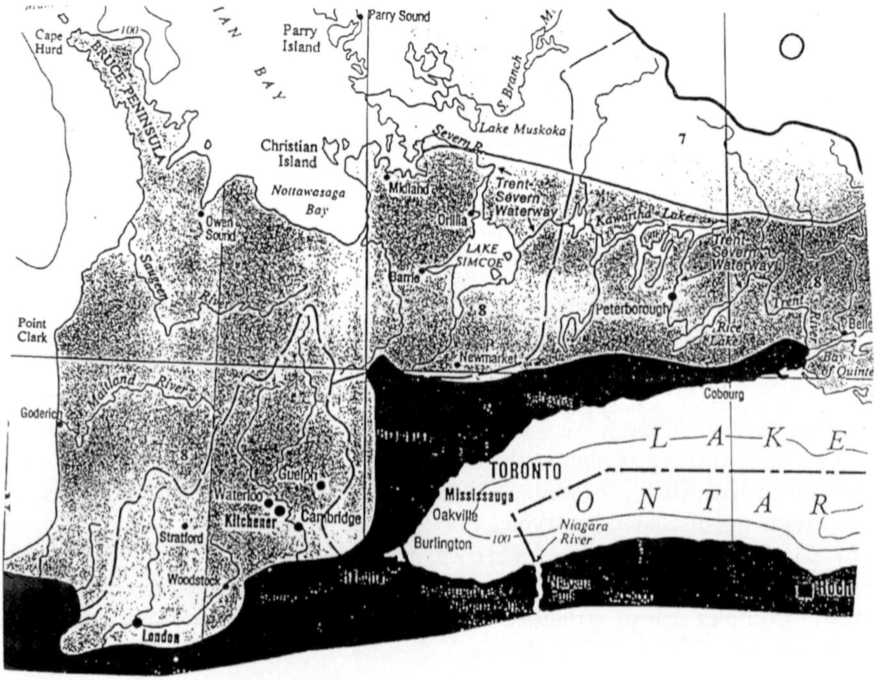

Fig. 9.70. Lake Simcoe study area (Tsanis and Wu, 1993).

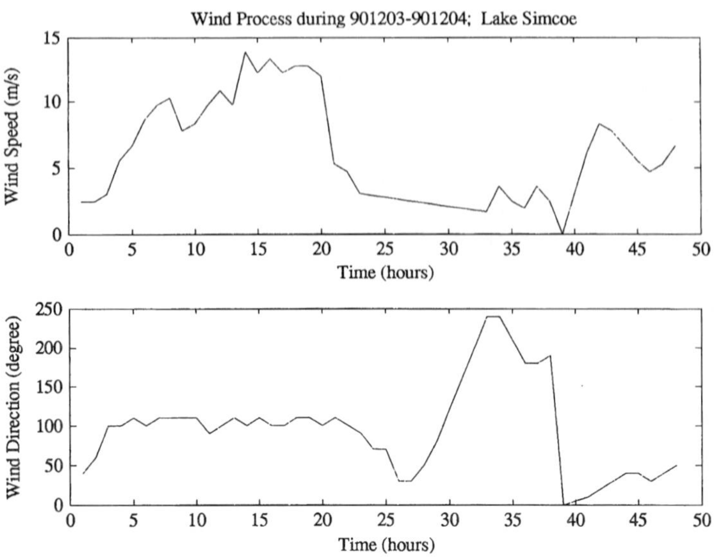

Fig. 9.71. Wind speed and direction during May 10–11, 1990 (Tsanis and Wu, 1993).

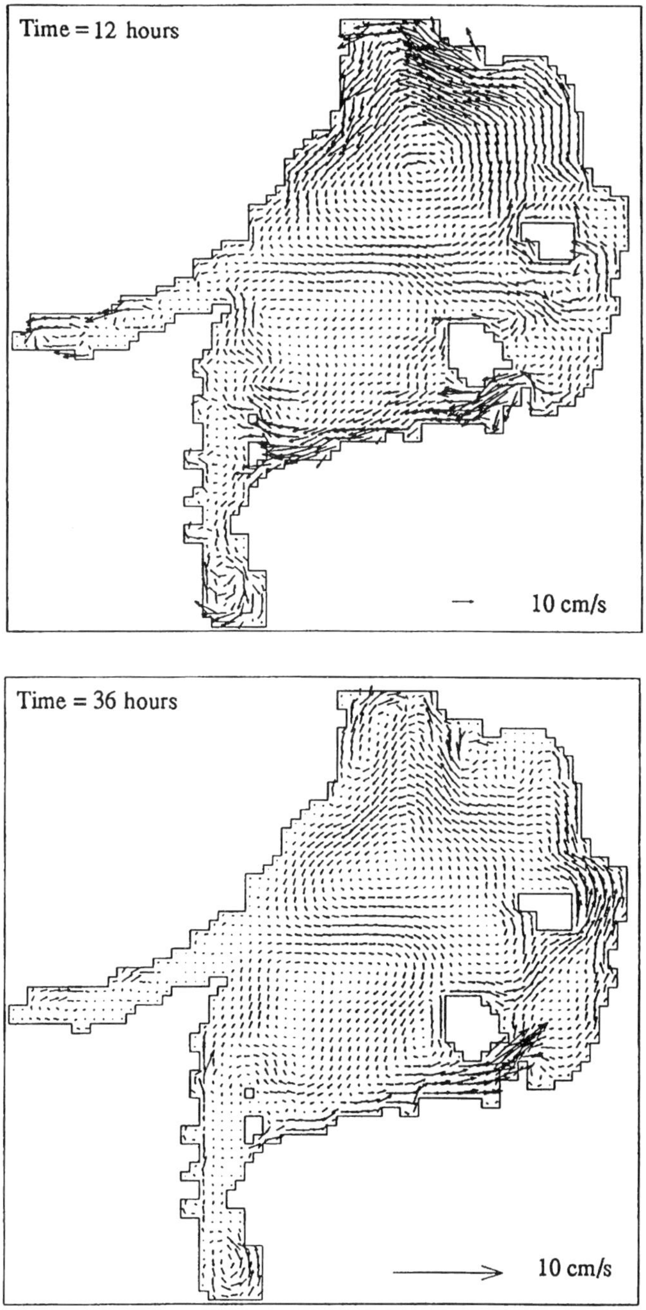

Fig. 9.72. Depth-averaged circulation in Lake Simcoe for a 10 m/s east wind (Tsanis and Wu, 1993).

averaged circulation for 12 and 36 h from the beginning of the last storm (May 17, 1990), for the east wind direction.

The currents close to the shore are in the wind direction, while the currents in the middle of the lake are in the opposite direction to the wind. The currents are stronger in the shallower parts of the lake.

Also simulated were the structure of the currents at the surface and the bottom. Figures 9.73(a) to (c) show the depth-averaged, surface and bottom circulation, 12 h from the beginning of the storm, in Kempenfelt Bay. Figures 9.74(a) to (c) show the depth-averaged, surface and bottom circulation for the same period in Cook's Bay. The changes in the wind direction are reflected in the circulation features in these bays.

The monthly mass volume flux between Kempenfelt Bay and Lake Simcoe was calculated: (a) based on monthly averaged currents; and (b) time-averaged flux over the month. The first method resulted in a flux of about 1000 m^3/s, while the second method resulted in a flux between 2000 and 2500 m^3/s. The mean current between station 1 (south station) was 0.81 cm/s from an NE direction, while the mean current from station 2 (north station) was 1.31 cm/s from a SW direction. This indicates that the mean monthly flow from Lake Simcoe is directed into Kempenfelt Bay from the north side and leaves the bay from the south side. The flow exchange

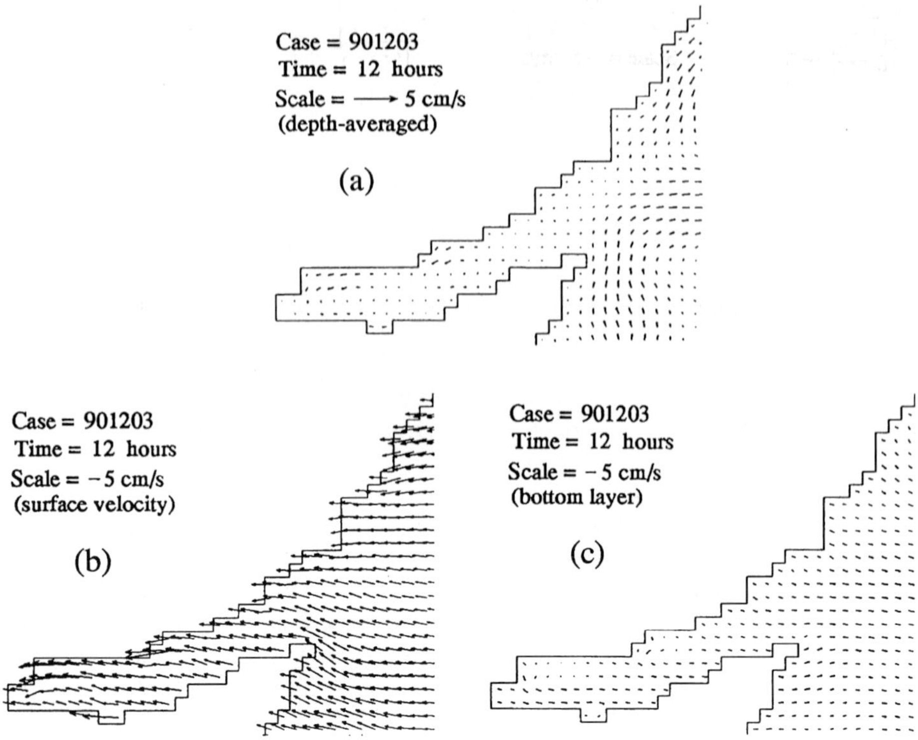

Fig. 9.73. (a–c) Depth-averaged circulation, surface currents, and bottom currents in Kempenfelt Bay, 12 h from the beginning of the storm (Tsanis and Wu, 1993).

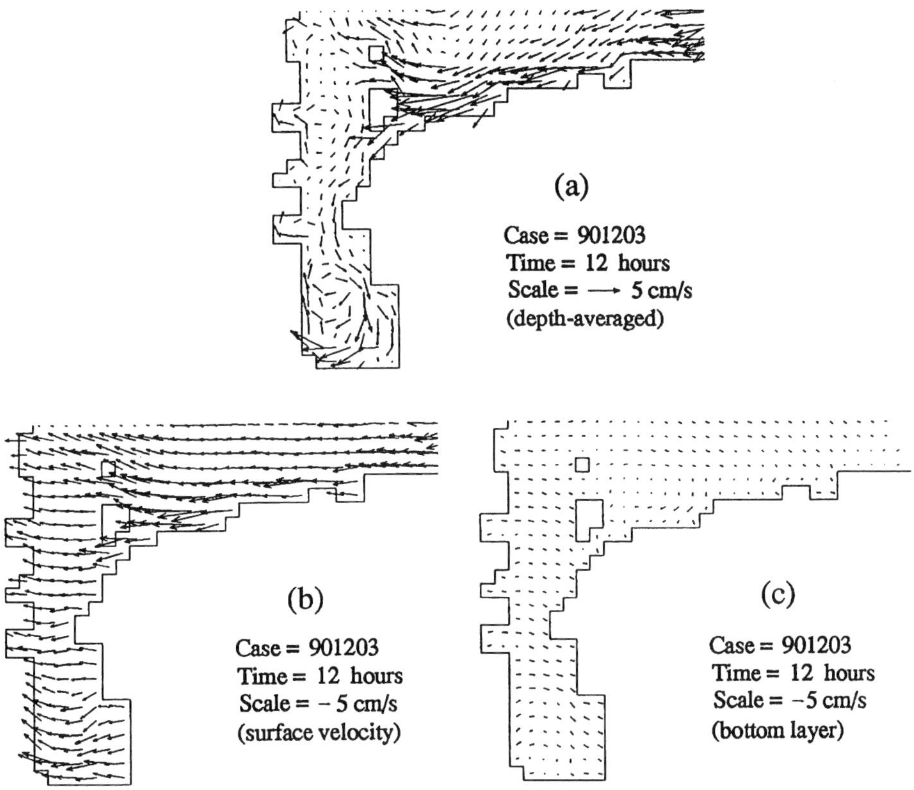

Fig. 9.74. (a–c) Depth-averaged circulation, surface currents, and bottom currents in Cook's Bay, 12 h from the beginning of the storm (Tsanis and Wu, 1993).

between Lake Simcoe and Kempenfelt Bay is approximately 2000 m^3/s (see Table 9.12).

9.8 LITTLE LAKE AND CRARY PARK MARINA STUDY[12]

This section addresses the need for remediation of contaminated sediments within Little Lake (see Fig. 9.75). This figure shows the location of a sewer pipe discharging contaminated sediments into the Crary Park Marina within Little Lake. It is suspected that, during periods of high flow, contaminated sediments are becoming dislodged and transported downstream. This scouring may be the primary factor in the movement of contaminated sediments from Peterborough to Rice Lake and eventually to the Bay of Quinte. Painter concluded that the sediments in the Crary Park Marina are probably highly susceptible to scouring

[12]Material in Section 9.8 is reproduced from Tsanis et al. (1996), copyright with permission from Environmental Technology.

Table 9.12. Mass flux between Kempenfelt Bay and Central Basin of Lake Simcoe.

(a) *Flux based on monthly averaged currents*	
Period:	13:00 2100992–07:00 181092
Data:	Total number of data = 1972
Mean currents:	Station 1–speed = 0.81 cm/s; direction = 58°
	Station 2–speed = 1.49 cm/s; direction = 251°
	(from the north)
	* $Um = u1*(h1+(h2-h1)/2)+u2*((h2-h1)/2$
	$+(h3-h2)/2)+u3*((h3-h2)/2+H-h3)$
Mass flux:	Flux = $(um1*3.28*0.65+Um2*33.0*0.35)*2700.0$
	= 932.365 m^3/s
(b) *Time-averaged mean monthly flux*	
Period = 210692–210792	Mean flux = 2497.7 m^3/s
Period = 210892–010992	Mean flux = 2031.3 m^3/s
Period = 210992–181092	Mean flux = 2008.0 m^3/s

Source: Tsanis and Wu (1993).

Fig. 9.75. Crary Marina and source of contaminants (Tsanis et al., 1994a).

during the spring freshet. He recommended further study to determine if scouring of the sediments has indeed occurred. Also, if movement of the deposit is observed, then removal of the contaminated sediment may be desirable, depending on the timetable of the clean-up at the source, even though the area may become contaminated again.

MacLaren Plansearch (1989) conducted a field study that examined current movement in Little Lake and in the Otonabee River, on August 1987 and May 1988. Survey stations, in conjunction with drogue releases, provided current data, meteorological data, and dispersion results for high and low flows typically experienced along the Otonabee River in the summer season. Also, Geo-Logic Inc. (1994) conducted a sediment survey of retrieved soil samples taken in the Crary Park Marina area of Little Lake. Their investigation identified lake and river bottom characteristics to provide baseline conditions prior to the spring freshet.

The objective of this study was to identify potential areas susceptible to scouring of contaminated sediments. The analysis was based only on the magnitude of the currents provided by the model, and the type of sediment. The study area consisted of Little Lake, from where the Otonabee River enters the Lake, downstream to Whitlaw Creek. The highest priority was the Crary Park Marina, located where the Otonabee River enters the Lake.

Little Lake was sub-divided into two sections. The first section involved the Otonabee River flowing into Little Lake near the Crary Park Marina. In this case a 4×4 m grid was established. The second section focused on the lake, excluding river inputs and for this case a 20×20 m grid was created. The effect of grid size on shoreline input information is illustrated in Figures 9.76A and B.

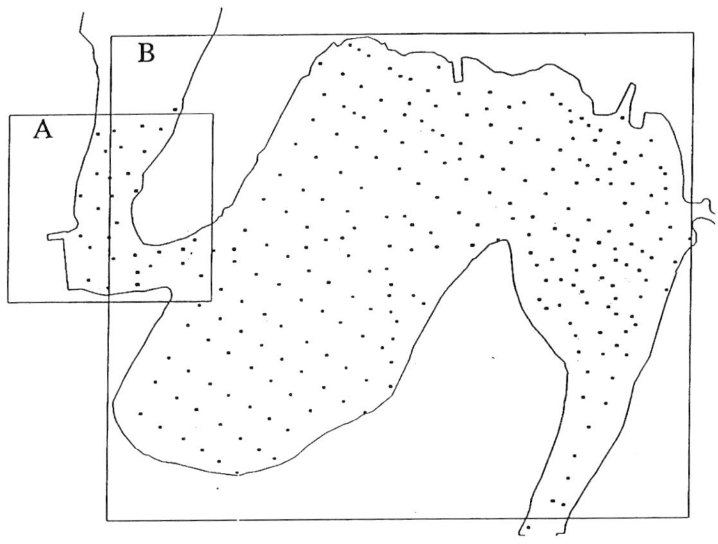

Fig. 9.76. (A, B) 4×4 and 20×20 m grids for Little Lake and Crary Marina (Tsanis, et al., 1994a).

The depth data from these grids and the gridded shorelines were used in the hydrodynamic model. The purpose of the hydrodynamic models was to simulate the current distribution within the Crary Park Marina, Otonabee River, and Little Lake. This information, coupled with information obtained by Geo-Logic Inc.'s field survey (1994), was used to determine the possibility and extent of scouring.

A 2D hydrodynamic model was used to simulate the combined hydraulic and wind-induced circulation in Little Lake and in the Crary Park Marina, for certain flows. However, the very high flows of the 100 year rain event on the Otonabee River produced certain problems for the quasi-3D circulation model. In this case, the hydrodynamic modelling necessitated the use of a less-sophisticated, but accurate depth-averaged, 2D circulation model.

9.8.1 Circulation Patterns in Little Lake

The finite difference method was used for the hydrodynamic models of Little Lake and the Crary Park Marina. High-resolution grids are used for both areas in order to represent the detailed shorelines and provide the detailed circulation pattern. The models were verified using data from the study conducted by MacLaren Plansearch (1989).

A 20 m grid was used for Little Lake, as shown in Figure 9.77. Figure 9.78 shows the depth contours in the model and a typical wind speed of 5 m/s wind was used in all the simulations. A mean annual flow condition for the Otonabee River was taken as 55 m^3/s.

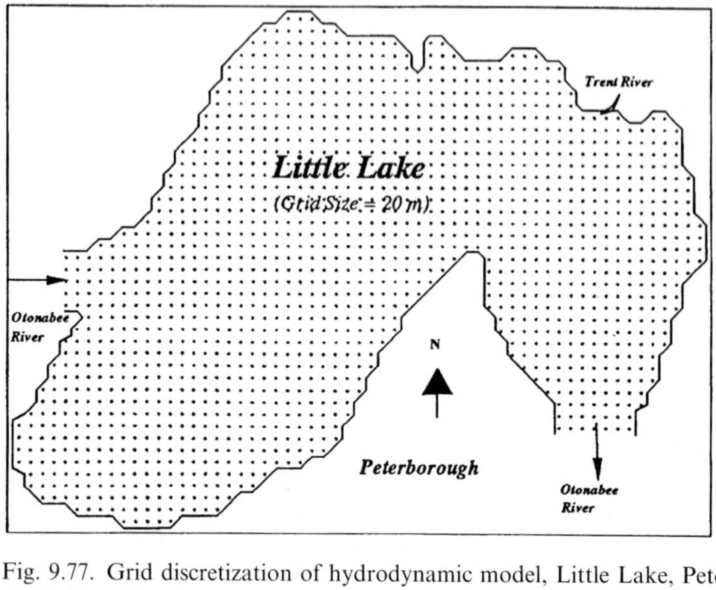

Fig. 9.77. Grid discretization of hydrodynamic model, Little Lake, Peterborough. The grid size is 20 m (Tsanis et al., 1996).

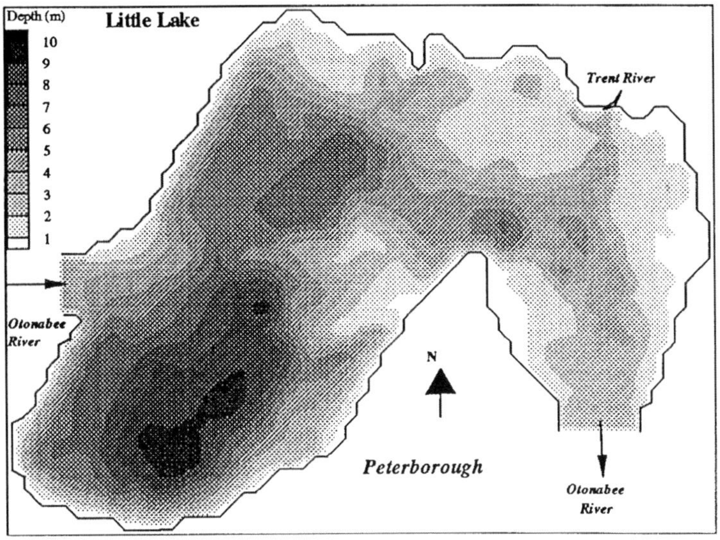

Fig. 9.78. Depth contours in the hydrodynamic model of Little Lake, Peterborough (Tsanis et al., 1994a).

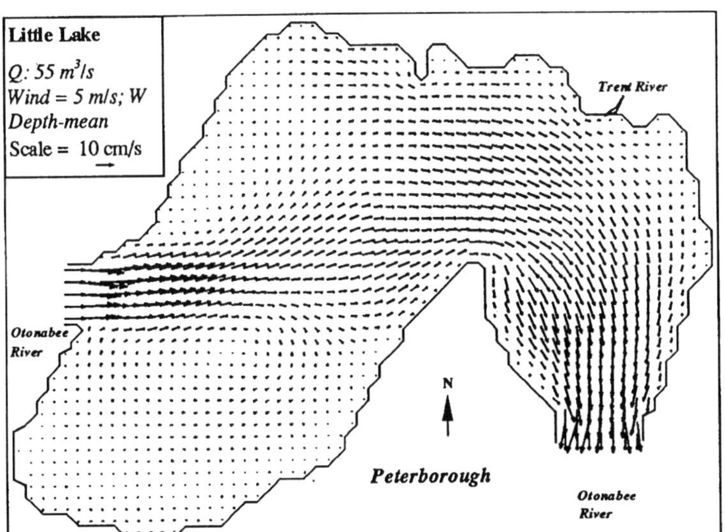

Fig. 9.79. Simulated depth-mean circulation pattern in Little Lake, Peterborough.

Figure 9.79 presents the depth-averaged circulation patterns in Little Lake under $55\,m^3/s$ flow conditions. An example of multi-layered circulation patterns for a 5 m/s west wind and $55\,m^3/s$ river flow is presented in Figures 9.80. The streamlines in Figure 9.81 demonstrate drifter movement under certain circulation pattern.

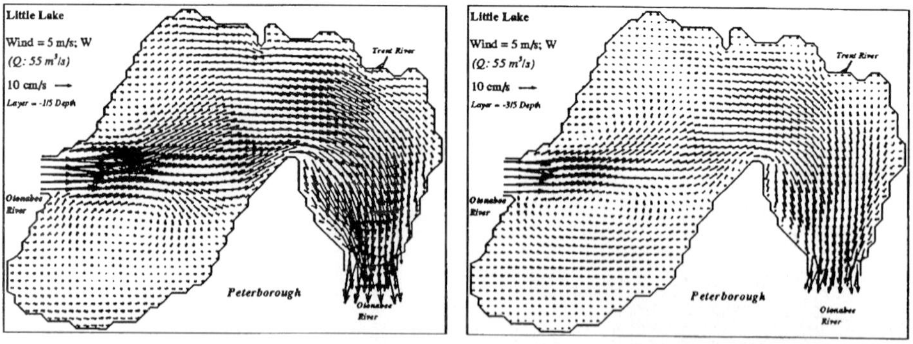

Fig. 9.80. Simulated circulation pattern at various layers in Little Lake, Peterborough. The discharge of Otanabee River is $55\,m^3/s$. The wind speed is 5 m/s from the west (Tsanis et al., 1994a).

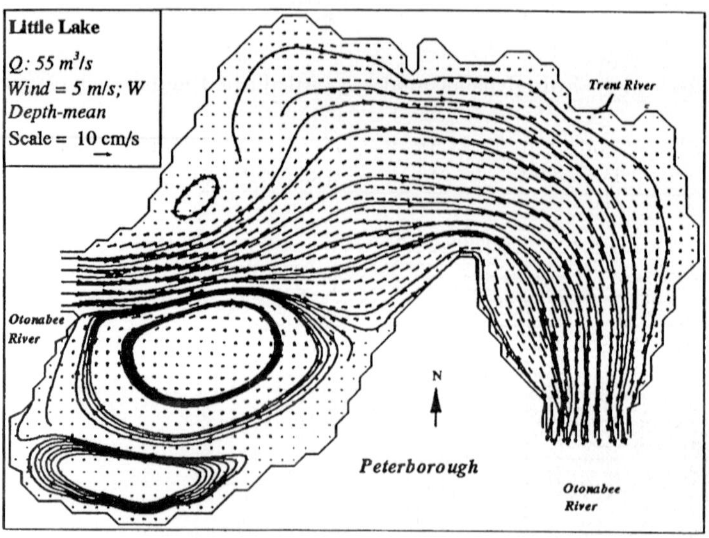

Fig. 9.81. Streamlines in the depth-mean circulation pattern in Little Lake, Peterborough. The discharge of Otonabee River is $55\,m^3/s$. The wind speed is 5 m/s from the west (Tsanis et al., 1994a).

9.8.2 Circulation Patterns in the Crary Park Marina

A 4 m grid was used for the Crary Park Marina, as shown in Figure 9.82, which also shows the station locations where sediment grain size analysis was conducted. Figure 9.83 shows the depth contours in the model. A typical wind speed of 5 m/s wind was used in all of the simulations and a $55\,m^3/s$ flow condition used for the Otonabee River. The discharge of Jackson Creek was taken as $20\,m^3/s$.

Figure 9.84 presents the depth-averaged circulation patterns in the Crary Park Marina. An example of multi-layered circulation patterns for a 5 m/s west wind and

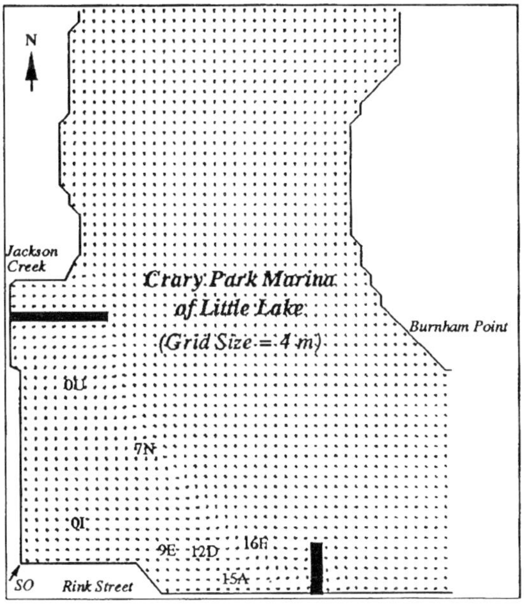

Fig. 9.82. Grid discretization of Crary Park Marina in Little Lake, Peterborough and sediment sampling locations. The grid size is 4 m (Tsanis et al., 1994a).

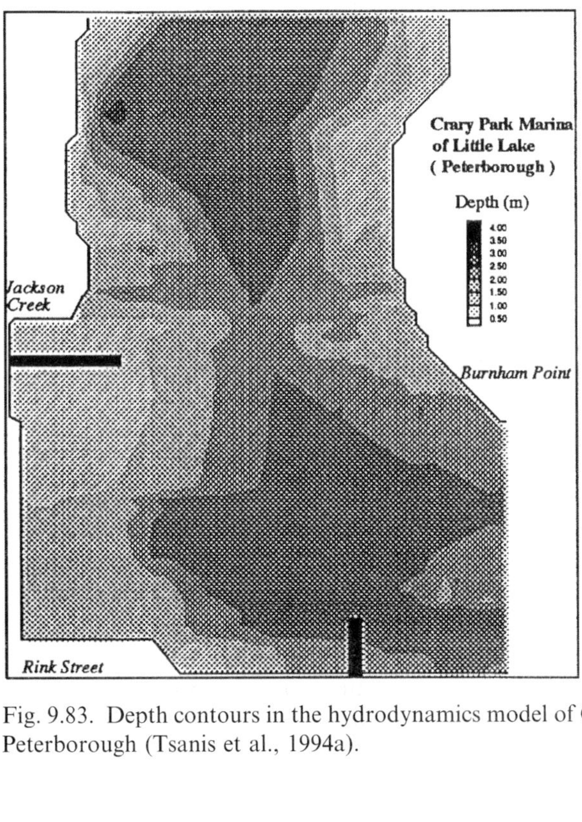

Fig. 9.83. Depth contours in the hydrodynamics model of Crary Park Marina of Little Lake, Peterborough (Tsanis et al., 1994a).

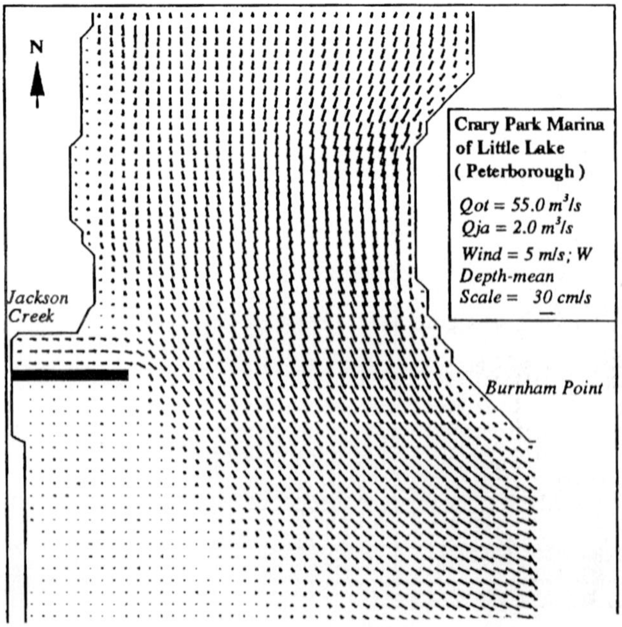

Fig. 9.84. Simulated depth-mean circulation pattern in Crary Park Marina in Little Lake, Peterborough. The discharge of Otonabee River is 55 m^3/s and the discharge of Jackson creek is 20 m^3/s. The wind speed is 5 m/s from the west (Tsanis et al., 1994a).

a 55 m^3/s river flow rate is presented in Figures 9.85. The streamlines in Figure 9.86 demonstrate drifter movement under a certain circulation pattern. Figure 9.87 presents the contours of bottom shear velocity under a 55 m^3/s flow condition.

Figure 9.88 presents the depth-averaged circulation pattern under a 10 m/s wind. The lower diagram in the figure demonstrates the detailed current structure in the southwest corner.

9.8.3 Analysis of Sediment Movement

The potential for sediment is primarily determined by the sediment size and the bottom shear velocity. The Shield's diagram is a useful tool for determining the onset of sediment motion. Its use, however, rests on the designation of a characteristic sediment size for a given sample. Simons and Senturk (1992) indicate that Shields used the median sediment size from a cumulative sediment size distribution curve, or D_{50}, when constructing the initiation of motion curve.

They also indicate that more recent investigators pinpoint D_{35} as the characteristic grain size of a sample. Since analysis of a smaller sediment size would produce more conservative results, both sizes are considered in this section. These characteristic sizes are shown below, in Table 9.13, for the seven stations where Geo-Logic Inc. (1994) conducted sediment grain size analysis (see Fig. 9.75).

The analysis from the hydrodynamic modelling provided the depth-averaged velocity components u and v for each of the seven stations. The total depth averaged

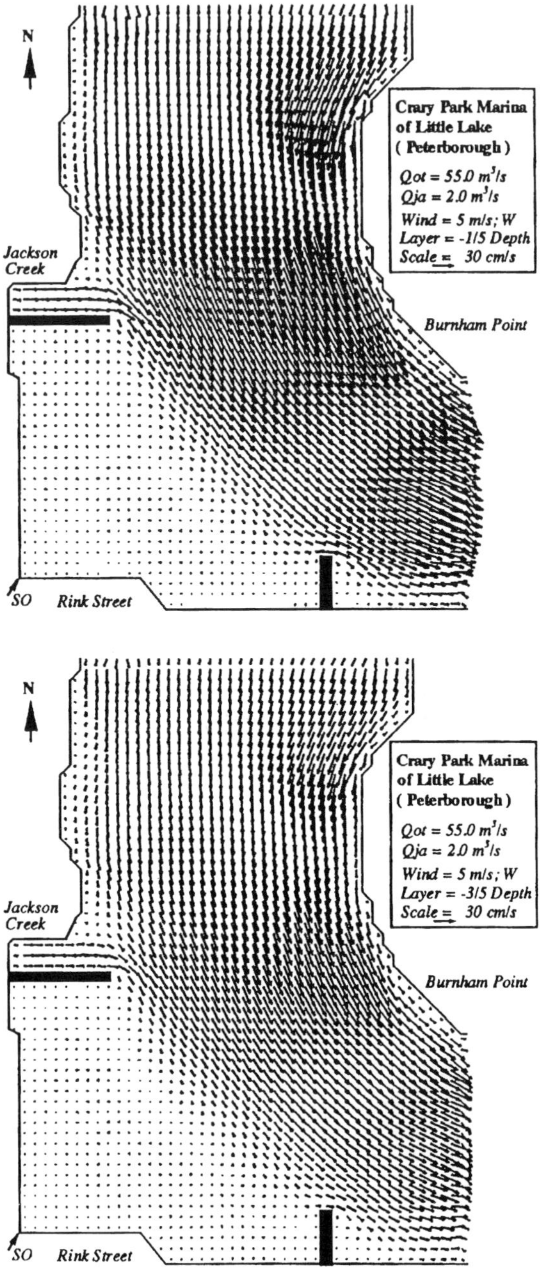

Fig. 9.85. Simulation circulation pattern at various layers, Crary Park Marina of Little Lake, Peterborough. The discharge of Otonabee River is $55\,m^3/s$ and the discharge of Jackson Creek is $2\,m^3/s$. The wind speed is 5 m/s from the west (Tsanis et al., 1994a).

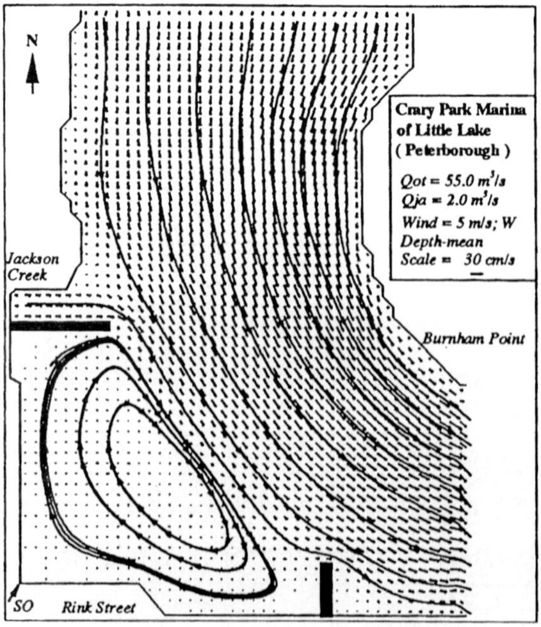

Fig. 9.86. Stream lines in the simulated depth-averaged circulation pattern, Crary Park Marina of Little Lake, Peterborough. The discharge of Otonabee River is $55\,m^3/s$ and the discharge of Jackson Creek is $2\,m^3/s$. The wind speed is 5 m/s from the west (Tsanis et al., 1996).

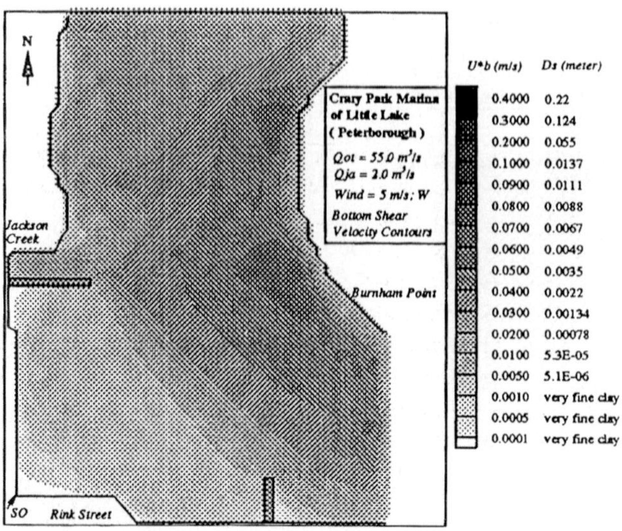

Fig. 9.87. Shear velocity contours near the bottom (U_*), Crary Park Marina of Little Lake, Peterborough. The discharge of Otonabee River is $55\,m^3/s$ and the discharge of Jackson Creek is $2\,m^3/s$. The wind speed is 5 m/s from the west (Tsanis et al., 1996).

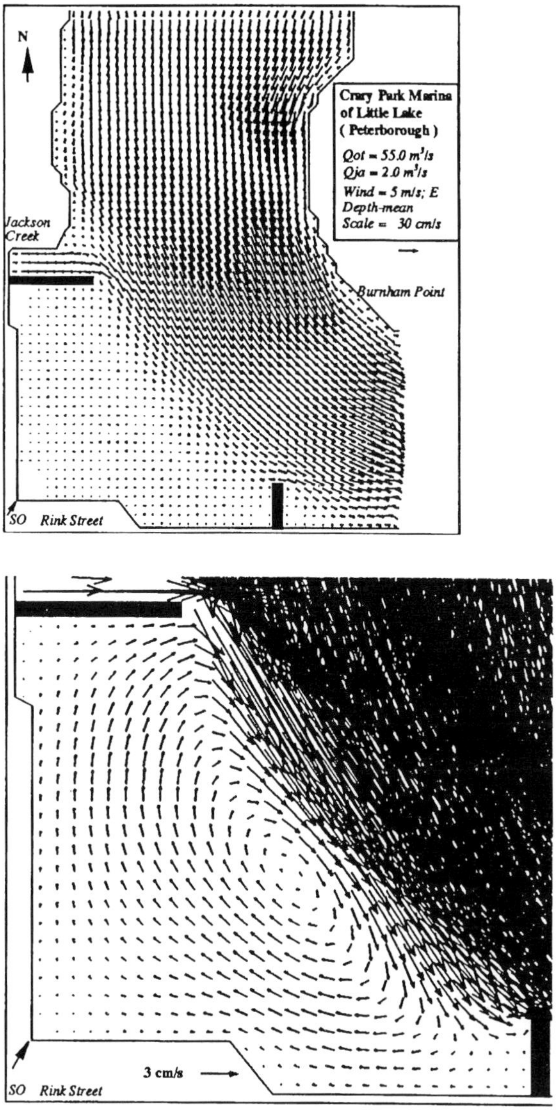

Fig. 9.88. Simulated depth-mean circulation pattern in Crary Park Marina in Little Lake, Peterborough. The discharge of Otonabee River is $55\,m^3/s$ and the discharge of Jackson Creek is $2\,m^3/s$. The wind speed is 5 m/s from the east. The lower figure shows the detailed current in the southwest corner of the Marina (Tsanis et al., 1994a).

velocity U is simply the square root of the sum of the squares of these components. The bottom shear velocity U^* was determined from the following equation:

$$U^* = \sqrt{\frac{U^2 n^2 g}{H^{1/3}}} \tag{9.16}$$

Table 9.13. Characteristic sediment sizes.

Grid	D_{35} (μm)	D_{50} (μm)
0I	300	450
0U	340	625
7N	190	260
9^E	150	200
12D	75	175
15A	850	2400
16F	350	425

Table 9.14. Average velocities for a 5 m/s wind.

Station	Depth (m)	u (cm/s)	v (cm/s)	U (cm/s)
0I	1.63	−0.14	0.50	0.5
0U	1.30	1.12	1.48	1.9
7N	2.49	0.65	2.02	2.1
9E	1.93	−1.84	1.33	2.3
12D	2.13	−3.09	1.50	3.4
15A	1.54	−1.49	0.16	1.5
16F	2.61	−2.99	1.18	3.2

The bottom shear stress can be determined from the bottom velocity with the following equation:

$$\tau_c = \rho U_*^2 \tag{9.17}$$

The density of fine sand is typically 2.64 g/cm^3 and the specific weight (γ_s) is approximately 26,000 N/m^3. If the point determined by the two equations

$$Y_{cr} = \frac{\tau_c}{(\gamma_s - \gamma)D_s} \tag{9.18}$$

$$X_{cr} = \frac{U_* D_s}{\nu} \tag{9.19}$$

lies above the Shields curve, then that sediment size will be scoured.

The analysis was conducted for the 55 m^3/s flow rate. Since the Manning's n used in Eq. (9.16) is unknown, a range of Manning's n from 0.035 to 0.02 was considered.

9.8.4 Results and Discussion

Table 9.14 below lists the average velocity components that were transferred to this analysis from the hydrodynamic modelling in Section 4. U is the average velocity over the entire depth for each of the four flowrates.

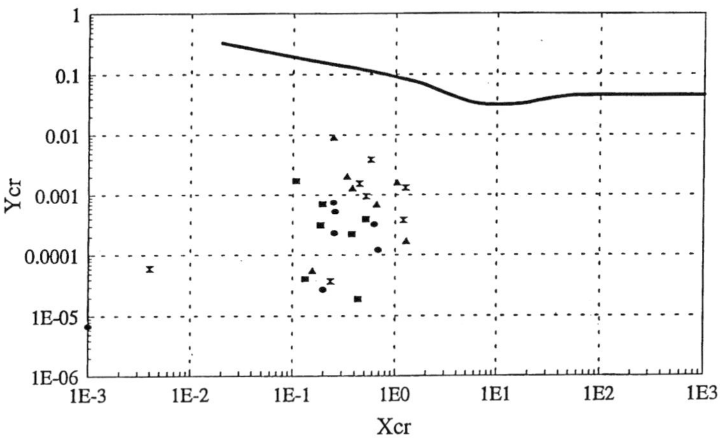

Fig. 9.89. Shield's diagram and placement of D_{35} and D_{50} characteristic sediment sizes. Manning's $n = 0.035$ (Tsanis et al., 1994a).

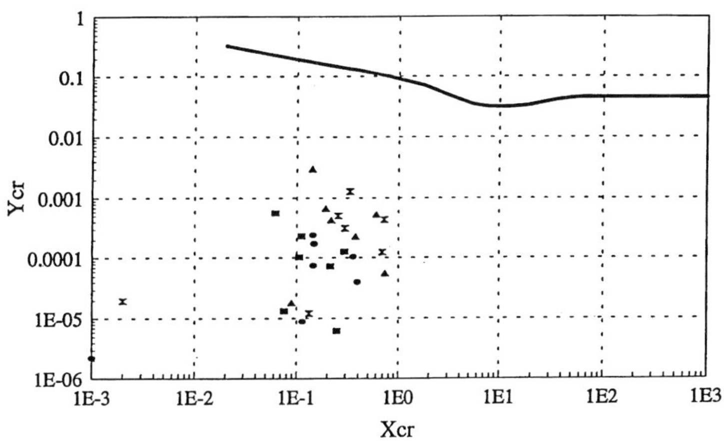

Fig. 9.90. Shield's diagram and placement of D_{35} and D_{50} characteristic sediment sizes. Manning's $n = 0.02$ (Tsanis et al., 1994a).

Figures 9.89 and 9.90 show the seven stations on the Shield's curve for the $55\,m^3/s$ flowrate and both characteristic sediment sizes, for the two values of Manning's *n*, respectively. The figures show that there is no movement within that range of Manning's *n* for the range of typical spring freshet flows in the Otonabee River. The values of ordinate and abscissa for each of the seven stations are, for all intents and purposes, far from the Shield's curve.

Figure 9.91 shows the effect of an increase in westerly wind from 5–10 m/s, used by the hydrodynamic model for a flow rate of $55\,m^3/s$. The figure plainly shows that the wind magnitude increase has only a marginal effect on the possibility of transport. The higher wind could possibly produce sediment movement at those stations which border the Shield's curve, for certain Manning's *n*.

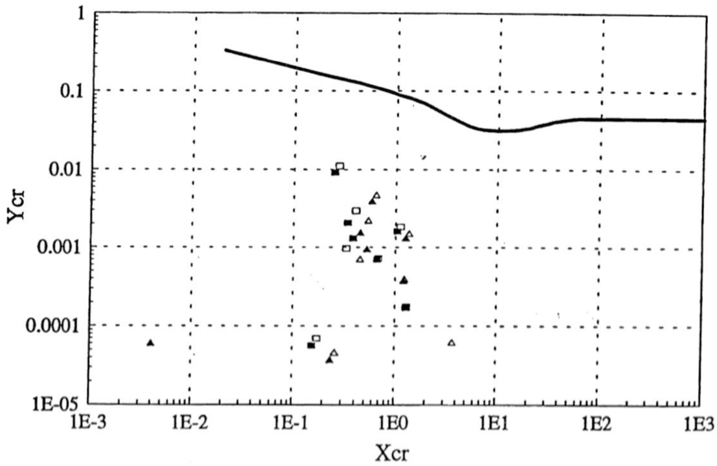

Fig. 9.91. Comparison between 10 and 5 m/s westerly wind rates on scouring (Tsanis et al., 1994a).

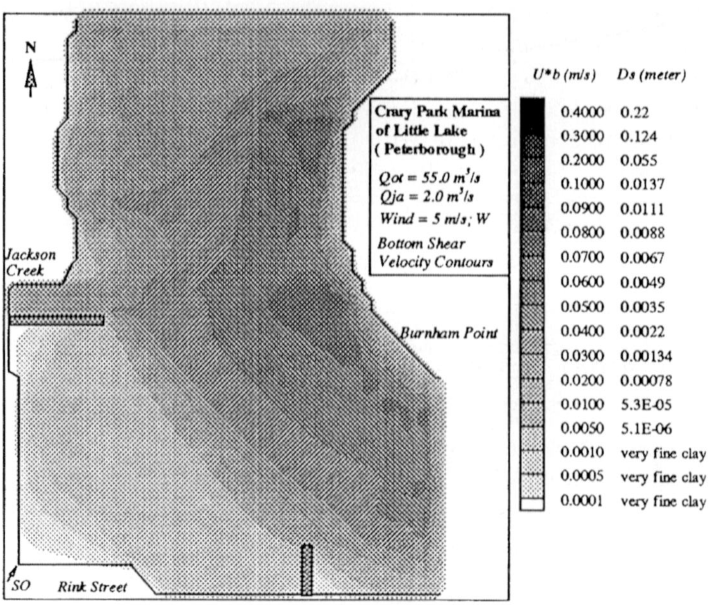

Fig. 9.92. Scouring contours for a flowrate of 55 m^3/s (Tsanis et al., 1994a).

The figures imply that, under typical annual conditions, there is no transport at any of the sampling station areas, which in turn implies that there is no scouring in the Crary Park Marina area under typical annual conditions.

A second approach considered the variation in sediment size in determining the potential for sediment transport. Figure 9.92 shows contours of sediment sizes that would be scoured under the flowrate detailed in the figure. For the two low flows, only the very fine sediment is re-suspended, and this sediment size is much smaller than any of the characteristic grain sizes. The figures show that the majority of sediment transport occurs in the trunk section of the river.

The bottom friction coefficient, or Manning's n, and horizontal eddy viscosity, are major parameters in the circulation model which will affect the magnitude of the current. Although model verification with field data is necessary to determine these parameters, those used in the present model are based on the literature. The bottom friction coefficient is given as 0.0025, which is broadly used for shallow waters. According to Fischer et al. (1979), the horizontal mixing coefficient varies from 0.12 for large smooth rivers to 1.1 m^2/s for large rough rivers. This coefficient for smooth laboratory open channel models is about 0.001 m^2/s, while for irrigation channels it is 0.01 m^2/s.

The Crary Park Marina area is a short reach in a small river. The horizontal eddy viscosity used in the model (0.16 m^2/s, where $t_h = 0.992\,m^2/s$) is reasonable for the river section. For the relatively closed southwest corner of the basin, this coefficient may be smaller; A test run is performed in which the horizontal eddy viscosity is 0.06 m^2/s ($t_h = 0.997\,m^2/s$). The current magnitudes increase by 1–1.5 cm/s in the corner and the influence of this increase on potential scouring is presented in Figure 9.93. For Figure 9.93, the flowrate is 55 m^3/s, the wind is 10 m/s, the Manning's n is 0.035, and the characteristic grain size is D_{35}. These choices of wind and Manning's n, etc., should produce the greatest possibility of scouring for the high annual flow. Both the table and the figure indicate that no sediment transport should occur with the decrease in horizontal eddy viscosity.

This study concludes that, based on the 3D and 2D models used in the hydrodynamic modelling of the Crary Park Marina, there is no sediment transport

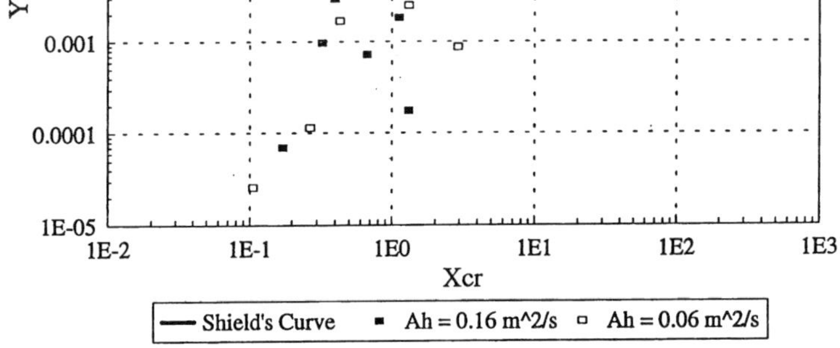

Fig. 9.93. Effect of varying the horizontal eddy viscosity on scouring. Flowrate = 55 m^3/s, west wind = 10 m/s, $n = 0.035$, and characteristic sediment size is D_{35} (Tsanis et al., 1994a).

in the immediate vicinity of the sewer pipe that discharges contaminated sediments, for typical annual flowrates of the Otonabee River (17–55 m^3/s).

For high flow rates, such as those associated with the 100 year return storm and the regional storm, some scouring may occur. In this case, it is recommended that the piers be retained in the marina since they impede fluid flow and prevent washout, which may increase the potential for scouring. Also, the potential for washout due to local inflows into the marina (from point and non-point sources other than the sewer outfall discharging the contaminants) should be investigated, since they may contribute to sediment transport.

Other manufactured phenomena may also contribute to scouring, such as the extensive action of boaters in the marina. Boat propellers can generate vortices that can propagate to the bottom and cause high enough shear stresses to re-suspend sediment. This phenomenon however is outside the hydrodynamic model's predictive ability. The model only considers fluid inflow, outflow, fluid input from the contaminant source, wind effects, and tributary inputs for its analysis.

Further verification of the hydrodynamic models is required in order to obtain more conclusive evidence concerning areas subject to scouring. The results obtained depend on the coefficients and parameters used in the model and these variables were chosen from the literature. The simple reason for this is that every situation is unique in hydrodynamic modelling, and it is only possible to verify the choice of parameters used by conducting a field study of the area in focus. Furthermore, the model at present cannot consider disturbances in the fluid flow created by boating activity, non-point sources and other outfall inflows which may affect fluid speeds close to the shore. A field study would collect information that could allow the model to incorporate the effects of these phenomena by adjusting parameter and coefficient values accordingly.

Chapter 10

MODEL APPLICATIONS TO OTHER LAKES AND COASTAL WATERS

10.1 MODEL APPLICATION IN GRETAN PELAGOS (MEDITERRANEAN SEA)

10.1.1 Northern Crete Waterfront

The Mediterranean Sea lies between Europe, Asia, and Africa, and, excluding the Black Sea, it covers about 2.5 million km^2, with an average depth of about 1.5 km and a volume of 3.7 million km^2 (Jeftic, 1990). It connects with the Atlantic Ocean through the Strait of Gibraltar. This narrow (14 km) and shallow (330 m) strait isolates the deeper waters from a direct exchange with the Atlantic (Jeftic, 1990).

The Mediterranean Sea consists of a series of interacting parts and adjacent seas, with two major basins, Western and Eastern. The Western Mediterranean (~0.85 million km^2) contains the Alboran Sea, the Algero-Provencal Basin, the Ligurian Sea, and Tyrrhenian Sea. The Eastern Mediterranean (~1.65 million km^2) includes the Adriatic Sea, the Ionian Sea, the Aegean Sea, and the Levant (Jeftic, 1990).

The physical aspects of the Mediterranean, that is, circulation, sea levels, tides, currents through straits, the dynamics of the Western and Eastern Mediterranean, etc. are given in a special issue of the Dynamics of Atmospheres and Oceans Journal (Malanotte-Rizzoli and Robinson, 1991).

The Eastern Mediterranean consists of two major basins, the Levant, which connects with the Aegean Sea in the north through a number of straits, and the Ionian Sea, which connects to the Adriatic Sea in the north through the strait of Otranto. The total water column below the seasonal thermocline is comparatively homogeneous, with temperatures between 13.3 and 13.7°C and salinity between 38.70 and 38.75%. A layer of greater salinity, 38.8–39%, extends from the entrance to the Sicilian passage, at depths of 400 m, and breaks out at the surface in the Levant near the island of Rhodes. This is called Levantine intermediate water. Atlantic water of salinity less than 38% moves eastward at the surface near the African coast and sinks beneath the surface after passing Libya. It continues to pervade the eastern waters past Libya, as a thin stratum at 100 m depth (Ketchum, 1983).

The Aegean Sea is characterized by a strong, deep seasonal overturn (Miller, 1972), that reduces the surface salinity from greater than 39% (winter stirring and convection) to 38.9% (a characteristic salinity at great depths). Water of lower salinity, 38.7%, enters the Aegean through the connecting straits at mid-depths. The deep-water oxygen content of the Aegean Sea is greater than 5 mL/L, indicating a rapid and deep convective phenomena.

Evidence of the effect of Aegean water on the deep water of the Eastern Mediterranean was not found in 1951 by Miller (1972), but in 1962, when the research vessel *Atlantis* made a section south of the Strait of Scarpanto, and observed Aegean water at depths of 1300 m. This water had the characteristics of deep Aegean water, with 14.3°C temperature, and a salinity of 38.9% found on the northern side of the Strait. Nearby stations supported this identification of an overflow.

An interdisciplinary program called the "Mediterranean Targeted Project" (MTP), is underway (period 1993–1995) to better the understanding of physical processes and biogeochemical cycles in the Mediterranean Sea. Tasks are distributed among three major research themes: (a) general circulation studies; (b) biogeochemical studies; and (c) ecosystem analysis. One of the sub-projects, with acronym "PELAGOS", will study the hydrodynamics and biogeochemical fluxes in the straits of the Cretan Arc (Aegean Sea, Eastern Mediterranean Basin). The general objectives of PELAGOS are to improve knowledge of the hydrodynamics of the South Aegean Sea and, in particular, the Straits of the Cretan Arc, and to estimate the exchanges of water and of dissolved and particulate matter between the Aegean Sea and the adjacent open sea region of the Eastern Mediterranean Basin.

The circulation in the nearshore of Northern Crete, which is the subject of this study, is determined by the wind, inflows and tides. The tidal effect is very small and can be neglected (Environmental impact of the sewage disposal in Iraklion Bay, 1993). The inflows from creeks and sewage treatment plants affect the circulation only within a small distance from the source. Their influence will be studied in high-resolution local circulation studies, that is, a grid size of 100 m. In the general circulation modelling studies that cover the northshores of Crete, a grid size is of the order of 2 km is selected. The only remaining significant factor is the wind. Measurements from the meteorological station of Iraklion indicate that the prevailing winds are from the north, northwest, and south directions.

The Princeton oceanic model (POM) (Blumberg and Mellor, 1987) and the hydrodynamic model IDOR (Tsanis and Shen, 1994a) are applied to the nearshore areas of Northern Crete. In this study, the 2D hydrodynamic components of the POM and IDOR models are used to describe the depth-averaged currents and transports in this nearshore area. The simulation run included a wind speed of 10 m/s, from two directions, west and northwest. This information is crucial to evaluating the impact on receiving water from four sub-merged outfalls discharging into the bays of Chania, Rethymno, Iraklion, and Agios Nikolaos.

10.1.2 Princeton Model versus IDOR Model

The common characteristics of the POM and IDOR models are that they use: (a) fully 3D primitive equations (non-linear) and depth-averaged equations (2D option); (b) shallow water assumption and Boussinesq approximation; (c) (k–l) turbulence model or constant eddy viscosity; and (d) Arakawa-C staggered grid and Cartesian coordinates. The POM model also uses σ coordinates, an implicit scheme for vertical mixing and a leap-frog scheme for time differencing. The IDOR model uses the control volume method, an implicit time scheme for the barotropic mode (SIMPLER method solved by SOR method), the Adams–Bashforth scheme for time differencing and the Donor-cell/Quadratic Upstream Interpolation for Convection Kinematics (QUICK) scheme for the advective terms. Although the two 3D models use different numerical schemes, the numerical schemes used for the 2D components (external mode) of the two models are the same. The only difference between the external modes of the POM and IDOR model is that the external mode equations of the POM and IDOR are expressed by vertically integrated velocity and water transport per unit width, respectively. This difference in formulation of governing equations will give different results when taking the average of velocity or flux in the discretization of the governing equations. When the bottom is flat, this formulation by velocity and flux gives the same results. The steeper the bottom topography, the larger the difference between the formulations by velocity and flux (Tsanis et al., 1994b).

Figure 10.1 shows a map of Greece indicating the study area. A morphological map of Crete is shown in Figure 10.2, in which the elevation contour lines are plotted in 200 m intervals. Figure 10.3 shows the map of Crete Island indicating the areas of environmental concern (Chania Bay, Rethymno Bay, Iraklion-Malia Bays, and Agios Nikolaos Bay). Figure 10.4 shows the spot depth location in Crete and Figure 10.5 shows the bathymetry contours in Cretan Pelagos. The grid discretization used was 1.2 ft. (1826 m in longitude and 2219 m in latitude) and there were $135 \times 44 = 5940$ grids for the study area. The 2D hydrodynamic components of the POM and IDOR model were used to simulate the wind-induced circulations under different environmental conditions. Three open boundaries were

Fig. 10.1. Map of Greece indicating the study area (Tsanis and Shen, 1994a).

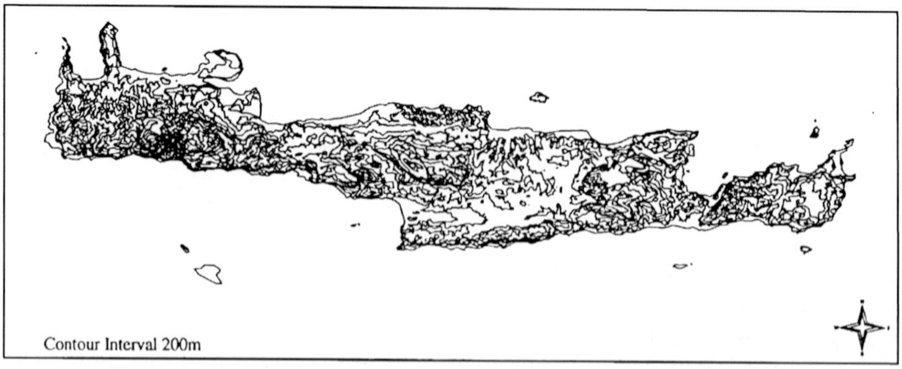

Fig. 10.2. Morphological map of Crete (Tsanis and Shen, 1994a).

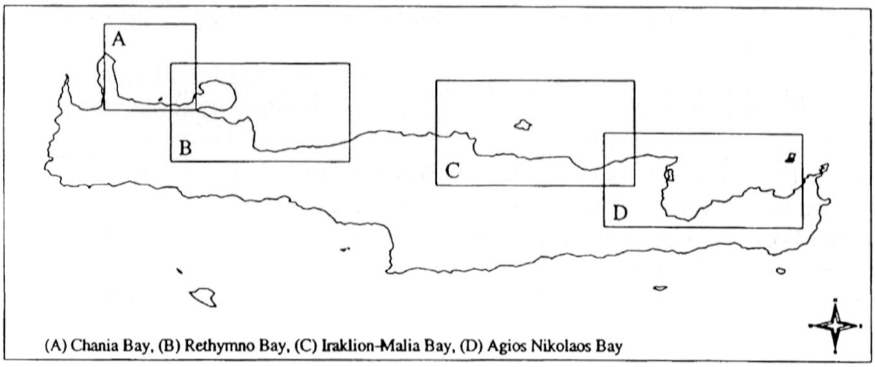

Fig. 10.3. Map of Crete indicating areas of environmental concern (Chania Bay, Rethymno Bay, Iraklion-Malia Bays, and Agios Nikolaos Bay) (Tsanis and Shen, 1994a).

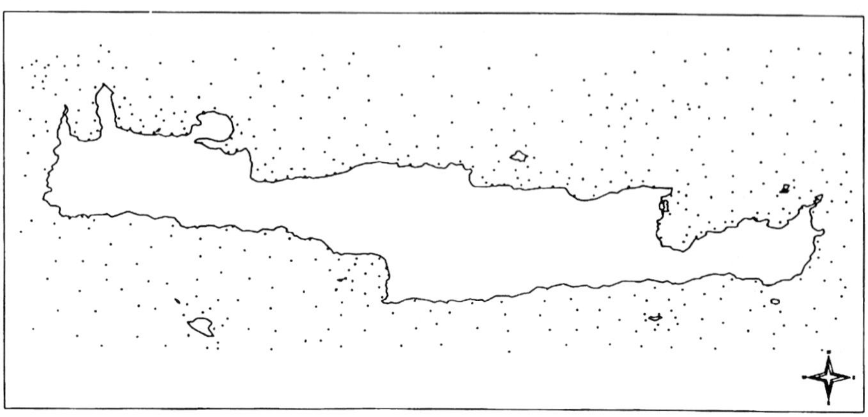

Fig. 10.4. Spot depth locations in Crete (Tsanis and Shen, 1994a).

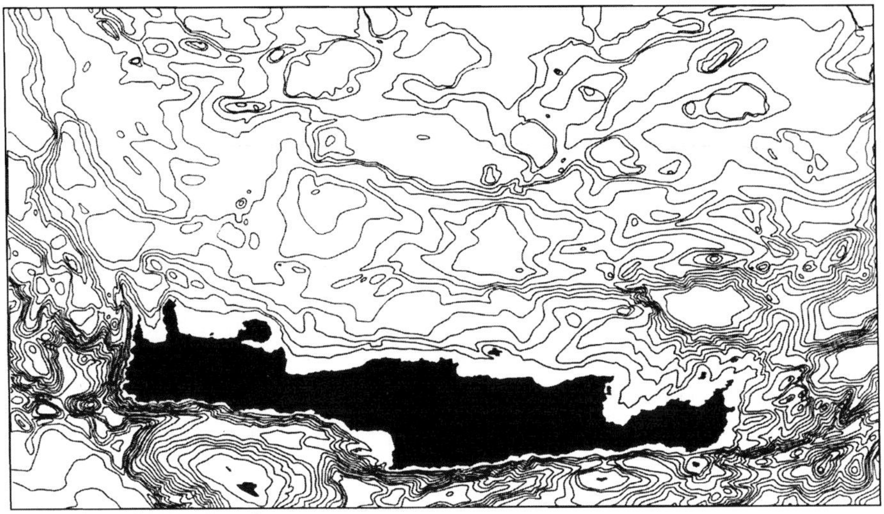

Fig. 10.5. Bathymetry contours of northern Crete (Tsanis and Shen, 1994a).

defined at the north, east, and west sides of the study area. Since no field data at the open boundaries were available, free radiation open boundary conditions were used. The horizontal eddy viscosity was chosen as $25\,m^2/s$ for both models. The simulated results for all cases are structured in different groups as follows:

(a) Depth-averaged velocities in Northern Crete for a 10 m/s west wind by the Princeton model.
(b) Depth-averaged velocities in Northern Crete for a 10 m/s west wind by the IDOR model.
(c) Total kinetic energies in Northern Crete for a 10 m/s west wind by the Princeton and IDOR models.
(d) Depth-averaged velocities in Area A for 10 m/s west wind by the Princeton and IDOR models.
(e) Depth-averaged velocities in Area B for 10 m/s west wind by the Princeton and IDOR models.
(f) Depth-averaged velocities in Area C for 10 m/s west wind by the Princeton and IDOR models.
(g) Depth-averaged velocities in Area D for 10 m/s west wind by the Princeton and IDOR models.

Figures from groups (a) and (b) (Figures 10.6 and 10.7) reveal that both models gave similar depth-averaged velocity patterns, but the magnitude differed significantly in the areas of steep bottom topography. For the west wind, strong currents along the shoreline occurred in the nearshore areas, and very small currents in the central areas.

In Figure 10.8, the total kinetic energy for the west wind obtained by the two models are compared. The total kinetic energy was found to be up to 30% different with the POM being higher and the IDOR model reaching steady state faster than

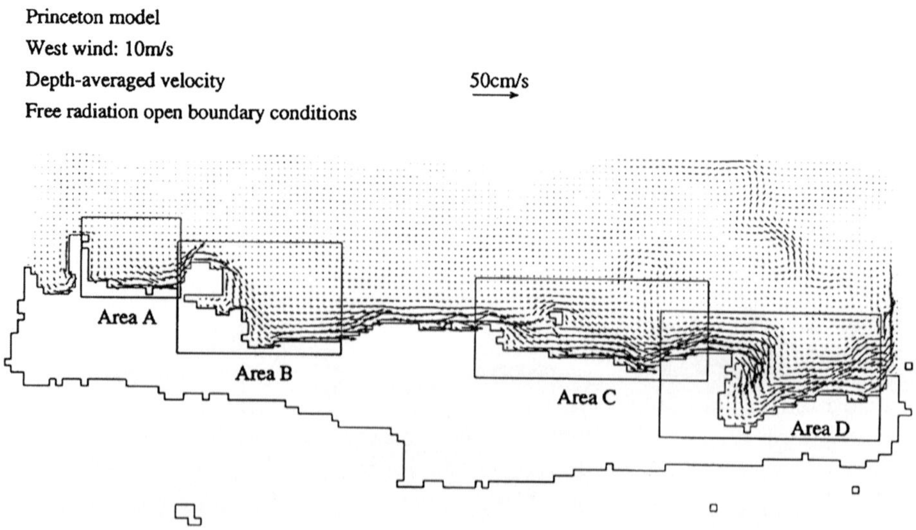

Fig. 10.6. Depth-averaged velocities in northern Crete for a 10 m/s west wind by the Princeton model (Tsanis and Shen, 1994a).

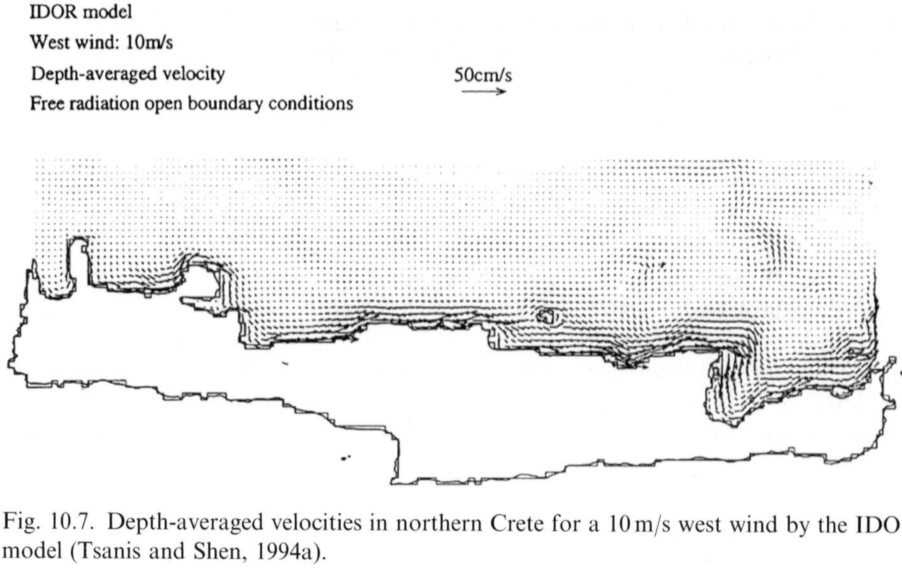

Fig. 10.7. Depth-averaged velocities in northern Crete for a 10 m/s west wind by the IDOR model (Tsanis and Shen, 1994a).

POM. The difference in total kinetic energy between the two models is largest for the west wind and increases with wind speed. This difference can be attributed to the different formulation of the governing equations. When the bottom topography is not steep, this difference is small. But in the case of very steep bottom topography, as in areas of the Cretan Pelagos, this difference is large, because there are areas with bottom slopes of up to 50%.

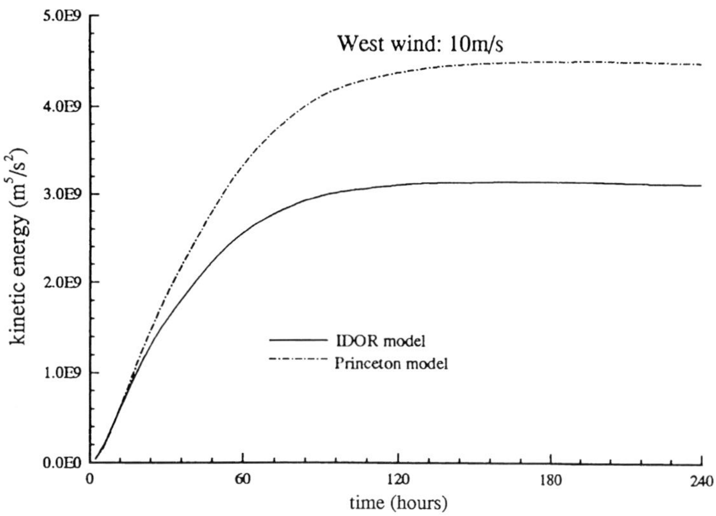

Fig. 10.8. Total kinetic energy in northern Crete for a 10 m/s west wind by the Princeton and IDOR models (Tsanis and Shen, 1994a).

Figure 10.9 illustrates comparisons in the simulated depth-averaged velocities in Area A (Chania Bay) of Northern Crete, between the POM and IDOR models. For the west wind case, a counterclockwise eddy could be seen in the eastern area of Chania Bay. It is evident that the resultant eddy in the area of steep bottom topography is smoother in the IDOR model, especially for a 10 m/s west wind case.

Figure 10.10 compares the simulated depth-averaged velocities in Area B (Rethymno Bay) of Northern Crete, in the POM and IDOR models. For the west wind case, strong currents occurred in the nearshore area of Rethymno Bay and a counterclockwise eddy could be seen in the central area, from the water transport patterns. Both the Princeton and IDOR models gave very similar depth-averaged velocity patterns because no steep bottom topography exists in Rethymno bay.

Figure 10.11 compares the simulated depth-averaged velocities in Area C (Iraklion-Malia Bay) of Nortern Crete, in the POM and IDOR models. For the west wind case, strong eastward currents occurred in the nearshore area and two clockwise eddies could be seen in the eastern area of Iraklion-Malia Bay. Both the Princeton and IDOR models show very similar depth-averaged velocity patterns for both cases.

Figure 10.12 compares the simulated depth-averaged velocities in Area D (Agios Nikolaos Bay) of Northern Crete, based on the POM and IDOR models. For the west wind case, similar eastward currents occurred in the nearshore area. The currents obtained by the two models could be seen in the northeast corner of Agios-Nikolaos Bay. This could be attributed to the existence of steep bottom topography in the area. The IDOR model was found to perform better than POM in this particular case, which had steep bottom topography. In addition, the IDOR model needs less CPU time than POM to reach to steady state for the same simulation (see Figure 10.8).

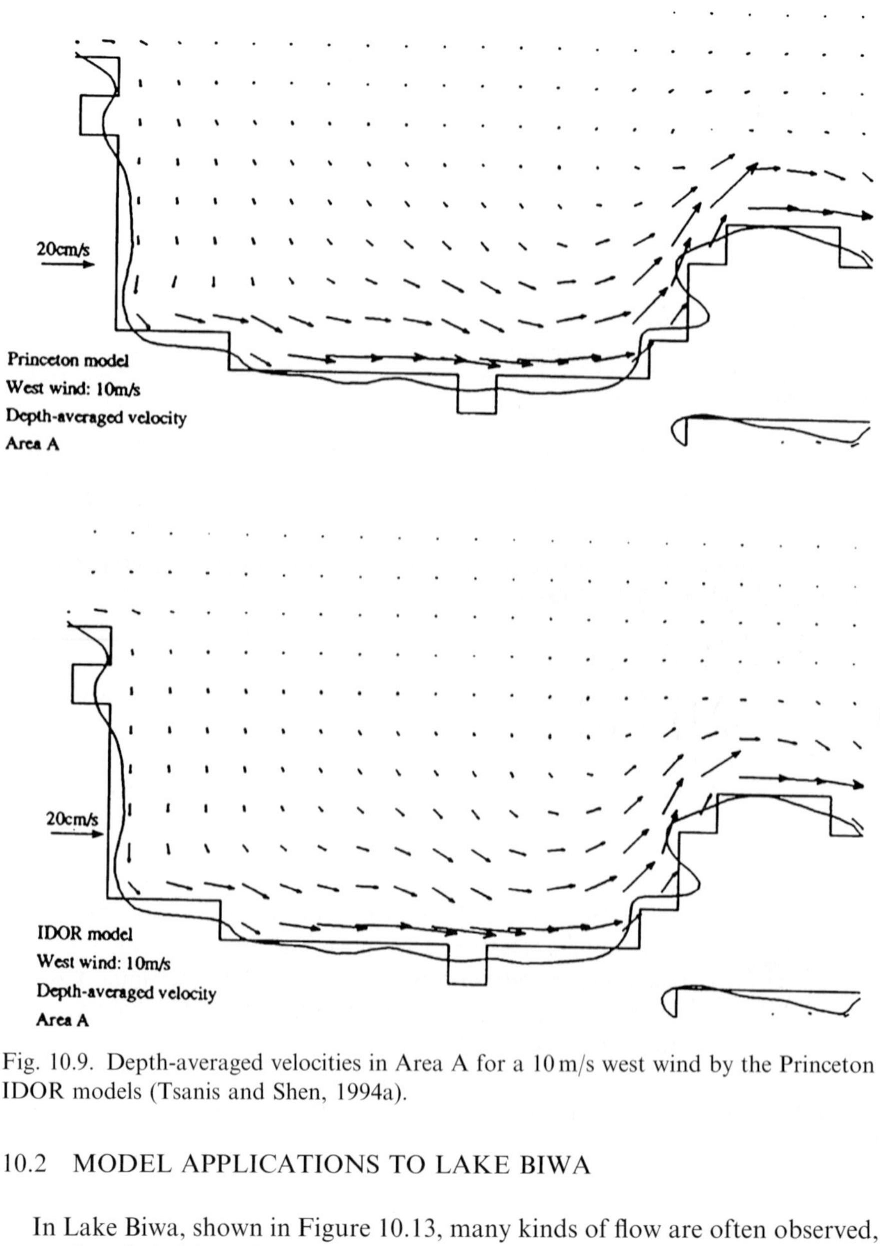

Fig. 10.9. Depth-averaged velocities in Area A for a 10 m/s west wind by the Princeton and IDOR models (Tsanis and Shen, 1994a).

10.2 MODEL APPLICATIONS TO LAKE BIWA

In Lake Biwa, shown in Figure 10.13, many kinds of flow are often observed, for example, wind-driven flow, circulation, internal Kelvin wave, and density current. In this section, some numerical methods are applied to Lake Biwa. Barotropic flow, baroclinic flow, and internal Kelvin wave are reproduced by simplifying the computational conditions, and then the flows during the destratification period are simulated. In the computations, Lake Biwa is divided into 891 columns (including 10054 cells) by mesh intervals $\Delta x = \Delta y = 1\,\text{km}$, $\Delta z = 2.5\,\text{m}$ for depths shallower

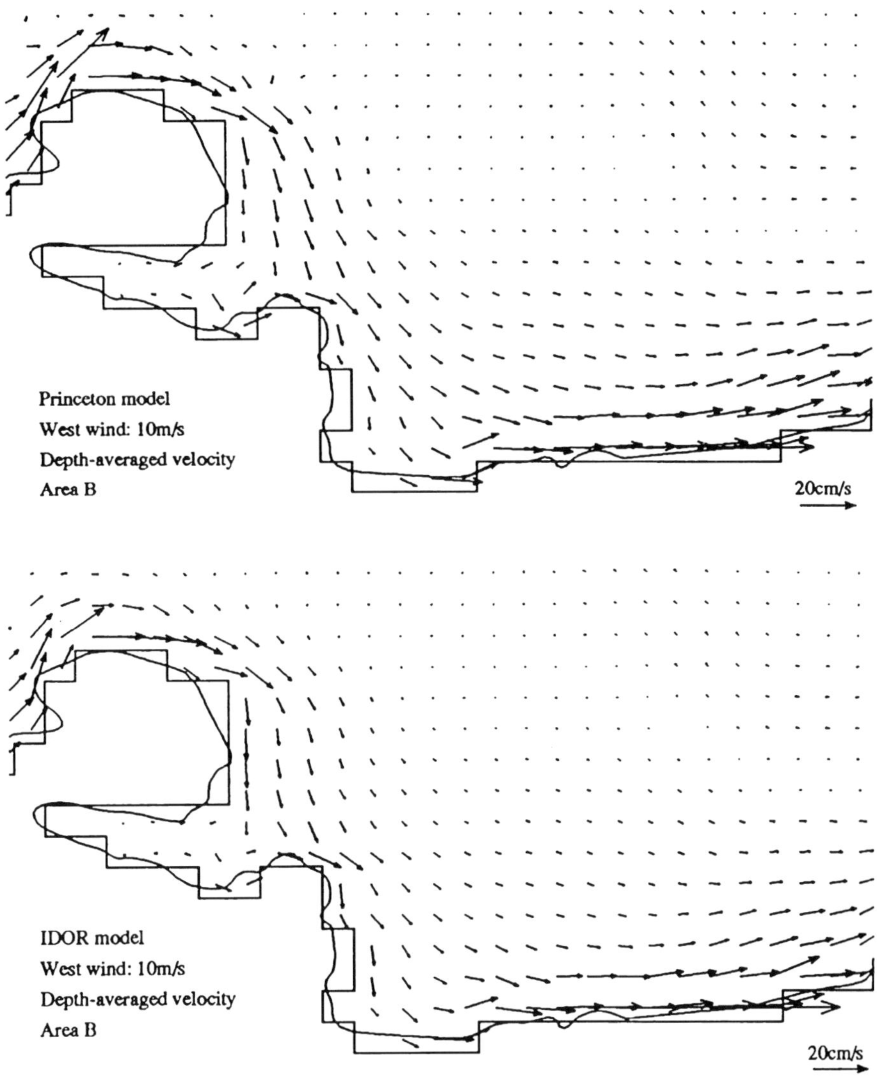

Fig. 10.10. Depth-averaged velocities in Area B for a 10 m/s west wind by the Princeton and IDOR models (Tsanis and Shen, 1994a).

than 50 m, and $\Delta z = 10$ m for depths deeper than 50 m. The parameters used in the calculations are given as follows: the eddy viscosity and diffusivity coefficients are $K_h = 30\,\text{m}^2/\text{s}$, $N_h = 10\,\text{m}^2/\text{s}$, $K_{v0} = 0.01\,\text{m}^2/\text{s}$, and $N_{v0} = 0.001\,\text{m}^2/\text{s}$; and the parameters of the upstream schemes are $\theta_h = \theta_v = 0.7$. Table 10.1 shows five cases (or methods) which combine different numerical schemes. When the barotropic component of the pressure term is treated implicitly, the time increment Δt is taken as 300 s; if it is treated explicitly, Δt is chosen as 20 s.

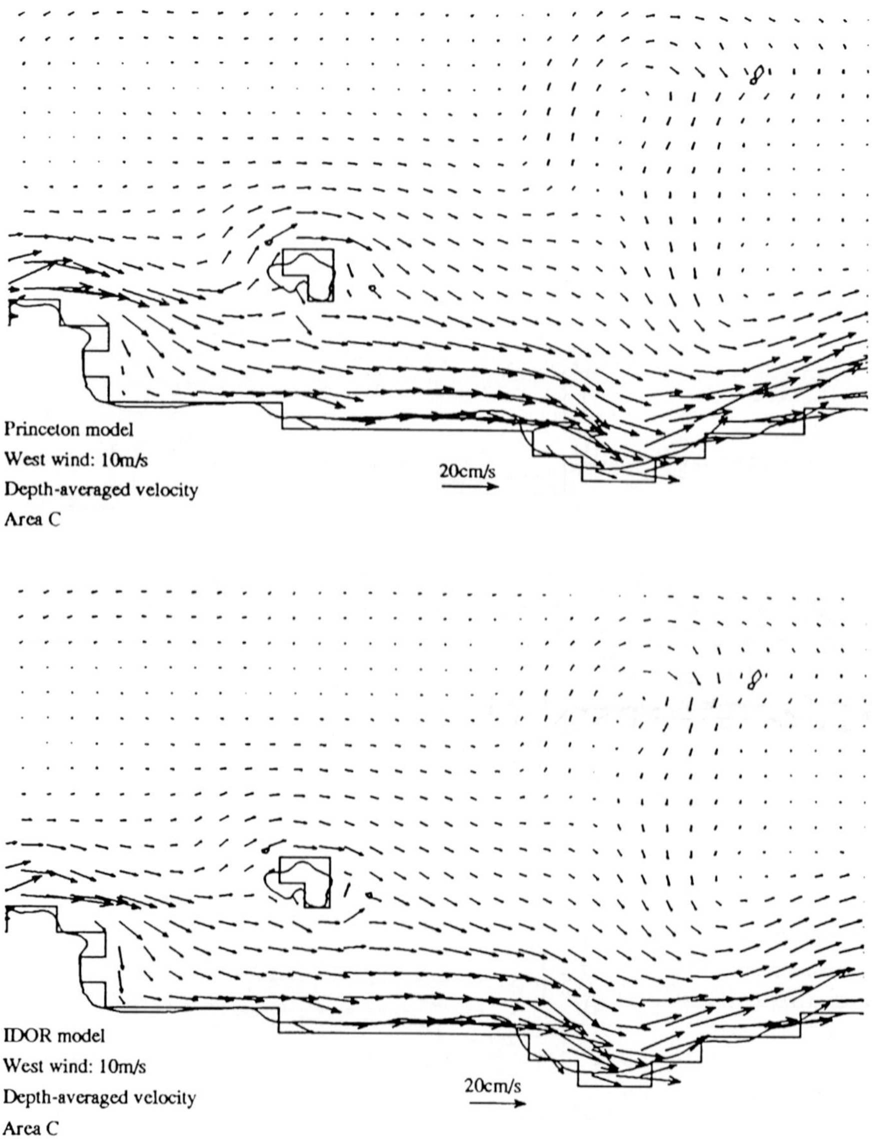

Fig. 10.11. Depth-averaged velocities in Area C for a 10 m/s west wind by the Princeton and IDOR models (Tsanis and Shen, 1994a).

10.2.1 Barotropic Flow

Barotropic flow has a great influence on the transport of mass and momentum and is a result of the variation in water stage, which is possibly wind-induced. It can also be approximately considered as wind-driven flow in the case of uniform

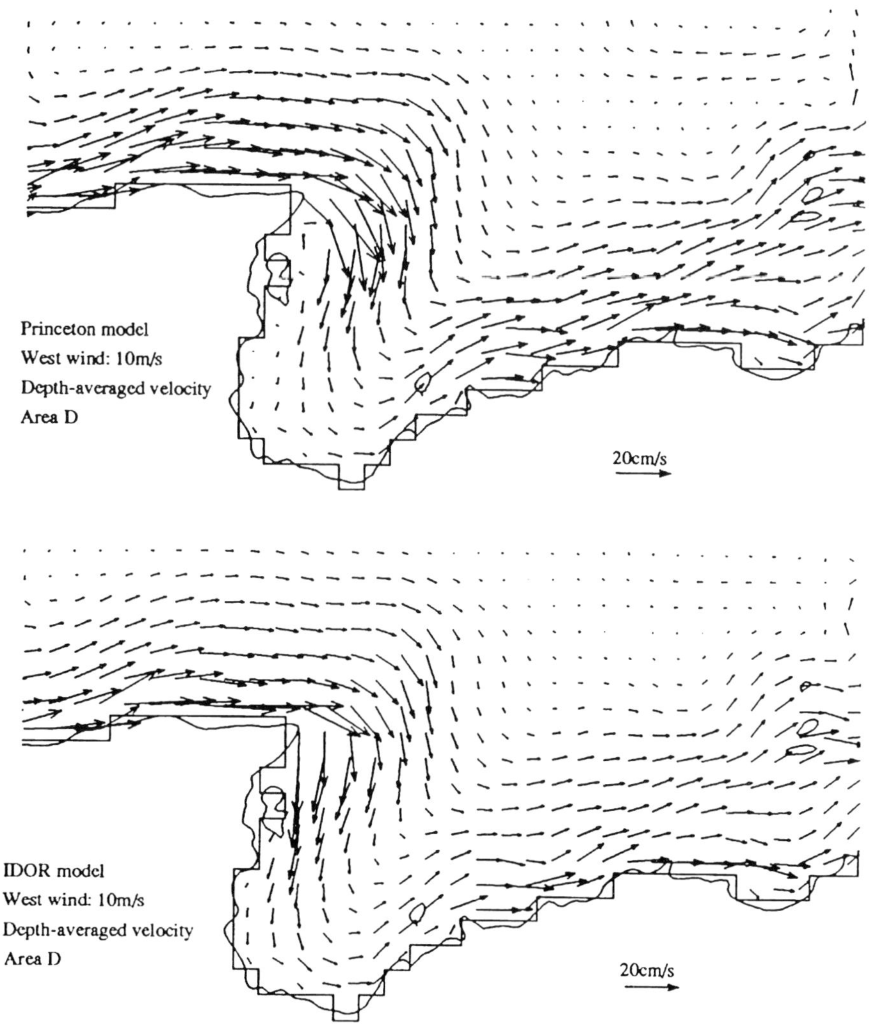

Fig. 10.12. Depth-averaged velocities in Area D for a 10 m/s west wind by the Princeton and IDOR models (Tsanis and Shen, 1994a).

temperature distribution. In order to investigate the barotropic flow (or wind-driven flow) in Lake Biwa, it is assumed that the temperature distribution is uniform over the whole lake and does not vary with time. The southwestern wind, with a velocity of 5 m/s, blows for 12 h, and no inflow and outflow rivers are taken into account. The computation period is chosen as 72 h and the method of Case 2 in Table 10.1 is adopted.

Figures 10.14 and 10.15 illustrate the velocity distributions of the surface layer and fifth layer at different times. It is observed from these figures that the water of the surface layer is driven by the wind, and the energy gradually transferred to the

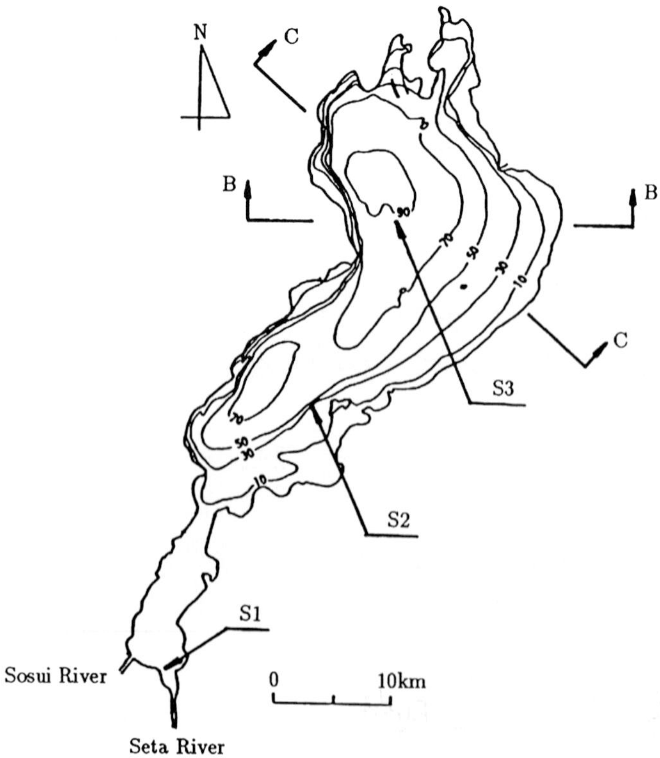

Fig. 10.13. Sketch map of Lake Biwa (Shen, 1991).

Table 10.1. Different computational cases for comparison.

Case	Time difference	Advective term	Pressure term	Δt (s)
1	Leap-frog	Donor cell	Implicit	300
2	Adams–Bashforth	Donor cell	Implicit	300
3	Adams–Bashforth	Donor cell	Explicit	20
4	Adams–Bashforth	QUICK	Implicit	300
5	Fractional step	Donor cell	Implicit	300

lower layers at the initial period of the wind blowing, after the wind calmed down. However, the velocities of the surface layer and the fifth layer tend to become uniform, while attenuating gradually. Figures 10.16 and 10.17, which are the velocity vectors in the vertical section, and the depth-wise distribution of velocity component u in the north basin, respectively, also verify the above characteristic.

In Figures 10.14 and 10.15, no horizontal circulation of large scale can be observed, and the currents in the central area of the north basin rotate clockwise with time, which can be considered as the inertia oscillation due to the Coriolis

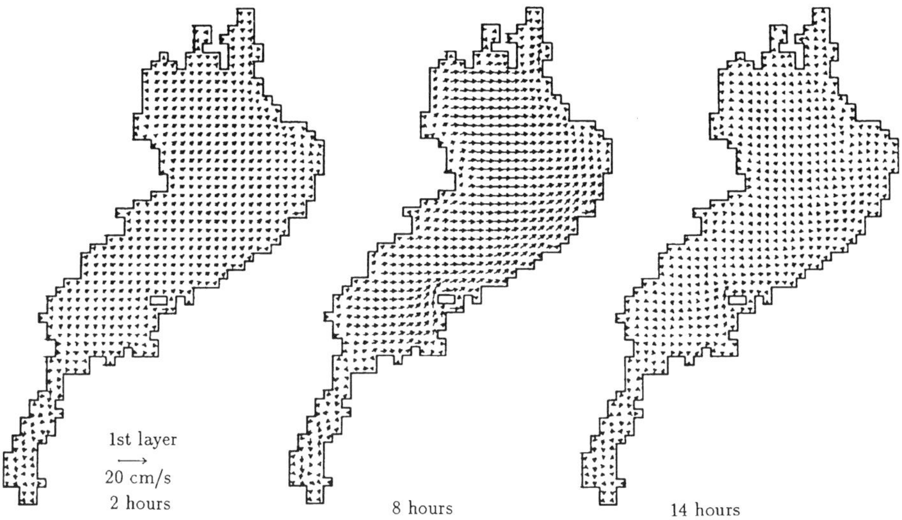

Fig. 10.14. Velocity distributions in the surface layer (Shen, 1991).

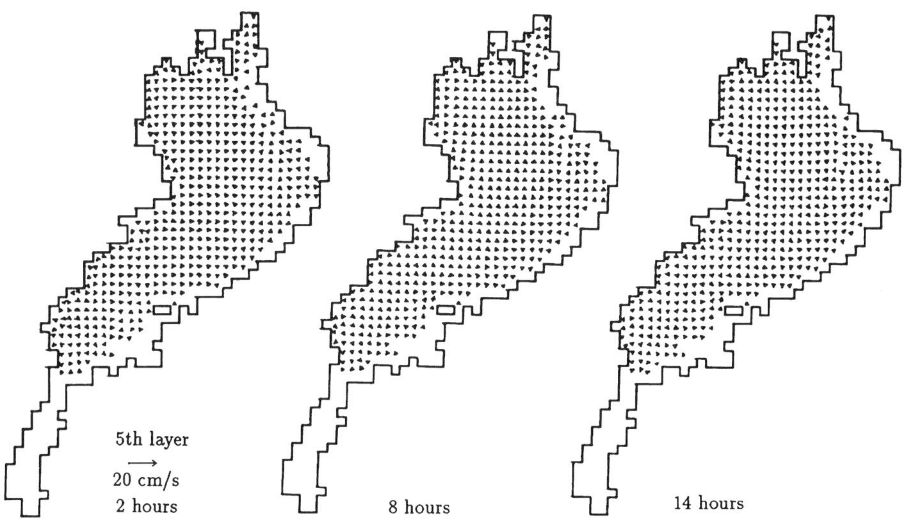

Fig. 10.15. Velocity distributions in the fifth layer (Shen, 1991).

force. Figure 10.18 is the temporal variation of the water stage at point S1 (see Fig. 10.13). In this figure, it is observed that the surface water is set up by the wind and the water stage at S1 falls down while the wind is blowing. After the wind stopped, the water surface is released and the water stage oscillates due to the seiche. The period of the surface seiche is estimated as 4 h, which is often observed in Lake Biwa.

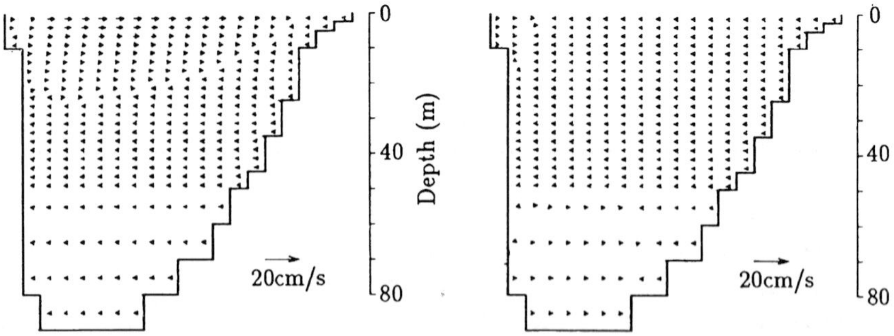

Fig. 10.16. Velocity vectors in the vertical plane of B–B section (Shen, 1991).

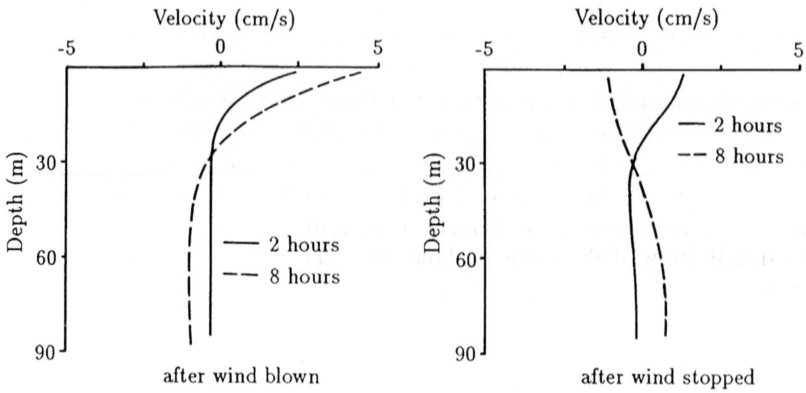

Fig. 10.17. Depth-wise distributions of velocity component *u* at S3 (Shen, 1991).

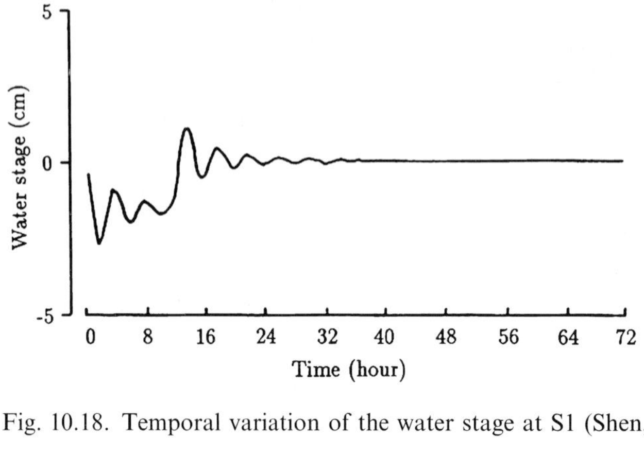

Fig. 10.18. Temporal variation of the water stage at S1 (Shen, 1991).

10.2.2 Baroclinic Flow

According to observations, large-scaled circulation is sometimes observed in the north basin of Lake Biwa. Though the mechanism of the circulation has not been clearly understood, there are two theories about the formation of the circulation, the so-called wind theory, and the heat theory. In the wind theory, the water of the lake is driven by wind, especially wind with velocity, and circulation occurs due to the topography of the lake. On the other hand, according to the heat theory, the circulation is possibly caused by non-uniform temperature distribution. This kind of circulation has been verified by laboratory experiments (Ookubo et al., 1984). In order to study the circulation in relation to the heat theory, it is assumed that a mass of water, with a temperature 4°C lower than the surroundings, exists in the central area of the north basin at the initial state, and the flows produced by it are calculated by using the method of Case 1 (Iwasa et al., 1985).

Figure 10.19 illustrates the velocity distributions of the various layers and Figure 10.20 shows the velocity and temperature distributions in the vertical plane of the C–C cross-section (see Fig. 10.13), after 3 h have passed. From the velocity distribution in the vertical plane, it is observed that the currents move toward the centre of the cool water mass at the surface layer, and outward at the lower layers. It is because the cooler water mass with a larger density moves downward that the water stage falls. As a result, the surrounding water with higher temperature concentrates to the central area. This kind of current can also be verified in the temperature distribution of the vertical plane, in which the isothermal lines in the lower layers have a concave shape. In the surface layer, strong circulation, rotating counterclockwise, occurs, because the Coriolis force acts on the currents concentrating toward the cooler water mass. Therefore, circulation due to a non-uniform temperature distribution could be verified in the calculation. However, the circulation attenuates quickly, whereas the circulation observed could last rather

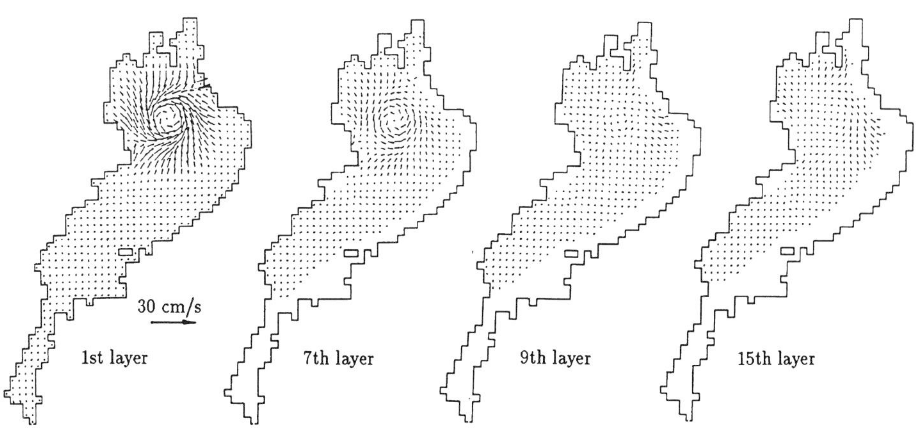

Fig. 10.19. Velocity distributions in the various layers (Shen, 1991).

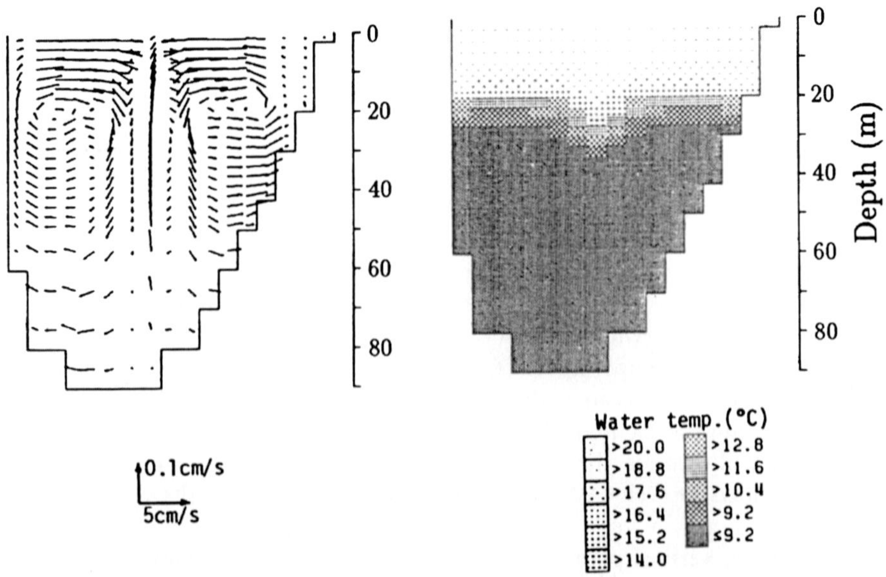

Fig. 10.20. Velocity and temperature distributions in the vertical plane of C–C section (Shen, 1991).

longer and therefore, further studies are required to clarify the mechanism of circulation.

10.2.3 Internal Kelvin Wave

It is well known that the internal Kelvin wave occurs in the north basin of Lake Biwa during the summer and plays a great role in the transport of momentum and mass. The condition which develops the Kelvin wave is a strong wind blowing for half the period of the internal Kelvin wave. In order to reproduce the internal Kelvin wave clearly, the computational conditions are simplified. The metalimnion is assumed to exist in the layer of water depth 20–22.5 m, at the initial state, and the water is set in motion by the wind. The method of Case 1 is adopted to simulate the internal Kelvin wave (Iwasa and Inoue, 1985).

Figure 10.21 shows the temperature distribution in the horizontal plane near the metalimnion, after the wind has calmed. In the figure, that portion of the west shore with the lower temperature rotates counterclockwise. In other words, the isothermal surface in the north basin is not horizontal but inclined, and the inclined surface rotates counterclockwise in the central portion of the north basin. Its period is estimated to be about 56 h. According to the two-layer model (Csanady, 1967), the internal Rossby radius R_2 of Lake Biwa and the speed 'c' of the internal Kelvin wave are given by Eqs. (10.1) and (10.2), respectively:

$$R_2 = \frac{1}{f}\sqrt{\frac{(\rho_2 - \rho_1)}{\rho_2}\frac{gh_1h_2}{h_1 + h_2}} \tag{10.1}$$

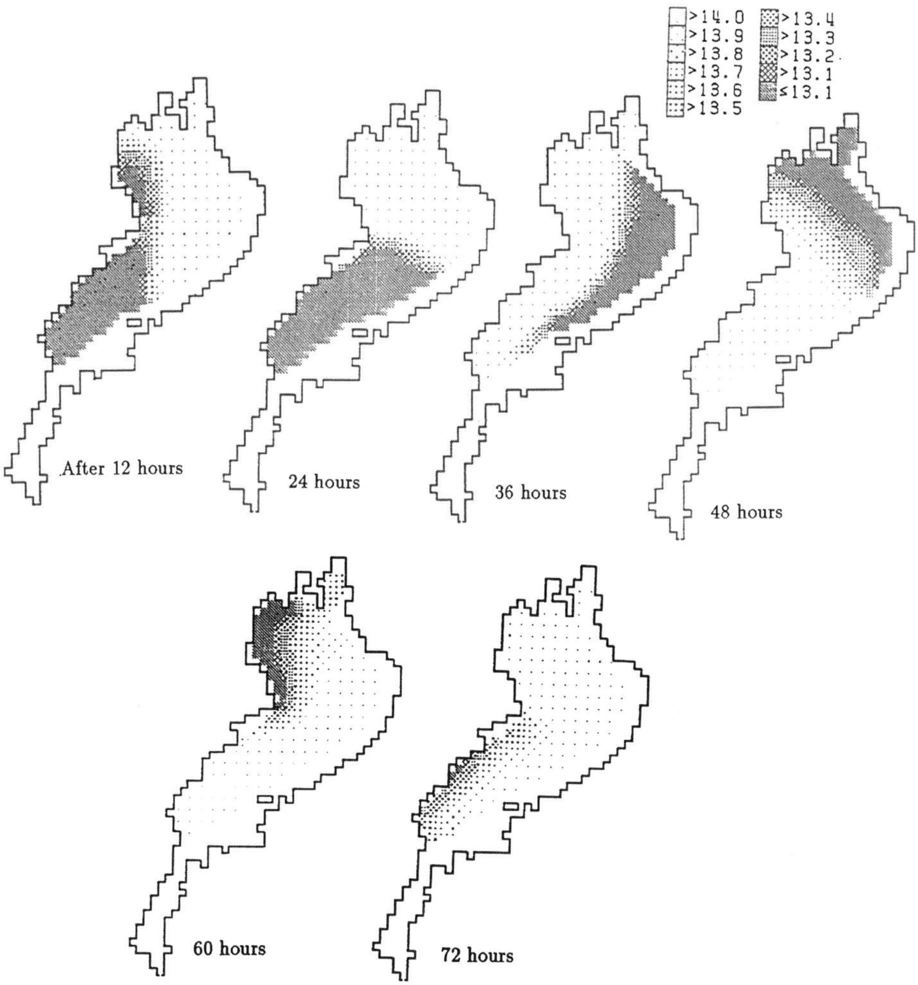

Fig. 10.21. Temperature distributions in the horizontal plane near the metalimnion (Shen, 1991).

$$R_2 = \frac{1}{f}\sqrt{\frac{(\rho_2 - \rho_1)}{\rho_2}\frac{gh_1h_2}{h_1 + h_2}} \tag{10.2}$$

If $\rho_1 = 998.234\,\mathrm{kg/m^3}$ (corresponding to 20°C), $\rho_2 = 999.848\,\mathrm{kg/m^3}$ (corresponding to 8°C), $h_1 = 20\,\mathrm{m}$, and $h_2 = 40\,\mathrm{m}$ are given, $R_2 = 4.13\,\mathrm{km}$ and $c = 0.458\,\mathrm{m/s}$ are calculated. Since the equivalent radius of Lake Biwa is calculated as 14.0 km (the area of the north basin is 620 km^2), the period of the internal Kelvin wave is 62.5 h. Therefore, the internal wave reproduced by the calculation can be considered as the internal Kelvin wave.

10.2.4 Flows during Destratification Period

In autumn, the water of the lake becomes cool with the falling of air temperature, and the stratification gradually collapses due to the mixing effect. The destruction of stratification exerts great influence over the flows in the lake. In this sub-section, the flows during the destratification period in Lake Biwa are simulated under actual meteorological conditions and the hydraulic properties of the flows are investigated. In the calculation, 19 main inflow rivers and two outflow rivers are taken into account and the observed meteorological data and inflow–outflow discharges are used. The lake flows are calculated from September 21 to October 5, 1982. The meteorological data and inflow–outflow discharges are shown in Figure 10.22. The initial condition is assumed to be the still state, with a smooth stratification, and the five numerical methods in Table 10.1 are adopted in turn (Shen et al., 1990a).

Figure 10.23 shows the depth-wise distribution of temperature at point S3. In this figure, the solid line shows the initial distribution of temperature and dotted lines express the temperature distributions on September 28 and October 5. It is clear that the temperature near the water surface is falling and becoming uniform, by the mixing. The stratification is gradually collapsing with time due to the cooling of the water surface during the destratification period. Figure 10.24 shows the velocity distributions in the surface layer, and from this figure it is found that the flows in the north basin, depending mainly on the magnitude and the direction of the wind, rotate counterclockwise in the central area, due to the Coriolis force. It

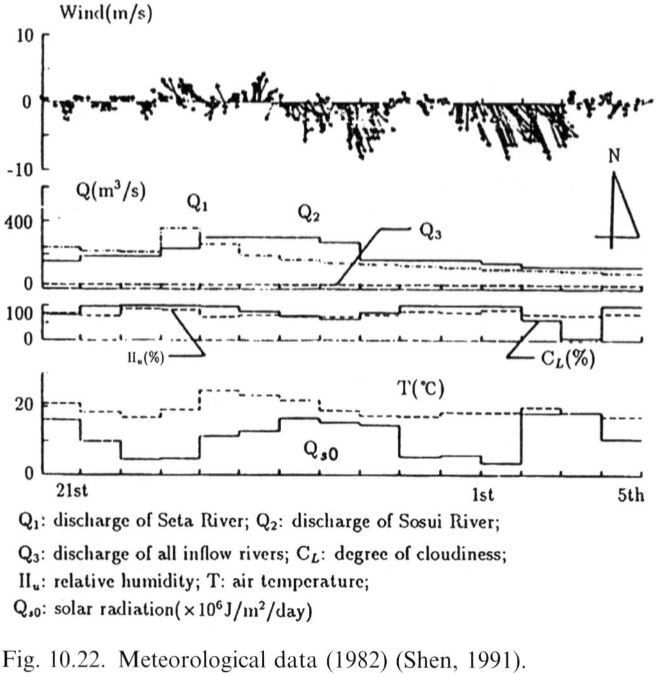

Fig. 10.22. Meteorological data (1982) (Shen, 1991).

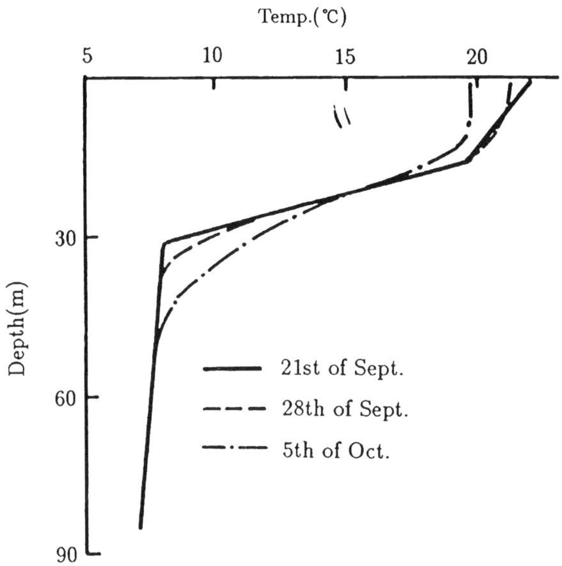

Fig. 10.23. Depth-wise distribution of temperature at S3 (Shen, 1991).

seems that the flows in the north basin are not greatly influenced by the inflow rivers; on the other hand, the flows in the south basin are mainly determined by the outflow discharge to the Seta River.

Figure 10.25 illustrates the water stage variations with time at points S1 and S3. The variations of water stage which are caused by the variation of inflow-outflow discharges are reproduced in both points, but the variation of high frequency can obviously be seen only in point S1. The period of this high-frequency variation is approximately 4 h, and it can be considered as the surface seiche, often observed in the south basin. Therefore, it is verified that the surface seiche is very weak compared to the strong wind-driven currents in the north basin, but is an extremely important flow in the south basin.

10.2.5 Comparison of Numerical Methods

Table 10.2 shows the time increments and CPU times of the five cases described, to simulate the flows during the destratification period. Figures 10.26 and 10.27 show the velocity variation with time at point S2; no great difference among the results can be seen, except in Case 3. From Table 10.2, it is seen that Case 2 spent the least CPU time, compared with the other cases.

In Case 1, more CPU time is required because Matsuno's scheme is involved for every six time steps, to avoid numerical oscillation. In comparing Case 4 with Case 2, it can be seen that the QUICK scheme has a higher spatial accuracy than the Donor-cell scheme, and the velocity of lake flows is usually not so large that convective terms have a small weight in the momentum equations; therefore, the

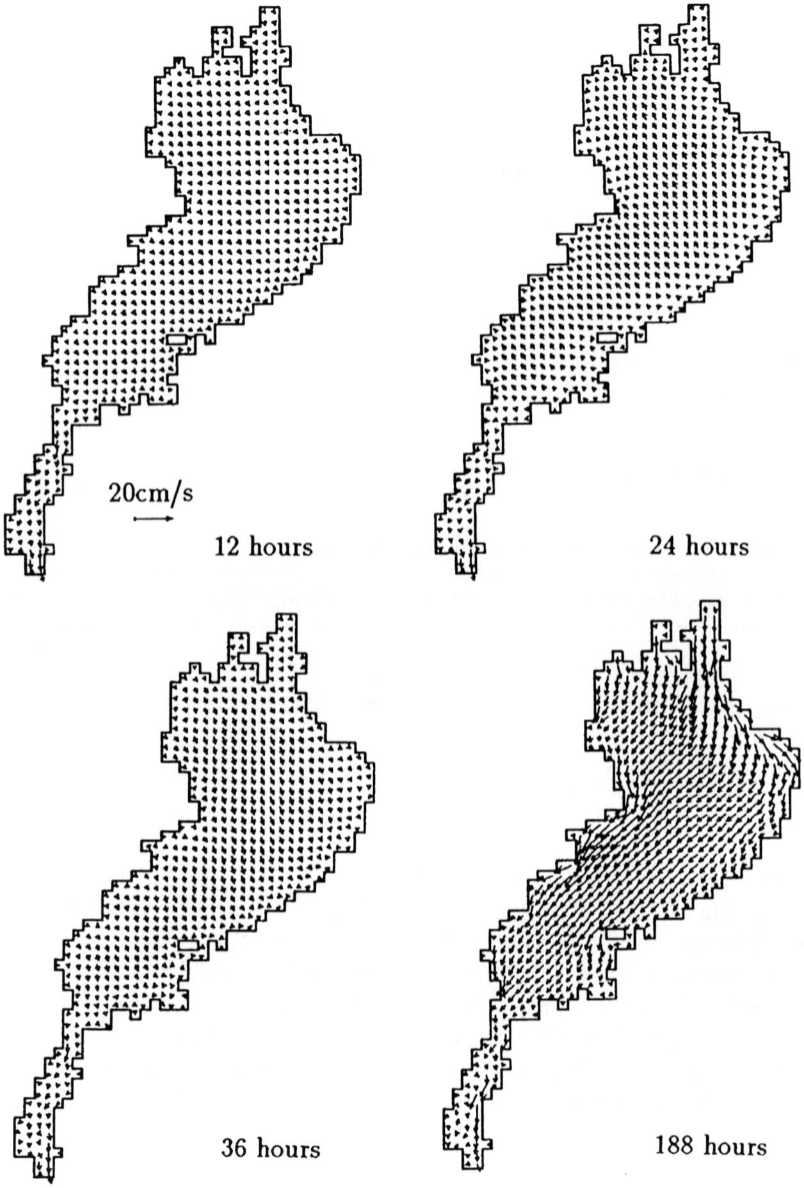

Fig. 10.24. Velocity distributions in the surface layer (Shen, 1991).

Donor-cell scheme is accurate enough to be applied to the advective term for lake flows and can save CPU time. As mentioned before, since the implicit scheme for the barotropic mode allows much larger time steps than the explicit scheme, it is natural that Case 3 consumed much more CPU time than Case 2. What has to be noticed is that the implicit scheme for the barotropic mode cannot reproduce the

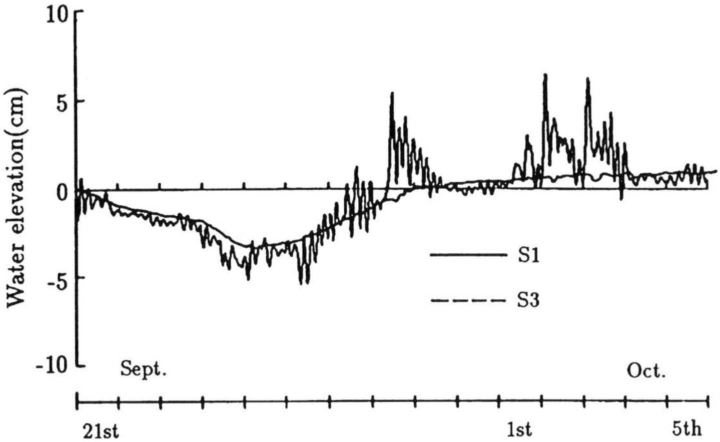

Fig. 10.25. Water stages with time at S1 and S3 (Shen, 1991).

Table 10.2. CPU times of the different computational cases.

Case	Δt (s)	Period (days)	CPU time (s)
1	300	15	1810
2	300	15	1442
3	20	2	2305
4	300	15	1761
5	300	15	1834

Source: Shen, 1991

waves of short periods, due to its long time step (see Fig. 10.25). Case 5 was expected to be able to save CPU time; unfortunately, it consumed more CPU time than Case 2. The reason is that part of the calculation performed by the double-sweep method cannot be vectorized (a FACOM VP–400 vector computer is used in this study) so that the advantages of the vector computer were not fully utilized. However, it is believed that the operator-splitting method will be an efficient and flexible method of promise if the computational technique is a little improved. In summary, the method used in Case 2 is considered the best method to simulate lake flows for long periods.

10.2.6 Density Current between the South and North Basins

Computational conditions

In order to investigate the basic characteristics of the density current and analyze the main factors which influence the flow, the computational conditions are simplified and the numerical simulations of the following two cases are carried out. It is assumed that the water surface of the whole lake coincides with the standard water surface and the water body is still at the initial state. The initial distribution of water

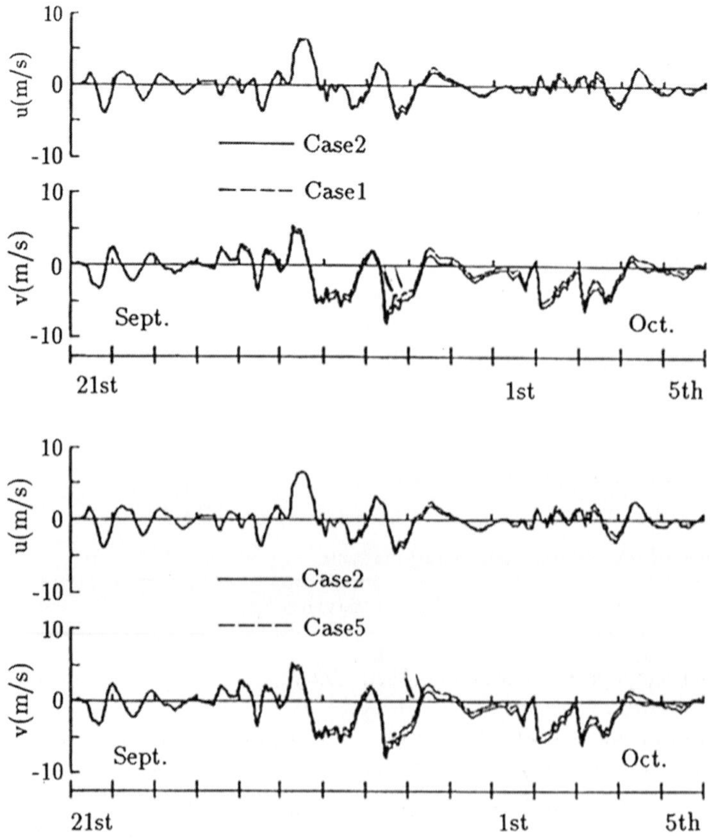

Fig. 10.26. Velocities at S2 obtained by different cases (Shen, 1991).

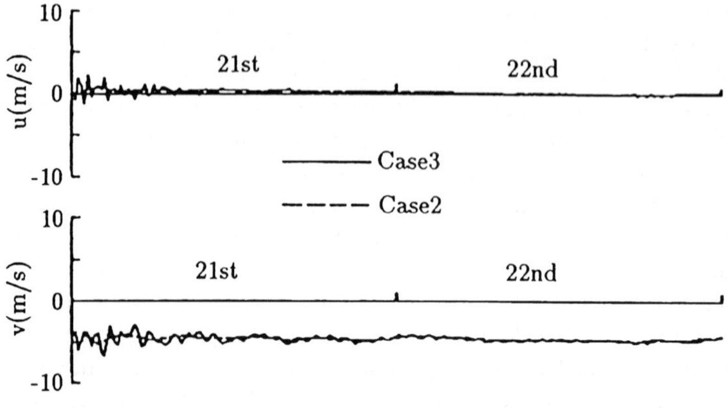

Fig. 10.27. Velocities at S2 by Cases 2 and 3 (Shen, 1991).

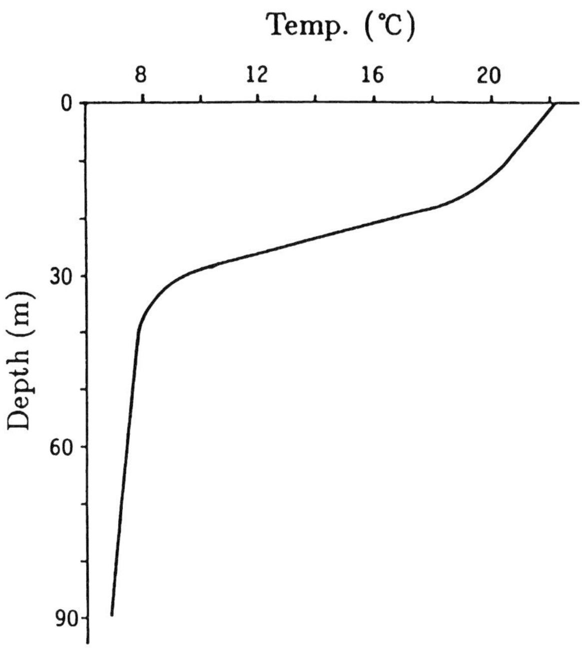

Fig. 10.28. Initial water temperature distribution (Shen, 1991).

temperature is uniform in the south basin at 8°C and is stratified in the north basin, as shown in Figure 10.28. Both inflow and outflow discharges are assumed to be zero, and wind is not taken into account. The purpose of this case is to investigate the properties of the density current produced only by the difference in density. The water of the south basin is assumed to have a tracer of concentration 100% and the temperatures of the inflow rivers are 13°C when they are taken into account.

Figure 10.29 illustrates the velocity distributions in the surface layer. It is observed from the figure that, initially, the flow is predominant only at the boundary of the two basins, and then it gradually rotates clockwise in the north basin, with time passing. Figure 10.30 illustrates the isothermal lines and velocity vectors of cross-section A–A (longitudinal direction of Lake Biwa) shown in Figure 7.14. This figure shows that the cooler water in the south basin moves into the north basin, and the warmer water in the north basin flows toward the south, which means a density current between the two basins occurs. This density current can be observed more clearly in Figure 10.31, which illustrates the depth-wise distribution of velocity component v at point B. It has been known that the influence of Coriolis force on the density current is large (Ookubo, 1988). For clarity, the isothermal lines are illustrated with the velocity vectors in Figure 10.32. In this figure, it can be observed that the current is almost moving toward the north, 12 h after the computation starts, but it gradually turns to the east and the isothermal lines also swell toward the east at the same time. This turn of direction can be considered an effect of Coriolis force.

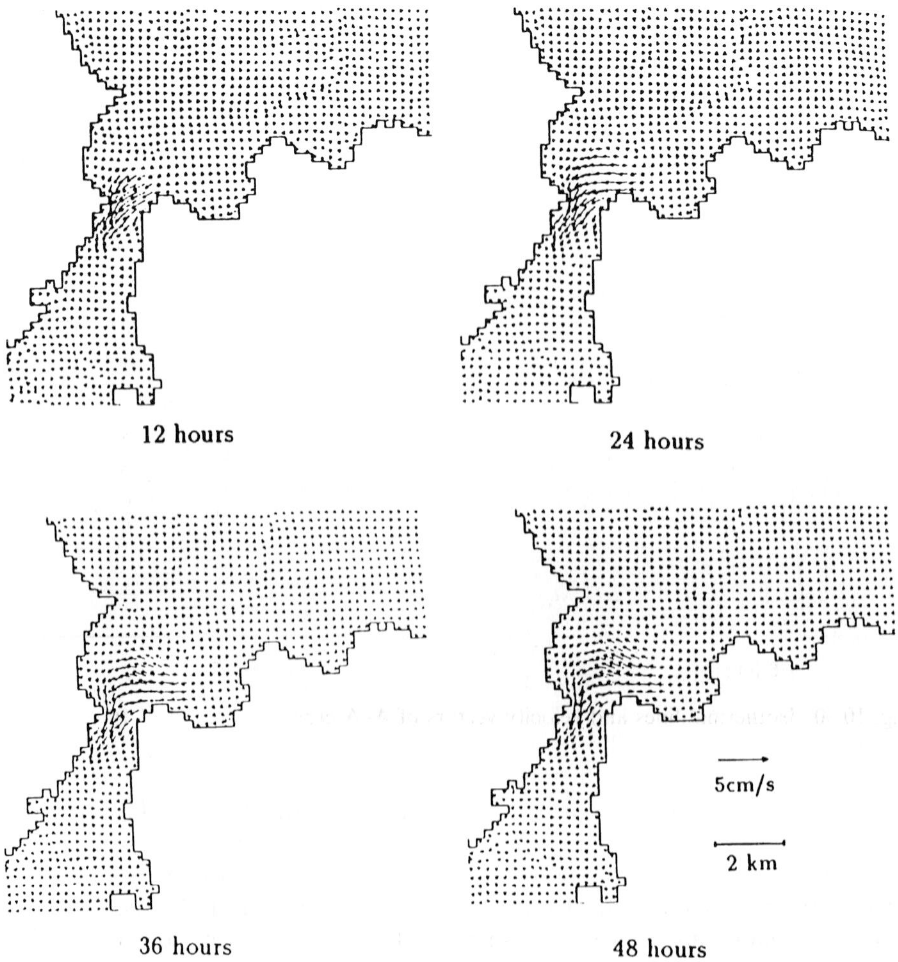

Fig. 10.29. Velocity distributions in the surface layer (Shen, 1991).

10.2.7 Flows Near the Inflow Estuaries

Yasu River

The Yasu River, whose catchment area is the second largest in Lake Biwa, next to the Ane River catchment, flows into the lake at the eastern shore from a southeastern direction, as shown in Figure 7.14. There is a great amount of sediment run-off in Yasu River and it has been considered a "suspended river", due to sedimentation. After the flood discharge improvement works in the river in 1981, the sediment run-off problem remained unsolved. In order to investigate the behavior of the inflow water and sediments in Lake Biwa, the coarse-fine grid model is applied to the inflow estuary.

Choosing the mesh sizes of the fine grid as $\Delta x_2 = \Delta y_2 = 250\,\text{m}$ is appropriate because the width of the river is 300 m. In this study, a simple case is assumed in

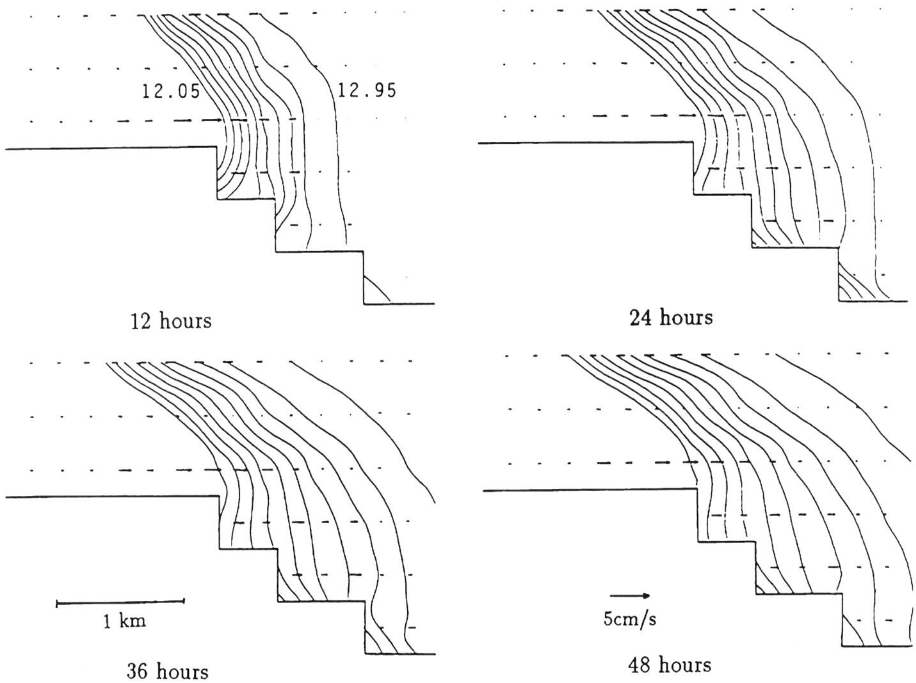

Fig. 10.30. Isothermal lines and velocity vectors of A–A cross-section (Shen, 1991)

which the thermocline exists in the 20 m depth below the water surface. The lake water is still at the initial state. As the external force, a northern wind with a velocity of 5 m/s is assumed to blow for 20 h and the outflow discharges of the Seta River and the Sosui River are given as 200 and 22 m^3/s, respectively. The same amount of inflow discharges are distributed to five rivers proportionally to their catchment areas, so that the discharge of Yasu River is approximately 50 m^3/s.

Furthermore, since the water temperature of the inflow river is usually lower than that of the lake water during a rainfall period, its temperature is given as 2°C lower than the surface layer of the lake.

In investigating the spread of the sediments in the lake, the settling velocity of sediments should be evaluated. Usually, the settling velocity is dominated by the grain size; however, it is neither easy nor necessary to assume the distribution of the grain size in this study. Here, the spread behavior of the river water is investigated by using the tracer without settling velocity and the sediments with a uniform settling velocity of 10^{-5} m/s, respectively.

First, the way in which the river water spreads (the case where settling velocity is zero, Case 1) is considered. The concentration distributions in the surface layer, third layer, and the lowest layer, are illustrated in Figures 10.33, 10.34 and 10.35, respectively. It is observed from these figures that the spread in the third layer was faster than that in the surface layer. This can also be observed from Figure 10.36, which illustrates the vertical concentration distribution at point C.

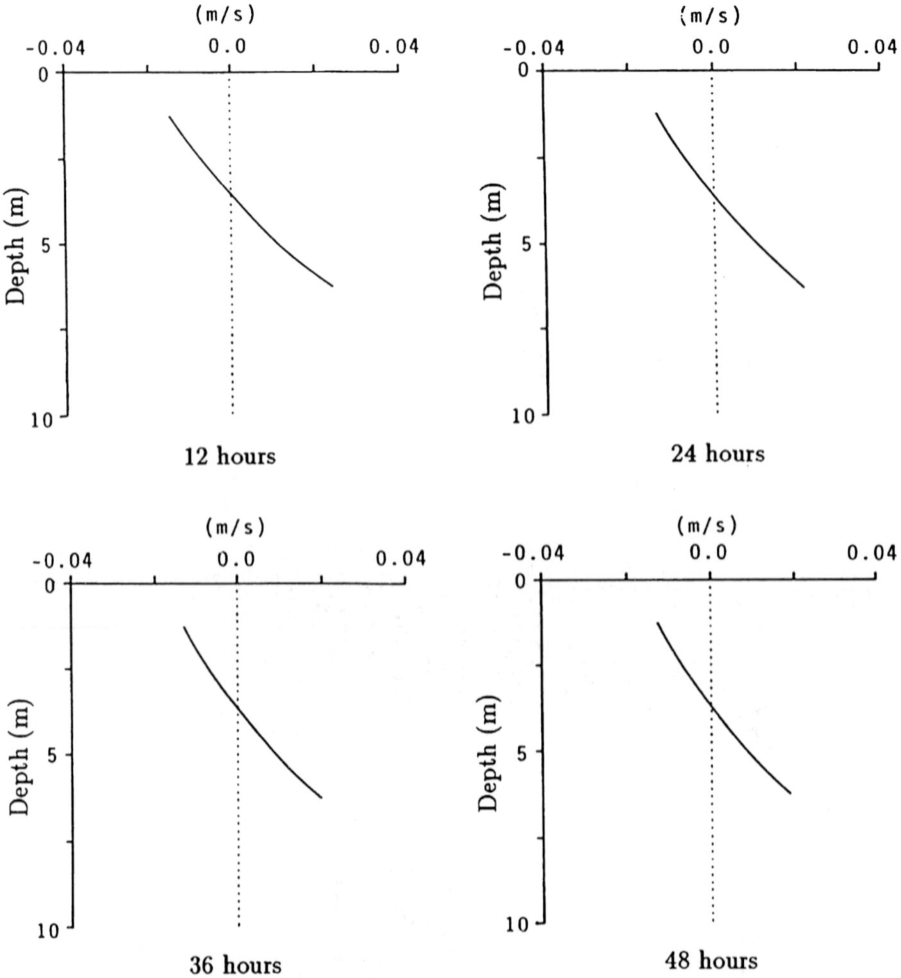

Fig. 10.31. Depth-wise distribution of velocity component v at point B (Shen, 1991).

The largest is approximately 21°C and higher than the water temperature of Yasu River. Comparing the temperature of the inflow water and the initial water temperature of the lake, it was expected that the water of the Yasu River would spread horizontally in the lower layer, but the simulation results show the spread in the upper layers. This is because the water from the river gave rise to a density inversion in the estuary and the instantaneous convection was assumed to happen in the calculation.

Considering the settling of sediments (Case 2), if the settling velocity is 10^{-5} m/s, the grain sizes are calculated as 10^{-4}–10^{-5} mm, from the Stokes formula, and can be considered as fine grains. Because the maximum value of the vertical velocity in this calculation is about 10^{-5} m/s, this settling velocity can be used to describe the settling process of sediments. Figures 10.37–10.39 illustrate the concentration

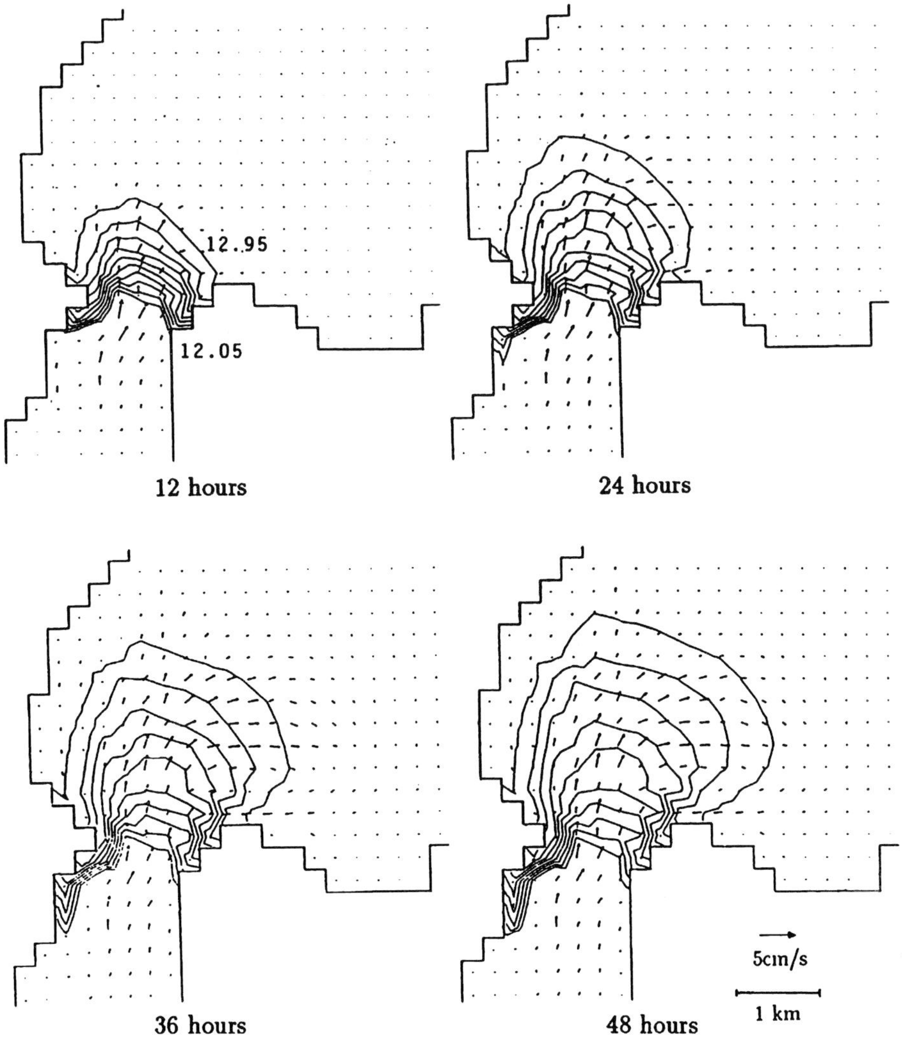

Fig. 10.32. Isothermal lines and velocity vectors near the boundary of the north and south basins (Shen, 1991).

distributions in the surface layer, third layer, and lowest layer, respectively. Compared to Figures 10.33–10.35, the spreads in the surface layer and third layer were a little smaller, while the spread in the lowest layer was larger. This difference is more remarkable between Figures 10.36 and 10.40. Therefore, when the settling velocity is zero, the sediments spread mainly due to horizontal convection, as a kind of tracer, while the sediments with settling velocity tend to spread toward the lower layers.

Since the estuary of Yasu River is close to Lake Biwa, sediments may possibly spread into the south basin. Figure 10.41 shows the proportion of tracer entered

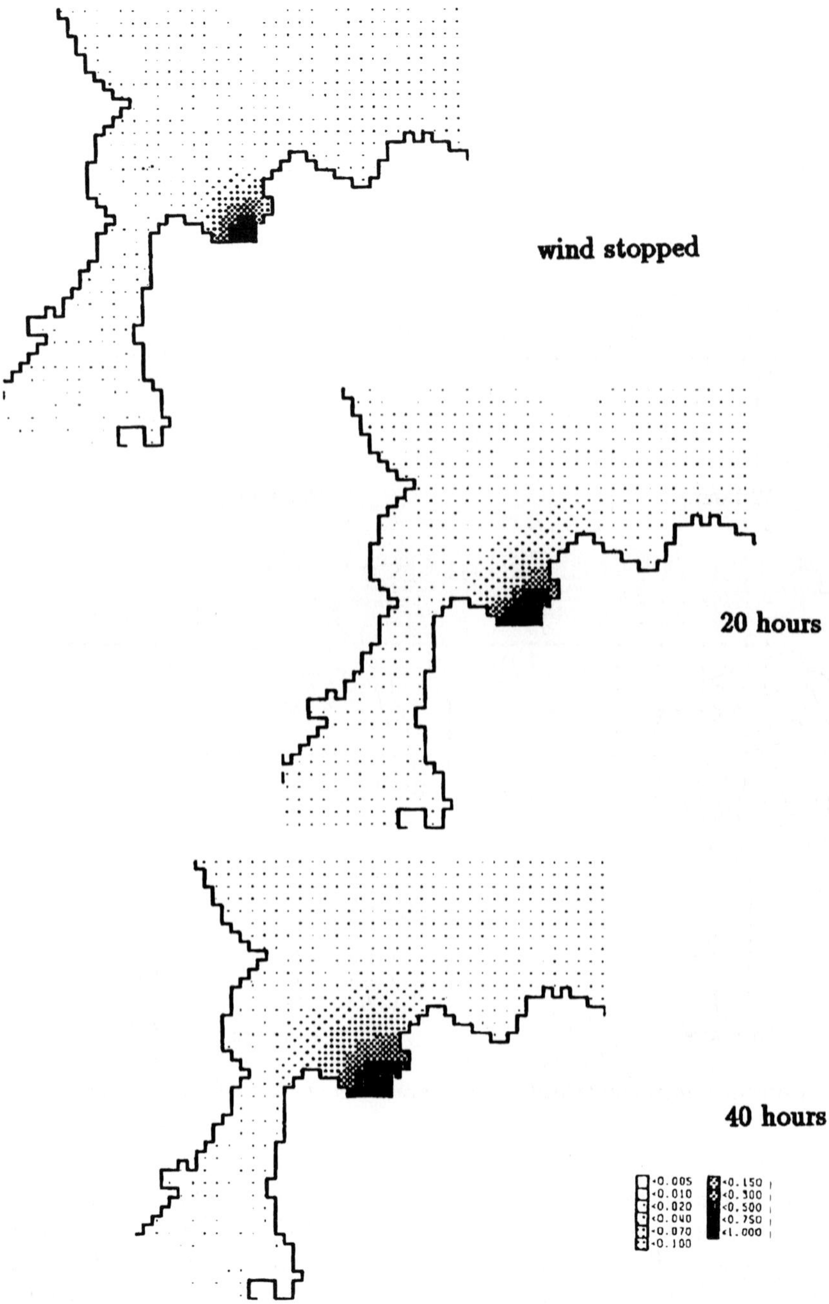

Fig. 10.33. Concentration distributions in the surface layer (Shen, 1991).

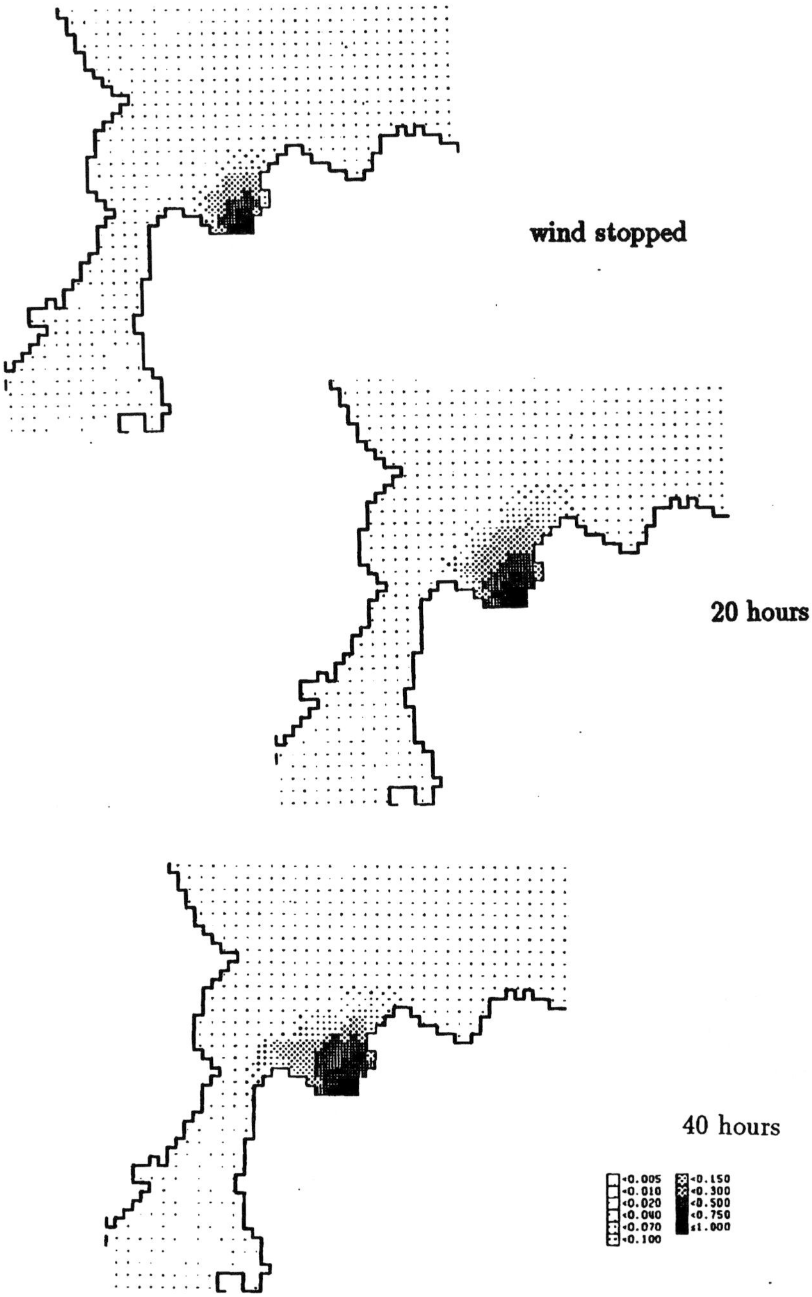

Fig. 10.34. Concentration distributions in the third layer (Shen, 1991).

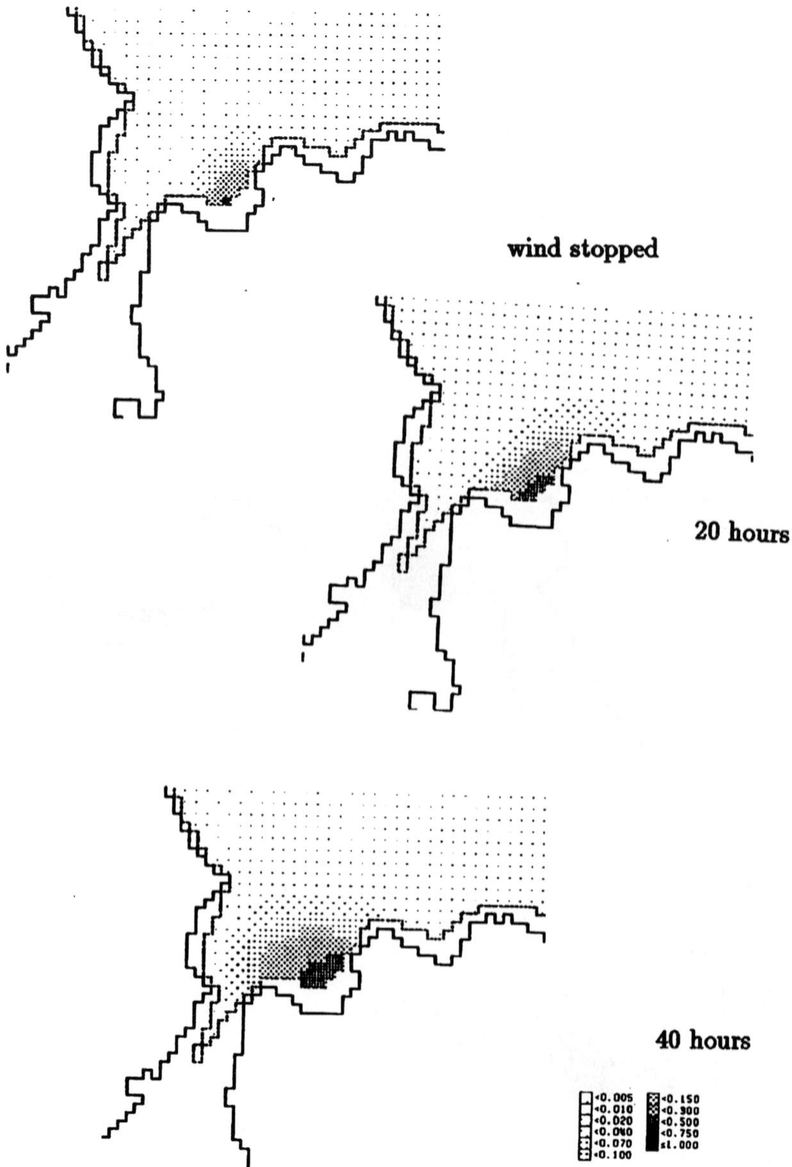

Fig. 10.35. Concentration distributions in the lowest layer (Shen, 1991).

into the south basin compared to the total inflow discharges; the solid line and dotted line express the results of inflow water (Case 1) and sediments (Case 2) from Yasu River, respectively. In this calculation, it is observed that the water from Yasu River took about 40 h to arrive at Lake Biwa Bridge, while the sediments with

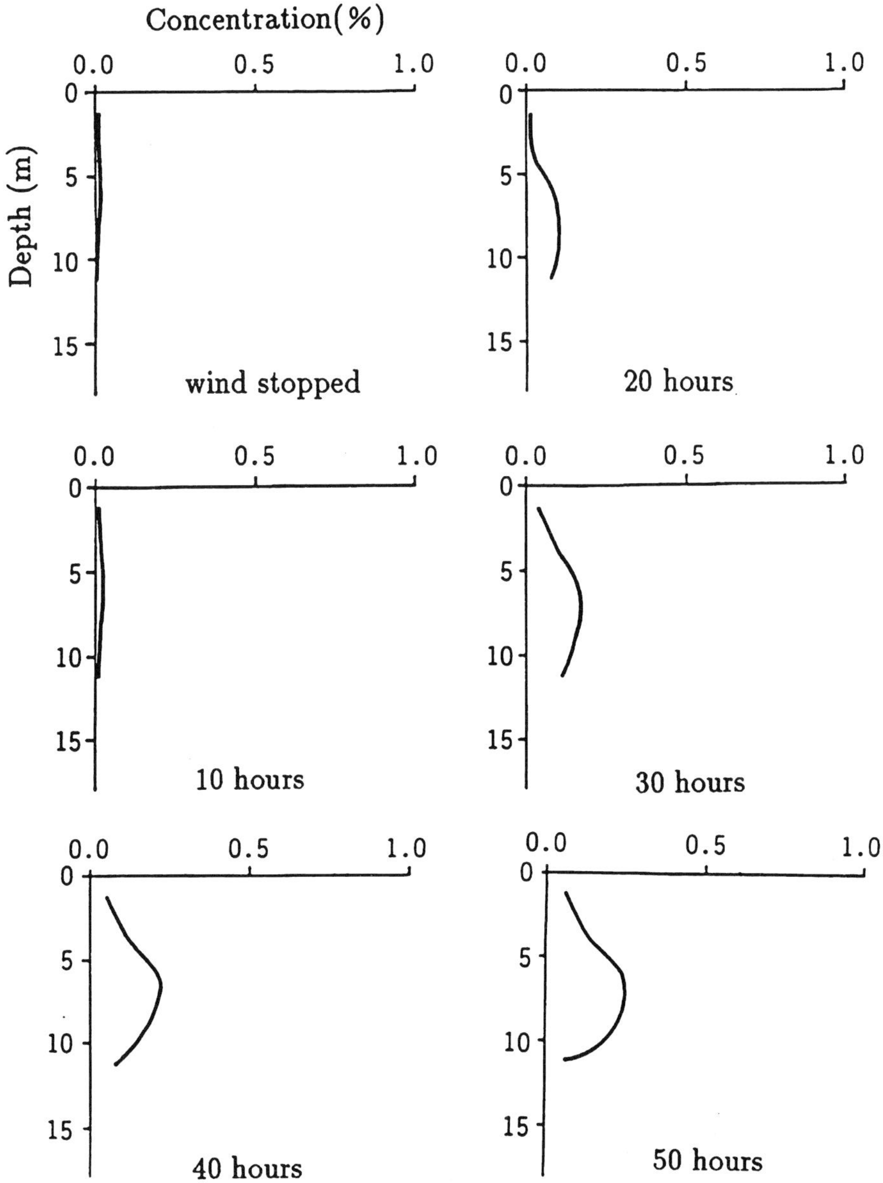

Fig. 10.36. Depth-wise concentration distribution at point C (Shen, 1991).

settling velocity moved into the south basin and gave rise to a turbid water problem. The spread of river water is influenced by the water temperature of the lake, while sediments, accepting little influence from water temperatures, spread toward the lower layers, because of settling velocity.

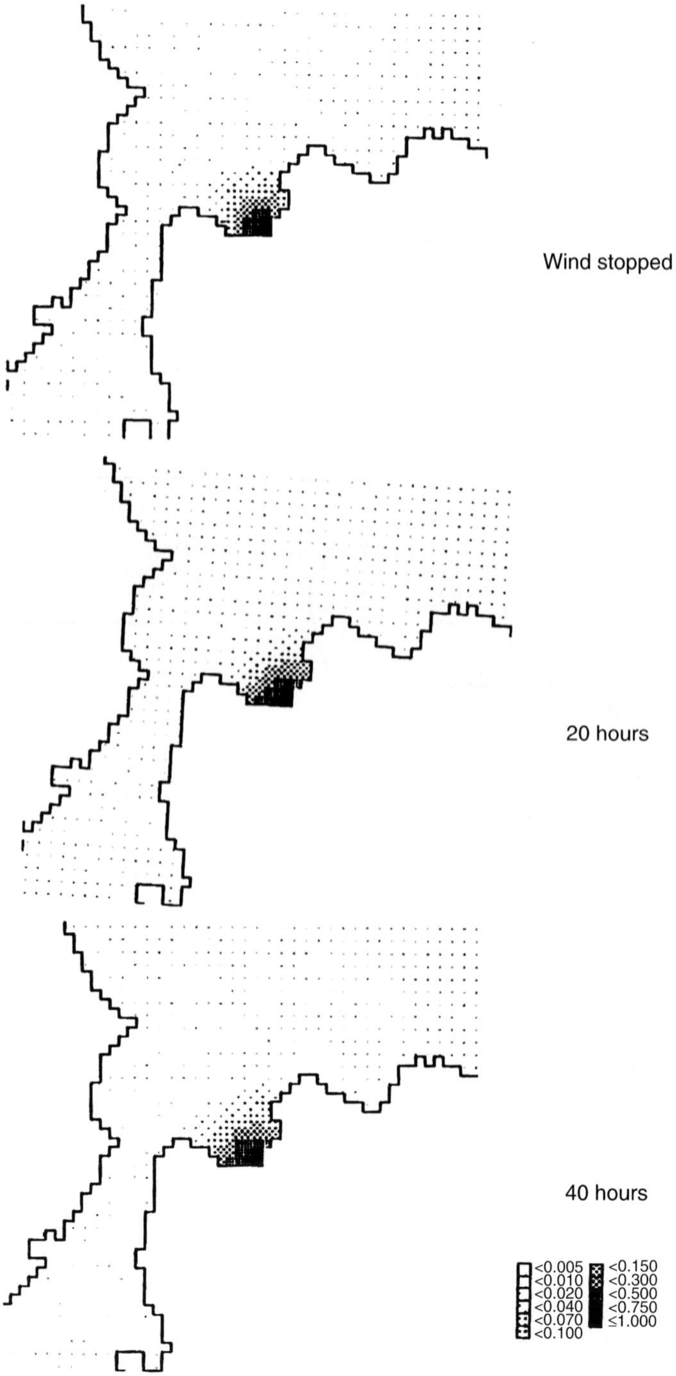

Fig. 10.37. Concentration distributions in the surface layer (Shen, 1991).

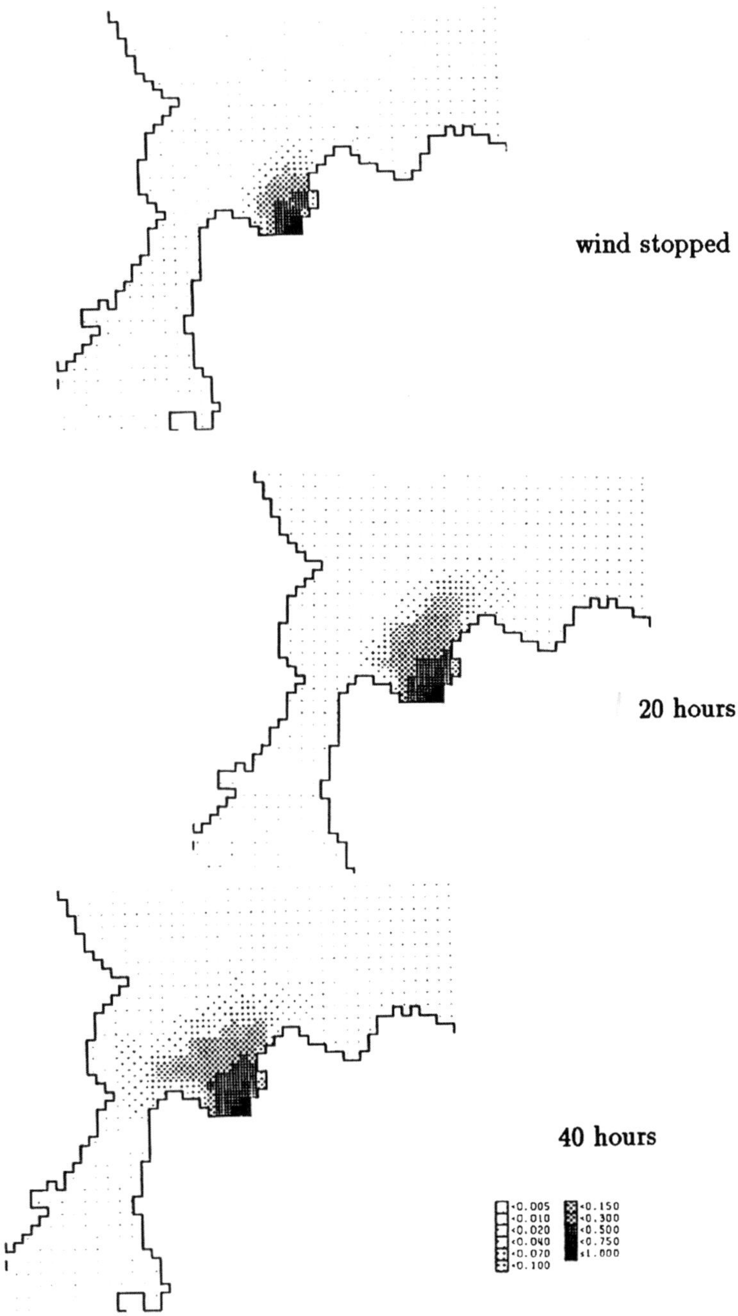

Fig. 10.38. Concentration distributions in the third layer (Shen, 1991).

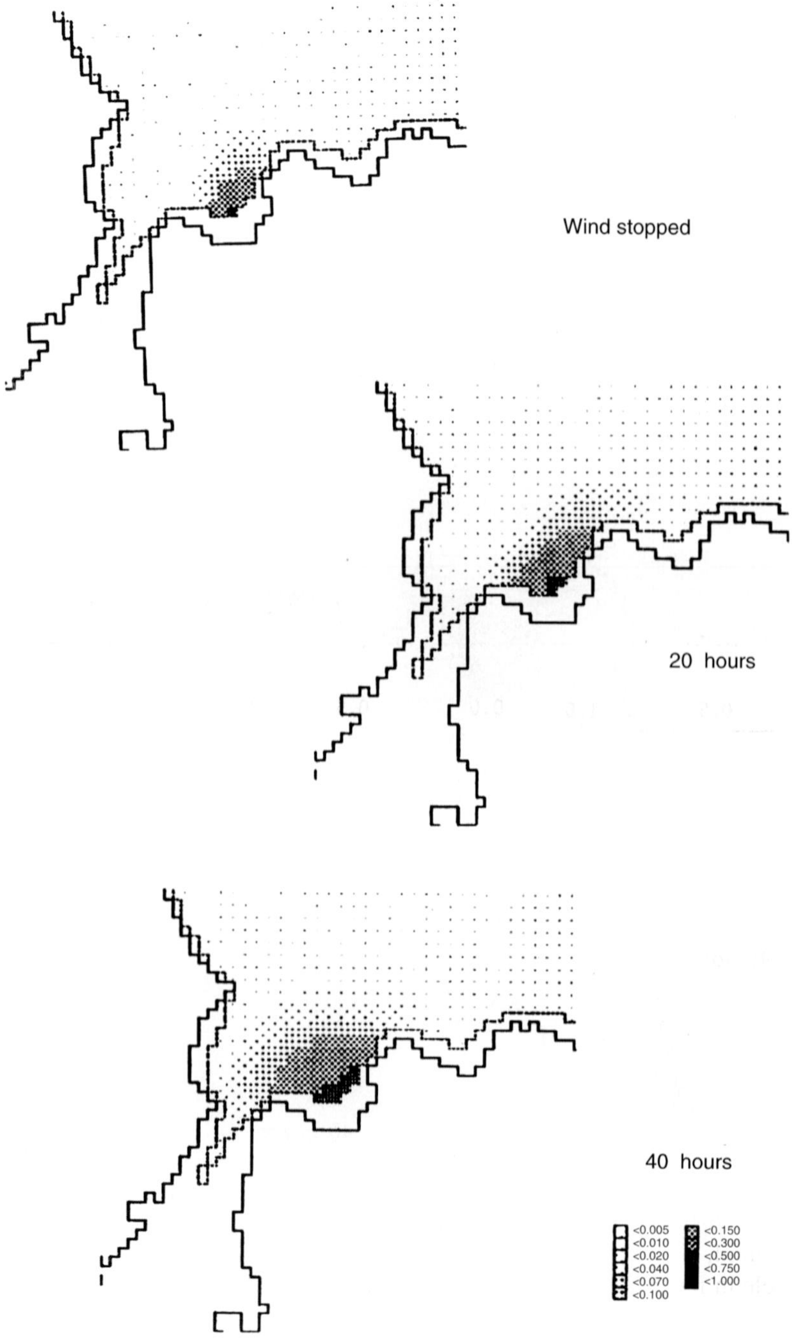

Fig. 10.39. Concentration distributions in the lowest layer (Shen, 1991).

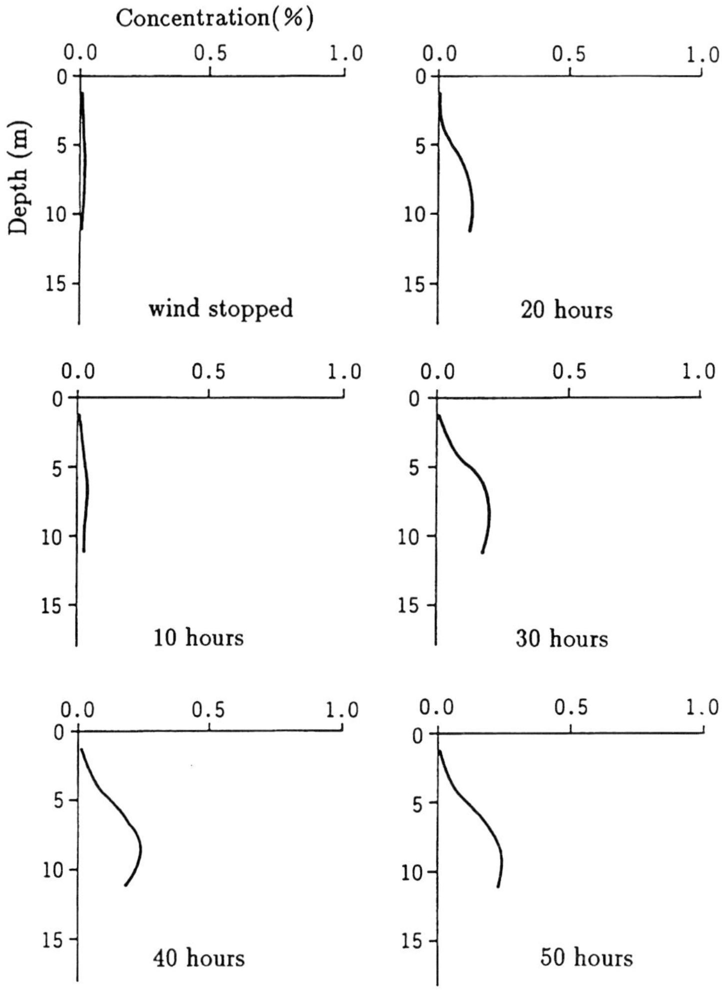

Fig. 10.40. Depth-wise concentration distributions at point C (Shen, 1991).

Ane River
Since Lake Biwa is located in the snow and cloud path from the Japan Sea to the Pacific Ocean, the northern area of the lake is among the few heavy snowfall areas in Japan. As the snow melts away and pours into Lake Biwa from the rivers and from underground, it plays an important role in water resources in the dry season, and a great role in improving the water quality of Lake Biwa. In this example, the Ane River, which is located in the heavy snowfall area, is examined, and the spreading process of the river water, due to the temperature difference between the river water and lake water, is analyzed.

In order to investigate the flows in the estuary of the Ane River in detail, the estuary is sub-divided, as shown in Figure 10.42. Since the Ane River pours into

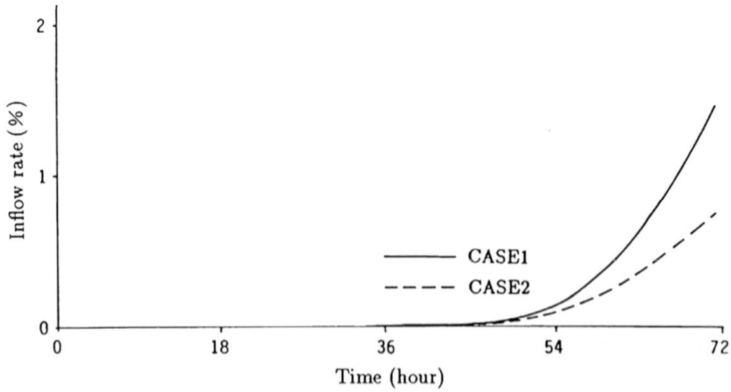

Fig. 10.41. Proportion of tracer entranced the south basin compared to the total inflow discharges (Shen, 1991).

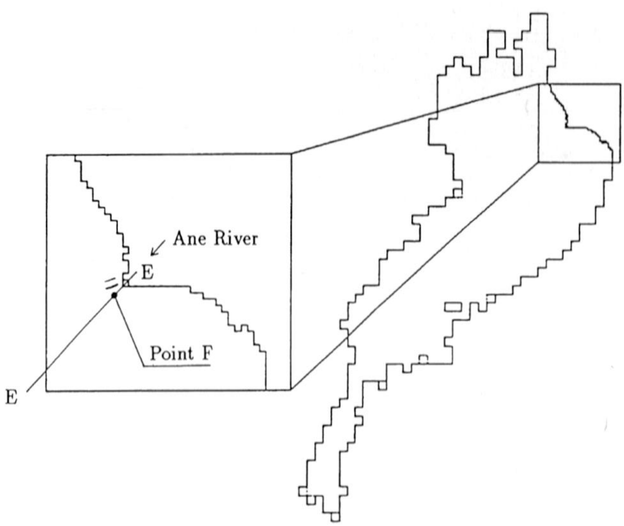

Fig. 10.42. Combined use of coarse-fine grid to the Ane River (Shen, 1991).

Lake Biwa from the northeastern direction, which is the diagonal direction to the mesh, the discharge of the Ane River must be divided into x- and y-wise components. However, this treatment makes the width of the Ane River 500 m, and therefore the velocity at the estuary of the Ane River is underestimated in the computation. Because the water temperature of the lake is quite different from that of the inflow river, and the density current in the river estuary is assumed to occur, it is problematic to assume the uniform distribution of velocity in the depth-wise direction at the river estuary. In this calculation, a 2 km river from the estuary toward the upstream direction is assumed and added to the computational domain. A uniform distribution is given at the upstream section, and the initial water

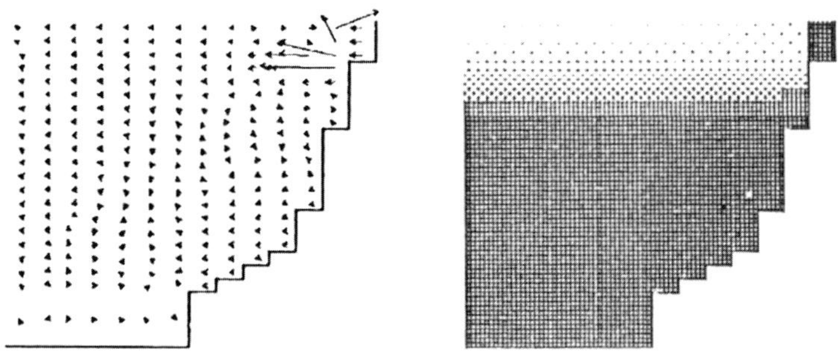

Fig. 10.43. Velocity and temperature distributions in E–E section (including the upstream river) (Shen, 1991).

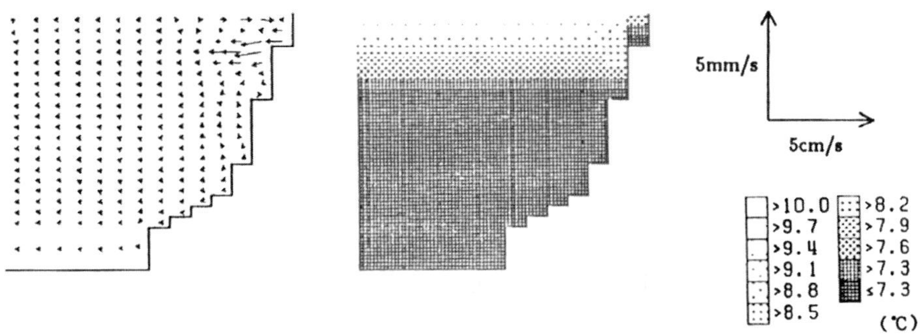

Fig. 10.44. Velocity and temperature distributions in E–E section (not including the upstream river) (Shen, 1991).

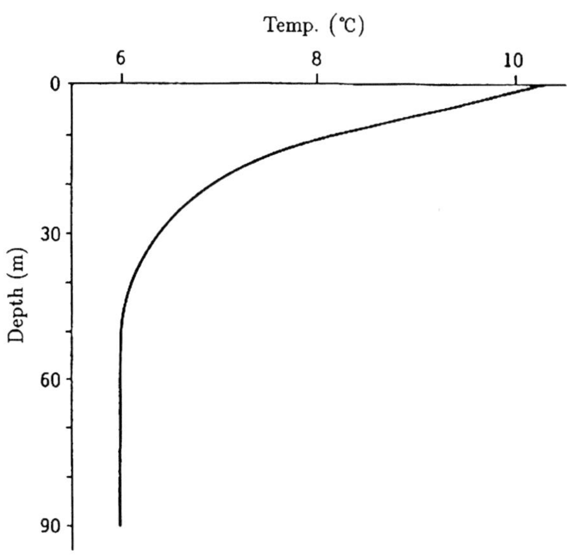

Fig. 10.45. Initial water temperature distribution (Shen, 1991).

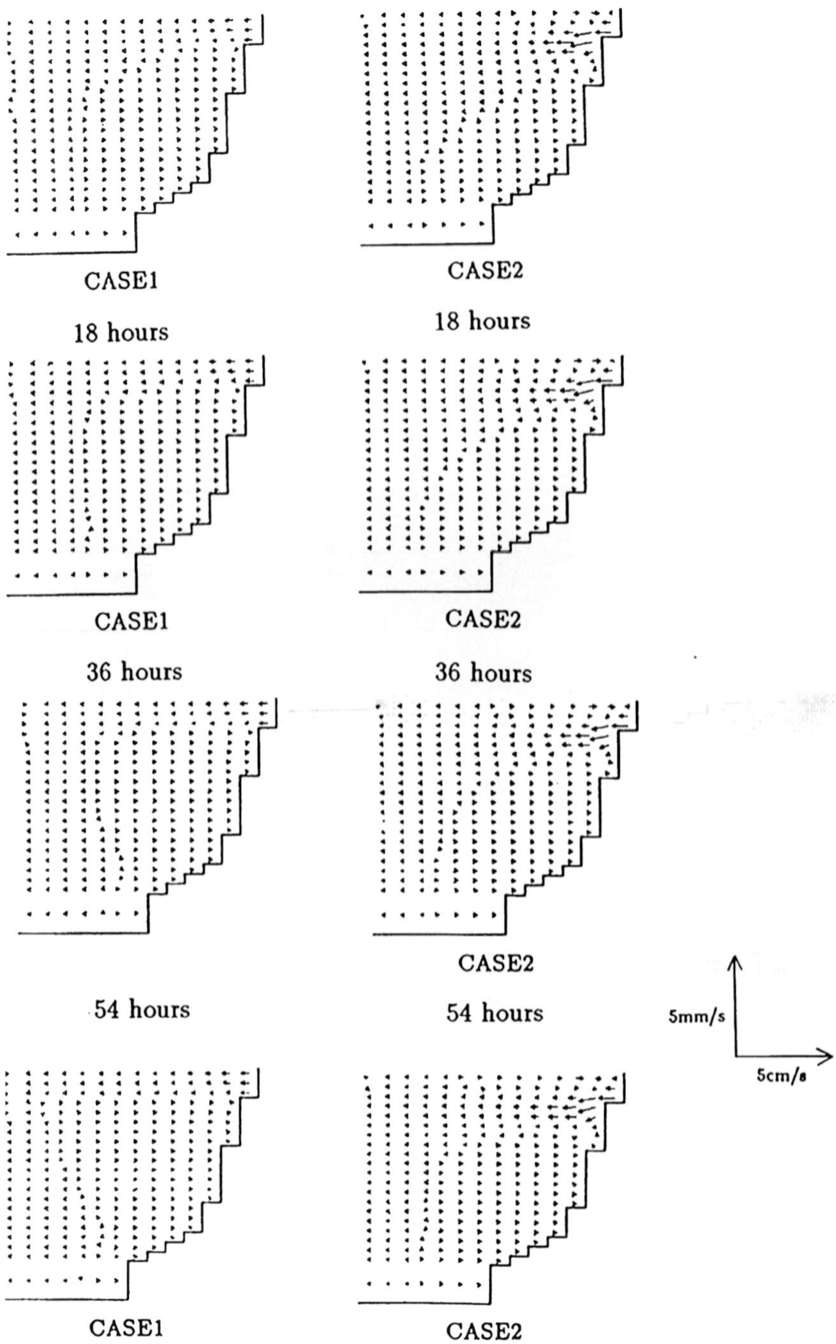

Fig. 10.46. Velocity distributions in E–E section (Shen, 1991).

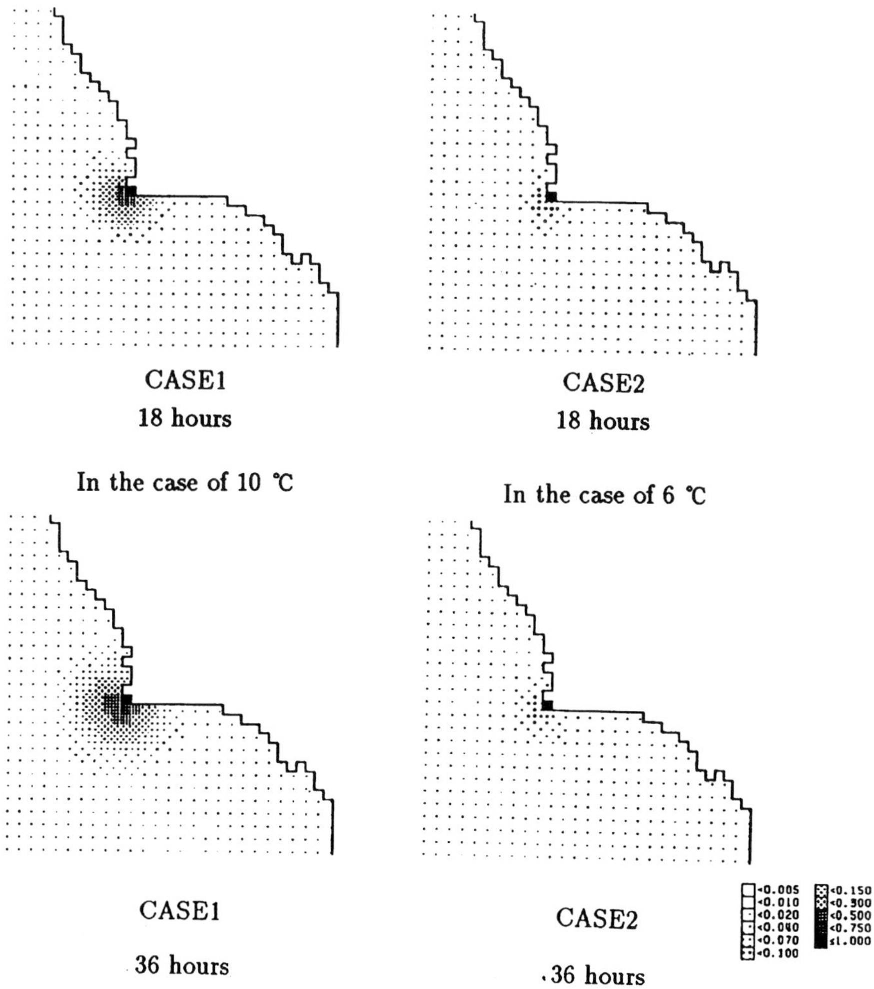

Fig. 10.47. Concentration distributions in the surface layer (Shen, 1991).

temperature of the river is given as 6°C. The calculated results are illustrated in Figure 10.43. In this figure, the predominant flow toward the river estuary, seen in Figure 10.44, cannot be observed, and the velocity is larger in the upper layers and smaller in the lower layers. Therefore, if the upstream river is included in the calculation, a fairly large velocity gradient in the vertical direction at the estuary results, but it must be verified by observation whether the velocity distribution in the vertical direction at the Ane River estuary actually exists. Moreover, though the river portion is assumed to be 2 km, by trial, in this calculation, there is little evidence to verify that this length is appropriate. At the initial condition, the water temperature distribution in the lake is shown in Figure 10.45. The water temperatures of the inflow river are given as the same as those in the lake, that is., 10°C at the surface layer (Case 1) and 6°C at the lowest layer (Case 2).

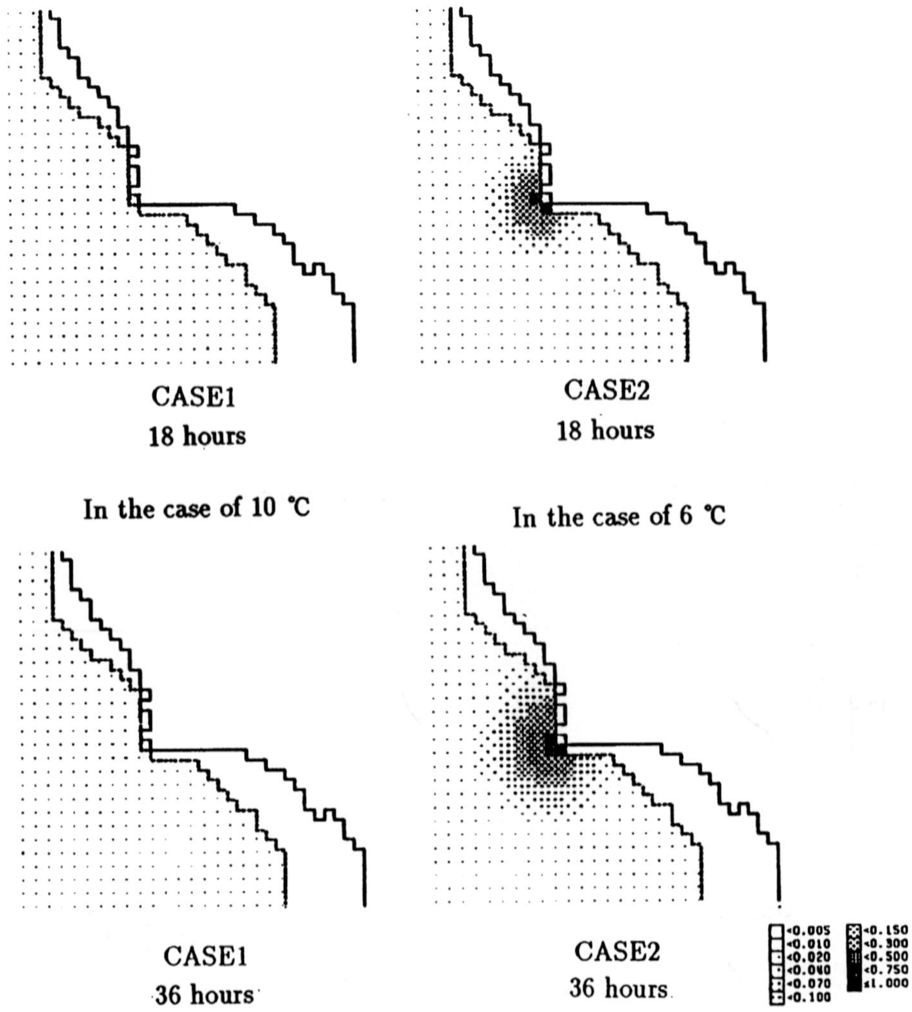

Fig. 10.48. Concentration distributions in the fourth layer (Shen, 1991).

From the velocity distributions of the E–E section, shown in Figure 10.46, it is observed that the velocity distribution at the estuary in Case 1 is almost uniform, while the current in Case 2 tends to move into the lower layers of the lake. In order to observe the spread of the river water, the tracer with concentration 100% is given as the boundary condition of all the cells included in the column of the river. Figures 10.47 and 10.48 present the concentration distributions in the surface layer and fourth layer, respectively (the diffusive equation of the tracer is solved only in the lake). In Case 1, the tracer spread greatly at the surface layer and did not reach the fourth layer, while in Case 2, the tracer spread greatly into the fourth layer. This difference can be seen more clearly in Figure 10.49 at point F. Ordinarily, when the water from a river whose density is relatively large flows into a stratified lake, the

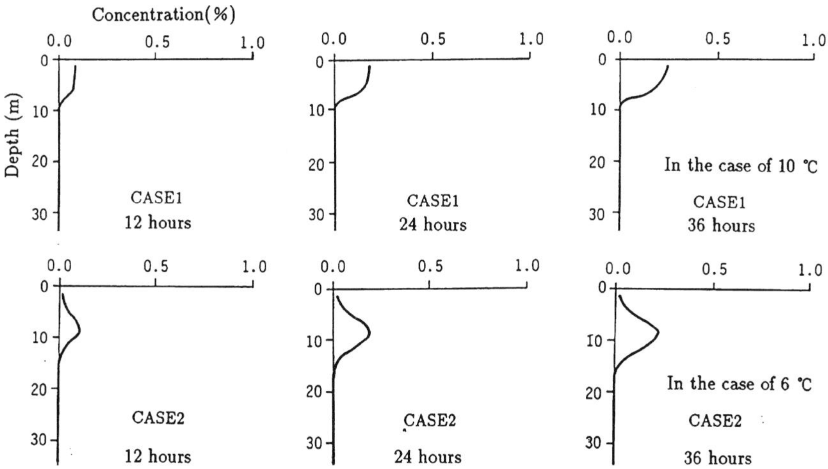

Fig. 10.49. Depth-wise concentration distributions at point F (Shen, 1991).

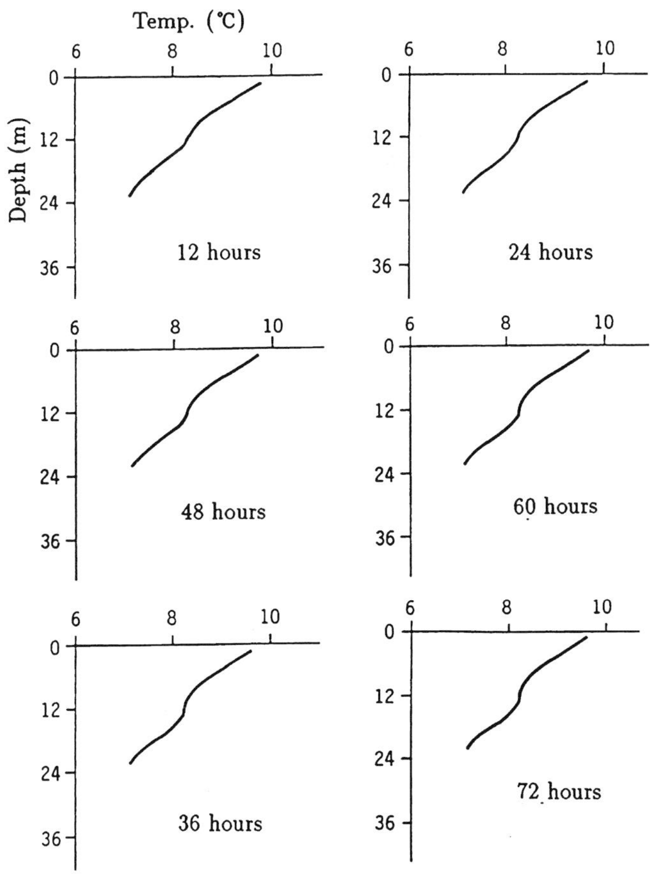

Fig. 10.50. Temporal variation of depth-wise water temperature distribution (Shen, 1991).

river water mixes with the lake water and advances along the bottom as density current; upon arriving at a layer which has the same density as its own, it will shift to a horizontal direction and spread toward the center of the lake. In this calculation, although the spread at the low layers in Case 2 is larger than in Case 1, the spread toward the deeper layers is not so obvious. As mentioned above, this is possibly because the density current along the lake bed has not been reproduced completely. In the grid system in this example, the bathymetry is expressed by the "stairs" topography, which deepens along the direction of the density current moving forward, and the gradient of the lake bottom at the Ane River estuary is steep; thus, the water depths of the columns along the density current are quite different. Therefore, the stratification will become unstable, due to the density inversion, and instantaneous convection will occur in these columns in computation. Figure 10.50 shows the temporal variation of depth-wise water temperature distribution at the neighborhood of the estuary. The portion where water temperature distribution becomes vertically uniform is the result of instantaneous convection, due to the computational reason mentioned earlier. Although the large vertical velocity along the inclined lake bottom could not be evaluated completely, due to the instantaneous convection, it was understood that the spread of river water depends strongly on the inflow water temperature, and the process of spread was well reproduced by this calculation.

Chapter 11

THE FUTURE IN HYDRODYNAMIC MODELLING

Three-dimensional hydrodynamic models have been used for the last 25 years to study the hydrodynamic properties of lakes, bays, and estuaries (Table 11.1, Tsanis and Wu, 1994; Tables 11.1 and 11.2, Cheng and Smith, 1990). These models are used as diagnostic and prognostic tools in lake and coastal water studies. A successful application of the 3D models in water bodies should be accompanied by adequate field data for proper calibration.

A proper simulation of the hydrodynamic behavior of a water body requires a large computational effort. Many applications of these models are executed in supercomputers with vector and parallel processing. A 3D code should be vectorized in order to reduce substantially the required computational time. The trend toward the use of supercomputers for execution of large 3D models will continue. Use of 3D models will also be increased in the coming years; one major reason being the faster PC computers that can allow execution of smaller-scale applications of these models in reasonable time, and medium-scale applications in multi-processing workstations.

Advanced Technologies such as Geographic Information Systems (GIS) will be used as a management tool for the pre-processing model inputs and the post-processing model results. GIS is an efficient database management system with exceptional graphic capabilities. Improved and versatile data management will aid in model calibration and verification. Use of advanced graphics will permit the visualization of large volumes of data in both space and time that represent different processes. User-friendly interfaces will help in disseminating research results to both technical and non-technical audiences.

Study of the ecological health of lakes and coastal waters requires hydrodynamic studies for longer time periods, that is, weeks or months. With the increasing computer speed hydrodynamic simulations longer than the short time span studies reported to the literature, will be possible and subsequently will improve the understanding of processes in ecosystems. Many field studies over the last 25 years have designed independent from any modelling, often leading to inadequate field program design and to incomplete results. Any field program designed in the future should be based on numerical modelling in order to achieve optimum results, that is, data collection and numerical modelling should be the two components of an integrated program. In addition, due to the complexity of the ecosystems, new technologies such as "Hydroinformatics" or in general terms, "Environmental Information Systems (EIS)," taking advantage of the rapid advancements of information technology, will be necessary to better manage complex ecosystems.

Table 11.1. Approximation and features used in 3D estuarine models.

Features or approximations*	Abbreviations
Advection	
Non-linear	AN
Neglected	AL
Bottom friction	
Quadratic	
Drag coefficient	BFD
Turbulent boundary layer	BFT
Linearized	BFL
Density	
Homogeneous (constant density)	CD
Stratified	
Baroclinic, diagnostically	BD
Baroclinic, coupled	BC
Turbulence models	
Constant or empirical	CE
One-equation model	T1
Two-equation model	T2
σ-Transformation	ZT
No transformation	ZNO
x,y transformation	
No transformation	XYNO
Algebraic	XYA
Curvilinear orthogonal	XYCO
Boundary fitted	XYBF
Treatment of z-variable	
Ekman dynamics	ZEK
Eigenfunction expansion	ZEF
Galerkin	ZG
Numerical, finite difference	ZNFD
Numerical, finite element	ZNFE
Treatment of x,y variables	
Finite difference	XYFD
Finite element	XYFE
Treatment of time variable	
Time stepping	TSP
Internal-external mode splitting	TMS
Spectral	SPL
Wetting and drying	
Allowed	YWD
Not allowed	NWD

Source: Adapted from Cheng and Smith (1990), with permission from American Society of Civil Engineers.

*Common approximations include incompressible flow, hydrostatic approximation, Boussinesq approximation.

Table 11.2. Examples of 3D numerical estuarine models.

Models	Features*
Backhaus (1980)	AN, BFD, BC, CE, ZT, XYNO, ZNFD, XYFD, YWD, TMS
Blumberg and Herring (1987)	AN, BFT, BC, T2, ZT, XYCO, ZNFD, XYFD, NWD, TMS
Blumberg and Mellor (1987), Oey et al. (1985a, b, c)	AN, BFT, BC, T2, ZT, XYNO, ZNFD, XYFD, NWD, TMS
Caponi (1976)	AN, BFD, BC, CE, ZNO, XYNO, ZNFD, TSP, XYFD, NWD
Davies (1980)	AL, BFD, CD, CE, ZT, XYNO, ZEK, XYFD
Davies (1983)	AL, BFL, BC, CE, ZT, XYNO, ZG, ZEF, XYFD, NWD
Feng (1977)	AL, BFT, CD, CE, ZNO, XYNO, ZEK, XYFD, SPL, NWD
Gordon and Spaulding (1987)	AN, BFD, CD, CE, ZT, XYNO, ZG, XYFD, TMS, NWD
Heaps (1972, 1973)	AL, BFL, CD, CE, ZNO, XYNO, ZEF, XYFD, TSP, NWD
Kawahara et al. (1983)	AN, BFT, CD, CE, ZNO, XYNO, ZNFE, XYFE, TSP, NWD
King (1985)	AN, CD, CE, ZT, XYNO, ZNFE, TSP, NWD
Leendertse and Liu (1975), Liu and Leendertse (1978)	AN, BFD, BC, T2, ZNO, XYNO, ZNFD, TSP, XYFD, NWD
Lynch and Werner (1987)	AL, BFL, CD, CE, ZT, XYNO, ZNFE, XYFE, SPL, NWD
Owen (1980)	AN, BFD, BC, CE, ZT, XYNO, ZG, XYFD, TSP, NWD
Sheng (1987)	AN, BFT, BC, CE, ZT, XYA, XYBF, ZNFD, XYFD, TMS, NWD
Stephens (1986)	AN, BFD, BC, BD, CE, ZT, XYNO, ZNFD, XYFD, TSP, YWD
Tee (1987)	AL, BFD, CD, CE, ZNO, XYNO, ZEK, XYFD, SPL, NWD

*See footnote in Table 11.1

11.1 ENVIRONMENTAL INFORMATION SYSTEMS (EIS)

Environmental Information Systems is a domain of interaction of Information Technologies such as GIS, artificial intelligence, and visualization systems with environmental models such as hydrodynamic, pollutant transport, ecological, and econometric models. With the rapid changes in information technologies, involving software as well as hardware developments, environmental information systems will play a significant role bridging information and fundamental and applied knowledge produced through research, engineering, and management practice. This interaction enables these systems to support decision making in environmental engineering and management. As a result, EIS will aid in sustainable development, protection of the ecosystems in a healthy environment.

Environmental Information Systems applications in the aquatic domain are usually referred to as *hydroinformatics systems*. These systems can handle the flow of information and knowledge within hydrology and hydraulics in all its aspects. Knowledge about the aquatic environment must be presented in a fast and flexible manner, but the speed and flexibility must not jeopardize the limitations and constraints built into the actual knowledge that is presented. Knowledge must be presented in framework systems that handle not only the hydroinformatics but also present information from other related sources. These can be achieved by introducing the types of electronic knowledge encapsulators that are currently being brought together and integrated in real-time-control, alarm-handling, computer-aided design, computer-aided management, and other systems (Abbott, 1991).

A decision-support system for the management of the North Sea is a successful application of hydroinformatics. This Database Management and Modelling System (DMMS) is a tool which encapsulates model and data and provides different users the means to use different components without the need to have knowledge about and technical expertise. The system's interface allows the user to see directly the links between model data and ecological or economical problems of the study domain (Van Zuylen et al., 1994). Another system dealing with scenarios concern hydrodynamic tide, wind and location of waste outfalls is the Coastal Modelling System (CMS). The different user tasks in CMS are (Van Zuylen et al., 1994):

(a) Project definition—selection of geographical area
(b) Scenario definition—selection of physical process, entry of parameters
(c) Scenario execution—hydrodynamics, mid-field water quality
(d) User interface—far field water quality, sediment transport, waves
(e) Visualization of results—screen and hard copy plots, animations, export of results
(f) Copy, archive, unarchive, remove, create—scenario or project results.

The goal of such a system is to minimize possible effects of waste water disposal by examining and comparing different options. The users can decide which options are acceptable, ensuring compliance to local laws and legislation.

An example of this type of integrated system may be found in the hydrodynamic/GIS interface system developed in the EIS Laboratory in the Department of Civil Engineering at McMaster University. This application makes use of GIS technology for the management of shoreline geometry, bathymetric data, and monitoring station data. Figures 11.1 and 11.2 show the Metropolitan Toronto waterfront area in Lake Ontario and the Thessaloniki Bay in the Aegean Sea. This menu-based system generates 3D models of water bodies, permits sub-areas to be extracted for analysis, provides surface quality control procedures, produces gridded shoreline files, and provides easy reporting, display, and plotting of database layers (Tsanis et al., 1996). In addition model output, such as pollutant concentrations and velocity vectors, may also be integrated. An example of a closely coupled hydrodynamic pollutant transport GIS model is the IDOR^{2D} GIS (Tsanis

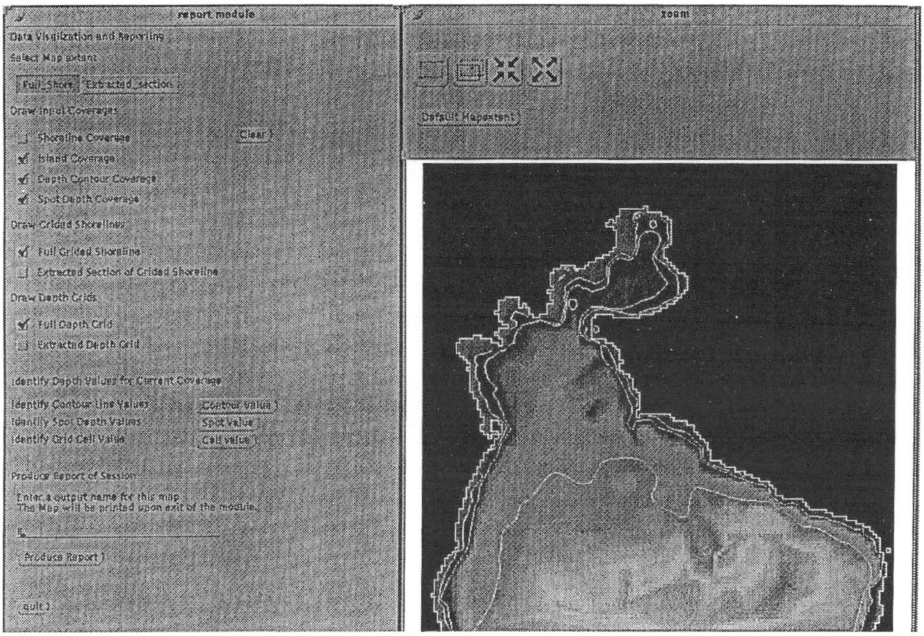

Fig. 11.1. GIS interface for the Thessaloniki Bay, Greece.

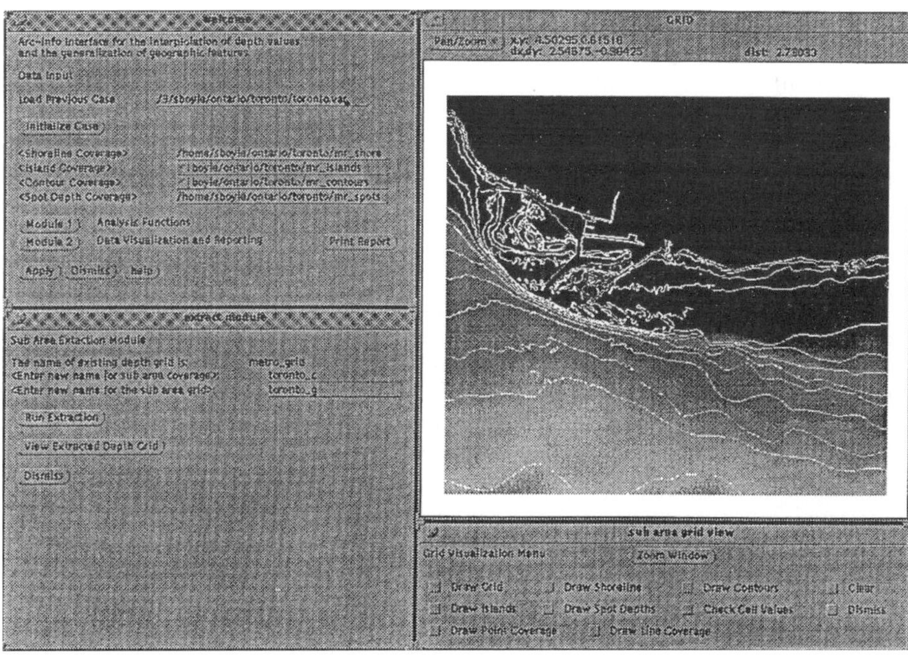

Fig. 11.2. GIS interface for the Metropolitan Toronto waterfront, Ontario, Canada.

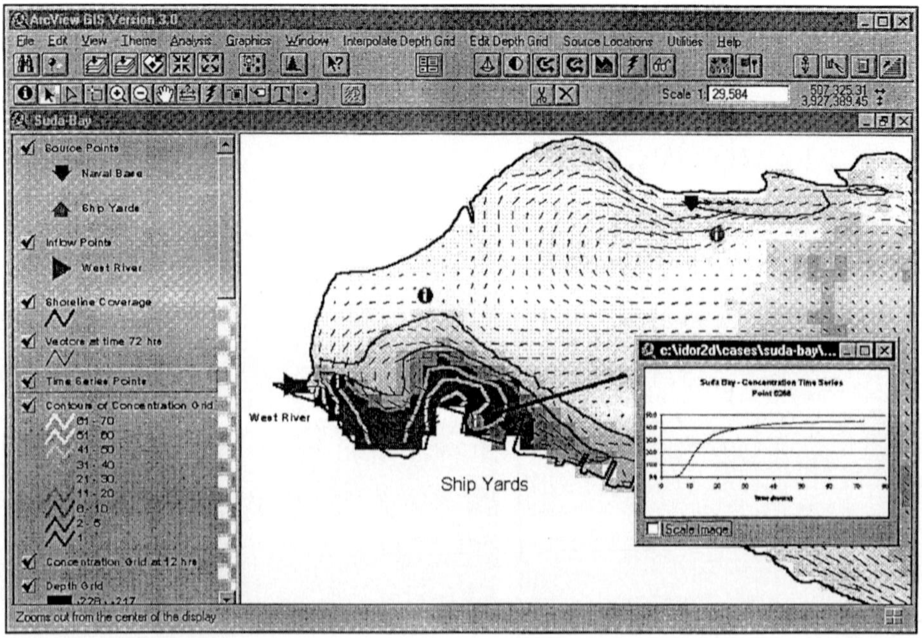

Fig. 11.3. Arc view GS interface for IDOR^{2D} GIS.

and Boyle, 1997). The arc view interface for IDOR^{2D} GIS is shown in Figure 11.3. This tool enables non-GIS users the ability to perform advanced tasks within this user-friendly environment.

The coupling of information technology tools and environmental engineering/water resources models will enhance decision making in engineering and management practice. Until the last decade the practice was to operate separate modelling systems by well-experienced users. The new integrated systems with state-of-the-art interfaces will allow all levels of users to apply them with a minimum of operational effort and generate high-quality output. Environmental problems are interdisciplinary in nature. The users do not need to have expertise in all kinds of technical issues, so water resource managers, engineers and scientists, decision makers and politicians that have the task to solve environmental problems, can use these systems. Possible alternative strategies, feasibility and impact of measures in solving an environmental problem can be evaluated by the users and solutions can be found that result in an acceptable environmental impact.

Appendix

DOCUMENTATION OF A TWO-DIMENSIONAL CIRCULATION MODEL

In this appendix, the FORTRAN code for a 2D circulation model is included. It also has an option for quasi 3D simulations. The model predicts the depth-averaged currents under steady state and variable wind conditions. Users can use the sample depth and wind data for demonstration runs. The program can be executed easily by simply following and responding to the programs' prompts. The user should use one of the commercial graphics packages to view and plot the simulation results.

INPUTS

Sample depth data (file name = depth.dat)

The model basin is a 10°km × 10°km rectangular lake with uniform depth along the west–east direction and shallow near the northern and southern shores. The center basin is 15deep. The grid size is 500 m.

```
22  22500.43.5
0.  0.  0.  0.  0.  0.  0.  0.  0.  0.  0.  0.  0.  0.  0.  0.  0.  0.  0.  0.  0.  0.
0.  5.  5.  5.  5.  5.  5.  5.  5.  5.  5.  5.  5.  5.  5.  5.  5.  5.  5.  5.  5.  0.
0.  7.  7.  7.  7.  7.  7.  7.  7.  7.  7.  7.  7.  7.  7.  7.  7.  7.  7.  7.  7.  0.
0.  9.  9.  9.  9.  9.  9.  9.  9.  9.  9.  9.  9.  9.  9.  9.  9.  9.  9.  9.  9.  0.
0.  11. 11. 11. 11. 11. 11. 11. 11. 11. 11. 11. 11. 11. 11. 11. 11. 11. 11. 11. 11. 0.
0.  15. 15. 15. 15. 15. 15. 15. 15. 15. 15. 15. 15. 15. 15. 15. 15. 15. 15. 15. 15. 0.
0.  15. 15. 15. 15. 15. 15. 15. 15. 15. 15. 15. 15. 15. 15. 15. 15. 15. 15. 15. 15. 0.
0.  15. 15. 15. 15. 15. 15. 15. 15. 15. 15. 15. 15. 15. 15. 15. 15. 15. 15. 15. 15. 0.
0.  15. 15. 15. 15. 15. 15. 15. 15. 15. 15. 15. 15. 15. 15. 15. 15. 15. 15. 15. 15. 0.
0.  15. 15. 15. 15. 15. 15. 15. 15. 15. 15. 15. 15. 15. 15. 15. 15. 15. 15. 15. 15. 0.
0.  15. 15. 15. 15. 15. 15. 15. 15. 15. 15. 15. 15. 15. 15. 15. 15. 15. 15. 15. 15. 0.
0.  15. 15. 15. 15. 15. 15. 15. 15. 15. 15. 15. 15. 15. 15. 15. 15. 15. 15. 15. 15. 0.
0.  15. 15. 15. 15. 15. 15. 15. 15. 15. 15. 15. 15. 15. 15. 15. 15. 15. 15. 15. 15. 0.
0.  15. 15. 15. 15. 15. 15. 15. 15. 15. 15. 15. 15. 15. 15. 15. 15. 15. 15. 15. 15. 0.
0.  15. 15. 15. 15. 15. 15. 15. 15. 15. 15. 15. 15. 15. 15. 15. 15. 15. 15. 15. 15. 0.
0.  15. 15. 15. 15. 15. 15. 15. 15. 15. 15. 15. 15. 15. 15. 15. 15. 15. 15. 15. 15. 0.
0.  15. 15. 15. 15. 15. 15. 15. 15. 15. 15. 15. 15. 15. 15. 15. 15. 15. 15. 15. 15. 0.
0.  11. 11. 11. 11. 11. 11. 11. 11. 11. 11. 11. 11. 11. 11. 11. 11. 11. 11. 11. 11. 0.
0.  9.  9.  9.  9.  9.  9.  9.  9.  9.  9.  9.  9.  9.  9.  9.  9.  9.  9.  9.  9.  0.
0.  7.  7.  7.  7.  7.  7.  7.  7.  7.  7.  7.  7.  7.  7.  7.  7.  7.  7.  7.  7.  0.
0.  5.  5.  5.  5.  5.  5.  5.  5.  5.  5.  5.  5.  5.  5.  5.  5.  5.  5.  5.  5.  0.
0.  0.  0.  0.  0.  0.  0.  0.  0.  0.  0.  0.  0.  0.  0.  0.  0.  0.  0.  0.  0.  0.
```

Sample wind data (file name = wind.dat)

The input wind can be steady (the wind speed and direction are required) or variable (hourly wind speed and direction for a period specified by the user are required). Here is an example of hourly wind data:

```
Title: sample wind data
48
1  2.5  245.0
2  2.5  250.0
3  4.5  300.0
4  6.0  355.0
5  7.0  331.0
6  5.6  301.0
7  6.0  266.0
8  8.0  188.0
9  2.0  155.0
10  2.5  245.0
11  2.9  244.0
12  2.5  250.0
13  4.5  300.0
14  6.0  355.0
15  7.0  331.0
16  5.6  301.0
17  6.0  266.0
18  8.0  188.0
19  2.0  155.0
20  3.4  160.0
21  2.5  245.0
22  2.5  250.0
23  4.5  300.0
24  6.0  355.0
25  7.0  331.0
26  5.6  301.0
27  6.0  266.0
28  8.0  188.0
29  2.0  155.0
30  2.9  180.0
31  2.5  245.0
32  2.5  250.0
33  4.5  300.0
34  6.0  355.0
35  7.0  331.0
36  5.6  301.0
37  6.0  266.0
38  8.0  188.0
39  2.0  155.0
40  8.0  177.0
41  2.5  245.0
42  2.5  250.0
43  4.5  300.0
44  6.0  355.0
45  7.0  331.0
46  5.6  301.0
47  6.0  266.0
48  8.0  188.0
```

The first line of the wind data file is reserved for title information such as station location, elevation, sample rate. The second line specifies the number of hours for the episode under consideration. Each of the remaining lines indicates the time, wind speed, and wind direction.

PROGRAM EXECUTION

Save the code as 2d.f90, compile the program using a FORTRAN compiler and create an executable file 2d.exe. Choose one parameter from the brackets just below the input prompts. The following example considers at a steady wind case with parameter choices shown within the brackets.

Wind condition? (1 = variable; 0 = steady)
0

Input wind components: wx = ? wy = ? (m/s)
(Use wind vector, for example: wx = 5., wy = 0. for a west wind)
5. 5.

Input the simulation time (hours)= ?
48.

Input a filename for the model output= ?
out.dat

The simulation begins with the following major parameters:

Time step	dt = 15. second
Wind velocity	wx = 5. wy = 0.
Coriolis coefficient	f = 0.00051
Surface drag coefficient	cs = 0.0000018
Bottom drag coefficient	cb = 0.025

***** The program is running *****

The program then displays the kinetic energy and the kinetic energy difference between two consecutive time steps every hour.

***** End of the Program *****

The following prompts and user responses are for a variable wind case.

Wind condition ?
1

Input wind data file (wind.dat) = ?
wind.dat

Four output times (t1, t2, t3, t4) = ?
12 24 36 48

Four output files = ?
out1.dat
out2.dat
out3.dat
out4.dat

SOURCE CODE

```
! ----------------- program 2d.f90 ------------------------------

! Two-Dimensional and Quasi Three-Dimensional Hydrodynamic Model

! imax=number of grids in x-direction (i=1,2, ...imax)
! jmax=number of grids in y-direction (j=1,2, ...jmax)
! is(j),ie(j)=the start and end grid index along each j
!             calculated from the depth in subroutine <domboun>
! d(i,j)=water depth in meters at center of each grid
! h(i,j)=free-water elevation above the bottom
! u,v(i,j)=velocity components along x and y directions
! ib,jb,nb=boundary grid types and numbering by the program
! dx,dt=grid size and time step

! wx,wy=wind speed
! cs=wind drag coefficient
! tcheck,tlast=K.E check and calculation time
! th=smooth factor (horizontal diffusivity)
! phi=mean latitude of the waterbody
! n_model=model options: n_model=1 for 2-D model;
!             n_model=2 for Q3D model

! ________________________________________________________________


dimension umn(100,100),um(100,100),vmn(100,100),vm(100,100)
dimension z(100,100),zn(100,100),is(100),ie(100),ib(1000)
dimension tws(100),twd(100),jb(1000),h(100,100),nb(1000)
character mac*60,fn1*20,fn2*20,f1*20,f2*20,f3*20,f4*20


write(*,'(/)')
write(*,'(" ********************************************" ,/)')
write(*,'(" Circulation Model ",//)')
write(*,'(" *********************************************" ,/)')

write(*,'(" Wind condition (1=variable; 0=steady) ",/)')
read(*,*) n_wind


if(n_wind.eq.0) then
   write(*,'(" Input wind components: wx=? wy=?(m/s)",/)')
   read(*,*) wx,wy
else
   write(*,'(/," Input wind data file (wind.dat)=? ",/)')
   read(*,'(a)') fn1
endif

write(*,'(/," Input the simulation time (hours)=",/)')
read(*,*) tlast
if(n_wind.eq.0) then
   write(*,'(/," Output filename (out.dat)?",/)')
   read(*,'(a)') fn2
else
   write(*,'(/," Four output times: ")')
   write(*,'("(For example: 12,24,36,48) ",/)')
   read(*,*) t1,t2,t3,t4
   write(*,'(/," Four output files:")')
   write(*,'("(example: out1.dat, out2.dat, out3.dat, out4.dat)",/)')
   read(*,'(a)') f1,f2,f3,f4
endif
```

```
! open depth file
  open (unit=20,file='depth.dat')

!--- n_model is set to 2 in this version
!

  n_model=2

  dt=15.0
  tlast=tlast*3600.00
  th=0.95
  cs=0.0000018
  cb=0.0025

! read in wind data
  if(n_wind.eq.1) then
    open (unit=10,file=fn1)
    read(10,'(a)') mac
    write(*,'(a)') mac
    read(10,*) npwind
    do i=1,npwind
       read(10,*) iq,tws(i),twd(i)
       write(*,*) iq,tws(i),twd(i)
    enddo
  endif
  read(20,*) imax,jmax,dx,phi
  do j=jmax,1,-1
    read(20,*)(h(i,j),i=1,imax)
  enddo
  call domboun(imax,jmax,nob,h,is,ie,ib,jb,nb)

  pi=3.14159/180.0
  cf=2.0*pi*180.*sin(phi*pi)/24/3600
  wx0=wx
  wy0=wy
  av=(1.-th)*(dx)**2./(4.*dt)

  write(*,'(/)')
  write(*,'(21x,"********** Major Parameters Used ****")')
  write(*,89) imax,jmax,dx,phi
  write(*,90) dt
  write(*,91) wx,wy
  write(*,'(/)')
  write(*,'(21x,"********** Program Is Running ******",/)')
89 format(/,20x,'imax,jmax,dx,phi =',2i5,2x,2f8.3)
90 format(/,20x,'time step dt =',f5.2,' (sec)')
91 format(/,20x,'wind velocity wx=',f8.4,' wy=',f8.4)

  en=0.
  e=0.
  time=0.
  100 n=n+1
  time=time+dt

! wind input
  if(n_wind.eq.1) then
     call wind(n,time,cs,tws,twd,wx,wy,tsx,tsy,ts,wxx,wyy)
  else if(n_wind.eq.0) then
     if(n.lt.300) then
     wx=n/300.0*wx0
     wy=n/300.0*wy0
```

```
   else
      wx=wx0
      wy=wy0
   endif
      tsx=cs*wx*sqrt(wx*wx+wy*wy)
      tsy=cs*wy*sqrt(wx*wx+wy*wy)
      ts=sqrt(tsx*tsx+tsy*tsy)
      tss=sqrt(ts)
endif

do j=2,jmax-1
   do i=2,imax-1
      if(h(i,j).le.0.0) cycle
      h1=0.5*(z(i+1,j)+z(i,j)+h(i+1,j)+h(i,j))
      h2=0.5*(h(i,j)+h(i-1,j)+z(i,j)+z(i-1,j))
      h3=0.5*(h(i,j+1)+h(i,j)+z(i,j+1)+z(i,j))
      h4=0.5*(h(i,j)+h(i,j-1)+z(i,j)+z(i,j-1))
      zn(i,j)=z(i,j)-dt*(h1*um(i+1,j)-h2*um(i,j)+h3*vm(i,j+1)&
            -h4*vm(i,j))/dx
   enddo ! i
enddo ! j

! ---------------------------------------------------
! Depth-Averaged Circulation Model (2DH)

if(n_model.eq.1) then
do j=2,jmax-1
   do i=is(j),ie(j)
            if(h(i,j).le.0.0) cycle
h0x=(h(i,j)+zn(i,j)+h(i-1,j)+zn(i-1,j))/2.
            v9=(vm(i,j)+vm(i,j+1)+vm(i-1,j)+vm(i-1,j+1))/4.
            u66=(um(i+1,j)+um(i-1,j)+um(i,j+1)+um(i,j-1))/4.
            bsx=cb*um(i,j)*sqrt(um(i,j)**2+v9**2)
            h0y=(h(i,j)+zn(i,j)+h(i,j-1)+zn(i,j-1))/2.
            u9=(um(i,j)+um(i,j-1)+um(i+1,j)+um(i+1,j-1))/4.
v66=(vm(i+1,j)+vm(i,j+1)+vm(i-1,j)+vm(i,j-1))/4.
            bsy=cb*vm(i,j)*sqrt(vm(i,j)**2+u9**2)
            u0=th*um(i,j)+(1.0-th)*u66
            u1=-um(i,j)*(um(i+1,j)-um(i-1,j))/(2.*dx)
            u2=-v9*(um(i,j+1)-um(i,j-1))/(2.*dx)
            u3=-9.81*(zn(i,j)-zn(i-1,j))/dx
            u4=cf*v9
            u5=(tsx-bsx)/h0x
            v0=th*vm(i,j)+(1.0-th)*v66
            v1=-vm(i,j)*(vm(i,j+1)-vm(i,j-1))/(2.*dx)
            v2=-u9*(vm(i+1,j)-vm(i-1,j))/(2.*dx)
            v3=-9.81*(zn(i,j)-zn(i,j-1))/dx
            v4=-cf*u9
            v5=(tsy-bsy)/h0y
            umn(i,j)=u0+dt*(u1+u2+u3+u4+u5)
            vmn(i,j)=v0+dt*(v1+v2+v3+v4+v5)
   enddo ! i
enddo ! j
endif
!---------------------------------------------------

! Quasi-Three Dimensional Parabolic Model (Q3D)
```

```
!
  if(n_model.eq.2) then
  alambda=0.020
  axx=1.0/alambda*tsx/sqrt(ts)
  ayy=1.0/alambda*tsy/sqrt(ts)
  do j=2,jmax-1
    do i=is(j),ie(j)
          if(h(i,j).le.0.0) cycle
          h0x=(h(i,j)+zn(i,j)+h(i-1,j)+zn(i-1,j))/2.
          v9=(vm(i,j)+vm(i,j+1)+vm(i-1,j)+vm(i-1,j+1))/4.
          u66=(um(i+1,j)+um(i-1,j)+um(i,j+1)+um(i,j-1))/4.
          bsx=3.0*alambda*um(i,j)*sqrt(ts)-0.5*tsx
          h0y=(h(i,j)+zn(i,j)+h(i,j-1)+zn(i,j-1))/2.
          u9=(um(i,j)+um(i,j-1)+um(i+1,j)+um(i+1,j-1))/4.
          v66=(vm(i+1,j)+vm(i,j+1)+vm(i-1,j)+vm(i,j-1))/4.
          bsy=3.0*alambda*vm(i,j)*sqrt(ts)-0.5*tsy
          u0=th*um(i,j)+(1.0-th)*u66
          u1=-(1.2*um(i,j)+axx/40.0)*(um(i+1,j)-um(i-1,j))/(2.*dx)
          u2=-(1.2*v9+ayy/40.0)*(um(i,j+1)-um(i,j-1))/(2.*dx)
          u3=-9.81*(zn(i,j)-zn(i-1,j))/dx
          u4=cf*v9
          u5=(tsx-bsx)/h0x
          v0=th*vm(i,j)+(1.0-th)*v66
          v1=-(1.2*vm(i,j)+ayy/40.0)*(vm(i,j+1)-vm(i,j-1))/(2.*dx)
          v2=-(1.2*u9+axx/40.0)*(vm(i+1,j)-vm(i-1,j))/(2.*dx)
          v3=-9.81*(zn(i,j)-zn(i,j-1))/dx
          v4=-cf*u9
          v5=(tsy-bsy)/h0y
          umn(i,j)=u0+dt*(u1+u2+u3+u4+u5)
          vmn(i,j)=v0+dt*(v1+v2+v3+v4+v5)
    enddo
  enddo
  endif

! prediction update
  do i=1,imax
    do j=1,jmax
      um(i,j)=umn(i,j)
      vm(i,j)=vmn(i,j)
      z(i,j)=zn(i,j)
    enddo
  enddo

! land boundary conditions
  do 200 k=1,nob
    i=ib(k)
    j=jb(k)
    nbb=nb(k)
      goto (200,205,210,215) nbb
        205 um(i,j)=0
        go to 200
        210 vm(i,j)=0.
        go to 200
        215 um(i,j)=0
        vm(i,j)=0
        go to 200
200 continue
```

```
! check kinetic energy
   e=en
   en=0.
   do j=1,jmax-1
     iss=is(j)
     iee=ie(j)
     do i=iss,iee
       ua=0.5*(um(i,j)+um(i+1,j))
       va=0.5*(vm(i,j)+vm(i,j+1))
       en=en+0.5*(ua*ua+va*va)*h(i,j)*dx*dx
     enddo
   enddo

   if(en.eq.0.0) go to 100
   de=abs((e-en)/en)
   if(mod(n,60).eq.0) then
                 write(*,390) time/3600.0,en,de
   endif
390 format(' time =',f7.0,2x,'en= ',e14.2,2x,'de =',f12.8)

! output velocity/elevation at four given times

   tm=time/3600.00
   n81=int(t1*3600.0/dt)
   n82=int(t2*3600.0/dt)
   n83=int(t3*3600.0/dt)
   n84=int(t4*3600.0/dt)

   if(n.eq.n81.or.n.eq.n82.or.n.eq.n83.or.n.eq.n84) then
     if(n.eq.n81) fn2=f1
     if(n.eq.n82) fn2=f2
     if(n.eq.n83) fn2=f3
     if(n.eq.n84) fn2=f4
     call outp(fn2,imax,jmax,axx,ayy,umn,vmn,um,vm,h,z)
   endif

   if(int(tm).le.npwind.or.time.le.tlast) go to 100

   if(n_wind.eq.0) then
     call outp(fn2,imax,jmax,axx,ayy,umn,vmn,um,vm,h,z)
   endif

   write(*,560)
560 format(///,'*** end of run ****',///)

stop
end
!==========================================================
!
! subroutine for model output
!
   subroutine outp(fn2,imax,jmax,axx,ayy,umn,vmn,um,vm,h,z)

   dimension umn(100,100),um(100,100),vmn(100,100),vm(100,100)
   dimension z(100,100),u3d(100,100,5),v3d(100,100,5),zh(5),h(100,100)
   character fn2*20
   open(30,file=fn2)

   do j=2,jmax-1
     do i=2,imax-1
       if(h(i,j).le.0.0) cycle
       umn(i,j)=0.5*(um(i,j)+um(i+1,j))*100.
```

```
      vmn(i,j)=0.5*(vm(i,j)+vm(i,j+1))*100.
    enddo
  enddo
  kmax=4
  zh(1)=.0
  zh(2)=-1/6.0
  zh(3)=-0.5
  zh(4)=-5/6.0

  do k=1,kmax
    do j=2,jmax-1
      do i=2,imax-1
      if(h(i,j).le.0.0) cycle
      uu3=umn(i,j)/100.0
      vv3=vmn(i,j)/100.0
      d2=zh(k)
      u3k=(3/4.*axx-3/2.0*uu3)*(d2**2-1.)+axx*(d2+1.)
      v3k=(3/4.*ayy-3/2.0*vv3)*(d2**2-1.)+ayy*(d2+1.)
      u3d(i,j,k)=u3k*100.0
      v3d(i,j,k)=v3k*100.0
      enddo ! i
    enddo ! j
  enddo ! k

  do j=jmax,1,-1
    do i=1,imax
      write(30,500) i,j,umn(i,j),vmn(i,j),z(i,j)
    enddo ! i
  enddo ! j

  do k=2,kmax
    write(30,600) k
    do j=jmax,1,-1
      do i=1,imax
        write(30,500) i,j,u3d(i,j,k),v3d(i,j,k),z(i,j)
      enddo ! i
    enddo ! j
  enddo ! k
500 format(2(1x,i3),3(1x,f9.4))
600 format('layer k=',i3)

  close(30)
  return
  end
!=====================================================================
!
! subroutine for determining the calculation domain
!
  subroutine domboun(imax,jmax,nob,h,is,ie,ib,jb,nb)

  dimension jb(1000),ib(1000),is(100),ie(100)
  dimension nb(1000),h(100,100)

  do j=1,jmax
    mm=0
    nn=0
    do i=1,imax
      if(mm.eq.1) cycle
      if(h(i-1,j).le.0.0.and.h(i,j).gt.0.0) then
        is(j)=i
```

```
        mm=1
      endif
    enddo ! i
    do i=imax,1,-1
      if(nn.eq.1) cycle
      if(h(i,j).gt.0.0.and.h(i+1,j).le.0.0) then
        ie(j)=i
        nn=1
      endif
    enddo ! i
  enddo ! j

! determine the boundary grid and boundary type
  kk=0
  do j=1,jmax
    do i=is(j),ie(j)
      if(h(i,j).le.0.0) cycle
      if(h(i-1,j).le.0.0) then
        kk=kk+1
        jb(kk)=j
        ib(kk)=i
        nb(kk)=2
      endif
      if(h(i,j-1).le.0.0) then
        kk=kk+1
        jb(kk)=j
        ib(kk)=i
        nb(kk)=3
      endif
      if(h(i-1,j).le.0.0.and.h(i,j-1).le.0.0) then
        kk=kk+1
        jb(kk)=j
        ib(kk)=i
        nb(kk)=4
      endif
    enddo ! i
  enddo ! j

  nob=kk

  return
  end
!==========================================================
!
! subroutine for time-varying wind data
!
  subroutine wind(n,time,cs,tws,twd,wx,wy,tsx,tsy,ts,wxx,wyy)
  dimension tws(100),twd(100)
! 3 hours warm up
  time99=time/3600.00
  ntime=int(time99)
  if(ntime.le.2) then
    ws0=tws(1)
    wd=twd(1)
  else
    ws0=(tws(ntime+1-2)-tws(ntime-2))*(time99-ntime)+tws(ntime-2)
    wd=(twd(ntime+1-2)-twd(ntime-2))*(time99-ntime)+twd(ntime-2)
  endif
```

```
! a step function for the first 100 time step for wind speed
  if(n.le.100) then
    ws=n/100.00*ws0
  else
    ws=ws0
  endif

  wx=-ws*cos((wd-90.0)*3.14159/180.0)
  wy=ws*sin((wd-90.0)*3.14159/180.0)
  tsx=cs*wx*sqrt(wx*wx+wy*wy)
  tsy=cs*wy*sqrt(wx*wx+wy*wy)
  ts=sqrt(tsx*tsx+tsy*tsy)

  wxx=0.415*tsx/sqrt(ts)
  wyy=0.415*tsy/sqrt(ts)

  return
  end
```

DISCLAIMER

This software are in the public domain and are furnished "as is". The authors of this book (1) make no warranty, express or implied, as to the usefulness of the software;or (2) to provide technical support to users. Permission to use or modify this software is hereby granted, provided that this disclaimer notice appears in all copies.All modifications to this software must be clearly documented, and are solely the responsibility of the agent making the modification.

REFERENCES

Abbott, M.B. (1991). *Hydroinformatics*, Avebury Technical, Aldershot.

Al-Rabeh, A.H. and Gunay, N. (1992). "On the application of a hydrodynamic model for a limited area", *Coastal Eng.*, Vol. 17, pp. 173–194.

Andreasson, P. (1991). *A numerical investigation of energy conversions in geophysical boundary layers*. Ph.D. Dissertation, Lulea University of Technology.

Backhaus, J.O. (1980). "On currents in the German Bight—A three-dimensional non-linear tidal model", Mathematical Modeling of Estuarine Physics, *Proc. Int. Symp.*, Hamburg, August 24–26, edited by J. Sundermann and K.-P. Holz, Springer, Berlin, pp. 102–132.

Baines, W.D. and Knapp, D.J. (1965). "Wind driven water currents", *J. Hydr. Div., ASCE*, Vol. 91, No. 2, pp. 205–221.

Bedford, K.W. (1981). "Spectral preservation capabilities of Great Lakes transport models", in *Transport Models for Inland and Coastal Waters*, edited by H.B. Fischer, Academic Press, London, UK, pp. 172–236.

Bedford, K., Yen, C.C., Kempf, J., Schwab, D., Marshall, R. and Kuan, C. (1990). "A 3D-stereo graphics interface for operational Great Lakes forecasts", in *Estuarine and Coastal Modelling*, edited by M.L. Spaulding, ASCE, New York, NY, pp. 249–257.

Bennett, J.R. and Clites, A.H. (1987). "Accuracy of trajectory calculation in a finite different circulation model", *J. Comp. Phys.*, Vol. 68, pp. 272–282.

Blaisdell, M.A. (1990). *Numerical model study of circulation in the Windermere Basin*, Master's Thesis, McMaster University, Hamilton, Ontario.

Blaisdell, M.A., Tsanis, I.K. and Krestenitis, Y. (1991). "Modelling the steady-stage circulation in a distorted physical model of the Windermere Basin", *J. Civil Eng.*, Vol. 18, No. 5, pp. 756–764.

Blumberg, A.F. (1986). "Turbulent mixing processes in lakes, reservoirs and impoundments", in *Physics Based Modelling of Lakes, Reservoirs and Impoundments*, edited by W.G. Gray, ASCE, New York, NY, pp. 79–104.

Blumberg, A.F. and Herring, H.J. (1987). "Circulation modeling using orthogonal curvilinear coordinates", in *Three Dimensional Models of Marine and Estuarine Dynamics*, edited by J.C.J. Nihoul, and B.M. Jamart, Vol. 45, Elsevier Oceanography Series, pp. 55–88.

Blumberg, A.F. and Kantha, L.H. (1985). "Open boundary conditions for circulation models", *J. Hydr. Eng.*, Vol. 111, No. 2, pp. 237–255.

Blumberg, A.F. and Mellor, G.L. (1987). "A description of a three-dimensional coastal ocean circulation model", in *Three-dimensional Coastal Ocean Models*, edited by N. Heaps, Vol. 4, American Geophysical Union, Washington, DC, p. 208.

Boyce, F.M. and Chiocchio, F. (1991). "Drogue measurements — Hamilton Harbour 1990", National Water Research Institute (NWRI), CCIW, Technical Note No. LRB-91-TN-08, Burlington, ON.

Brissette, F.P. (1992). *Estimation of wave directional spectra and applications to the study of surface gravity water waves*, Ph.D. Dissertation, McMaster University, Hamilton, Ontario.

Brissette, F.P., Tsanis, I.K. and Wu, J. (1993). "Wave directional spectra and current interaction in Lake St", *Clair. J. Great Lakes Res.*, Vol. 19, No. 4, pp. 553–568.

Caponi, E.A. (1976). "A three-dimensional model for the numerical simulation of estuaries", *Adv. Geophys.*, Vol. 19, pp. 189–310.

Chamberlain, C.E. and Mitchell, R. (1978). "A decay model for enteric bacteria in natural waters", in *Water Pollution Microbiology*, edited by R. Mitchell, Vol. 2, Wiley, New York, pp. 325–348.

Cheng, R.T. (1976). "Numerical models of wind driven circulation in lakes", *Appl. Math. Model.*, Vol. 1, pp. 141–158.

Cheng, R.T. and Smith, P.E. (1990). "A survey of three-dimensional numerical estuarine models", in *Estuarine and Coastal Modelling*, edited by M.L. Spaulding, ASCE, New York, NY, pp. 1–15.

Cheng, R.T., Powell, T.M. and Dillon, T.M. (1976). "Numerical models of wind-driven circulation in lakes", *Appl. Math. Model.*, Vol. 1, pp. 141–159.

Chow, V.T. (1959). *Open Channel Hydraulics*, McGraw-Hill, New York, NY.

Crean, P.B., Murty, T.S. and Stronach, J.A. (1988). *Mathematical Modelling of Tides and Estuarine Circulation*, Springer-Verlag, New York, NY.

Csanady, G.T. (1967). "Larger scale motion in great lakes", *J. Geophys. Res.*, Vol. 72, No. 16, pp. 4151–4162.

Csanady, G.T. (1975). "Hydrodynamics of large lakes", *Ann. Rev. Fluid Mech.*, Vol. 7, pp. 357–386.

Csanady, G.T. (1982). *Circulation in the Coastal Ocean*, D. Reidel, Dordrecht, Holland.

Davies, A.M. (1980a). "Application of the Galerkin method to the formulation of a three-dimensional numerical sea model", *Appl. Math. Model.*, Vol. 4, pp. 245–256.

Davies, A.M. (1980b). "On formulating a three-dimensional hydrodynamic sea model with arbitrary variation of vertical eddy viscosity", *Comput. Methods Appl. Mech. Eng.*, Vol. 22, pp. 187–211.

Davies, A.M. (1983). "Numerical modeling of stratified flow: a spectral approach", *Continental Shelf Res.*, Vol. 2, No. 4, pp. 275–300.

Davies, A.M. (1985). "Application of the Dufort–Frankel and Saul'ev methods with time-splitting to the formulation of a three-dimensional hydrodynamic sea model", *Int. J. Numer. Methods Fluids.*, Vol. 5, pp. 405–425.

Davis, J.M. (1991). "The finite element method. An alternative subdomain method for modeling unsteady flow in coastal waters and lakes", *Proc. Int. Symp. Unsteady Flow in Open Channels*, IAHR, Newcastle-upon-Tyne, UK, pp. B41–B54.

Deardorff, J.W. (1970). "A numerical study of three-dimensional turbulent channel flow at large Reynolds numbers", *J. Fluid Mech.*, Vol. 41, No. 2, pp. 453–480.

Dewey, R.J. and Palmer, M.D. (1984). "St. John's Harbour water quality model", *Can. J. Civil Eng.*, Vol. 11, pp. 65–73.

Donelan, M.A. (1980). "Similarity theory applied to forecasting of wave heights, periods and directions", *Proc. Can. Coastal Conf.*, pp. 47–61.

Donelan, M., Edler, F.C. and Hamblin, P.F. (1974). "Determination of the aerodynamic drag coefficient from wind set-up (IFYGL)", *Proc. 17th Conf. Great Lakes Res.*, Hamilton, Ontario, pp. 778–788.

Draper, D., Henry, D., Engler, F., Singer, S., Antonszek, J., Batten, S. and Walters, M. (1985). "Phosphorus modelling and control options", *Technical Report Imp. A.6*, Lake Simcoe Environmental Management Strategy (LSEMS), Lake Simcoe, Ontario.

Eid, B.M.F. (1981). *Investigation into interfacial transports and exchange flows for lake models*, Ph.D. Dissertation, McMaster University, Hamilton, Ontario.

Elder, J.W. (1959). "Dispersion of marked fluid in turbulent shear flow", *J. Fluid Mech.*, Vol. 5, pp. 544–560.

Elzawahry, A.E. (1985). *Advection, dispersion and settling in the coastal zone of Lake Erie*, Ph.D. Dissertation, McMaster University, Hamilton, Ontario.

Engel, P. (1988). "Hydraulic model study of the new Windermere Basin", *Research and Applications Branch Report*, National Water Research Institute, Canada Centre for Inland Waters, Burlington, Ontario.

Environment Canada, Environment Ontario, Ministry of Natural Resources, Metropolitan Toronto and Region Conservation Authority (1988). "*Metropolitan Toronto Remedial Action Plan—Environmental Conditions and Problem Definition*", Toronto, Ontario.

Feng, Shih-zao. (1977). "A three-dimensional nonlinear model of tides", *Scientia Sinica*, Vol. XX, No. 4, pp. 436–446.

Fischer, H.B., List, E.J., Koh, R.C.Y., Imberger, J. and Brooks, N.H. (1979). *Mixing in inland and coastal waters*, Academia Press, New York, NY.

Forssblad, L. (1947). "Effects of wind, waves and currents on floating timber", Institute of Hydraulics, Royal Institute of Technology, *Bulletin*, No. 7, Stockholm.

Ganoulis, J., Koutitas, C. and Tolikas, D. (1980). "Review of in situ measurements and analytical methods for the prediction of the water movement in Thessaloniki Bay," *Greek-U.S. Working Conference on Oceanography*, 7–11 July, Egina, Greece.

Geo-Logic Inc. (1994). "Report of baseline conditions of the PCB contaminated area prone to scouring", Little Lake, Peterborough, Project No. 94-G-105.

Goossens, L.T., Van Pagee, H.J.A. and Tessel, P.J. (1982). "Vertical diffusion in air driven water flows", *J. Hydr. Div., ASCE*, Vol. 108, No. 2, pp. 995–1009.

Gordon, R.B. and Spaulding, M.L. (1987). "Numerical simulations of the tidal- and wind-driven circulation in Narragansett Bay", *Estuarine, Coastal Shelf Sci.*, Vol. 24, pp. 611–636.

Halfon, E.H., Simons, T.J. and Schertzer, W.M. (1990). "Modelling the spatial distribution of seven halocarbons in Lake St. Clair in June 1984 using the TOXFATE model", *J. Great Lakes Res.*, Vol. 16, No. 1, pp. 90–112.

Haney, R.L. (1991). "On the pressure gradient force over steep topography in sigma coordinate ocean models", *J. Phys. Oceanogr.*, Vol. 21, pp. 610–619.

He, S. and Lin, B. (1984). "Applications of the operator splitting method to the calculation of two-dimensional tides", *Acta Oceanol. Sin.*, Vol. 6, No. 2, (in Chinese).

Heaps, N. (Editor) (1987). "Three-dimensional coastal and ocean models", Vol. 4, American Geophysical Union, Washington, DC, p. 208.

Heaps, N.S. (1972). "On the numerical solution of the three-dimensional equations for the tides and storm surge", *Mem. Soc. R. Sci. Liege*, 6e series, tome I, pp. 143–180.

Heaps, N.S. (1973). "Three-dimensional numerical model of the Irish Sea", *Geophy. J.R. Astron. Soc.*, Vol. 35, pp. 99–120.

Heaps, N.S. (1984). "Vertical structure of current in homogeneous and stratified lakes", in *Hydrodynamics of Lakes*, edited by K. Hutter, Springer, pp. 153–202.

Hinz, S.C., Katopodes, N., Freedman, P., Sullivan, M. and Freuberg, S. (1990). "Modelling total residual Chlorine in the upper potamic estuary", in *Estuarine and Coastal Modelling*, edited by M.L. Spaulding, ASCE, pp. 229–238.

Hirose, M. (1990). *Numerical simulation of the flow in lakes by a nesting method*, Master's thesis, Kyoto University, Japan (in Japanese).

Hirt, C.W. and Cook, J.L. (1972). "Calculating three-dimensional flows around structures and over rough terrain", *J. Comput. Phys.*, Vol. 10, pp. 324–340.

Ibrahim, K.A. and McCorquodale, J.A. (1985). "Finite element circulation model for Lake St. Clair", *J. Great Lakes Res.*, Vol. 11, No. 3, pp. 208–222.

Inoue, K. (1986). *Studies on the numerical methods of the unsteady flows in open channels and their applications in hydraulic engineering*, Ph.D. Dissertation, Kyoto University, Japan (in Japanese).

Irvine, D.E. (1987). "Extreme waves in the Agulhas-a case study in wave–current interaction", *J. Hopkins APL Tech. Digest*, Vol. 8, No. 1, pp. 100–106.

Ivetic, M.V. (1988). *Numerical simulation of turbulent flow in shallow flow domain by SGS modeling*, Doctoral Thesis, Kyoto University, Kyoto, Japan.

Iwasa, Y. and Inoue, K. (1985). "Mathematical simulation of plane and multi-layer flows in a large and deep lake", *Proc. 21st Congr. IAHR, Melbourne*, Vol. 2, pp. 225–259.

Iwasa, Y. and Inoue, K. (1989). "Large eddy simulation for large water bodies", *Proc. Comput. Model. Exp. Methods Hydraul.* (*Suppl.*), Dubrovnik, Yugoslavia, pp. 1–12.

Iwasa, Y., Inoue, K. and Matsuoka, T. (1985). "Some considerations on the flows in Lake Biwa by means of three-dimensional analysis", *Proc. 32nd Japanese Conference on Hydraulics*, Tokyo, Japan, pp. 751–756 (in Japanese).

Iwasa, Y., Inoue, K. and Shen, H. (1988). "The hydrodynamics of lake flows in view of numerical simulation methods", *Proc. 6th Congress of APD-IAHR*, Kyoto, Japan, pp. 217–224.

Iwasa, Y., Inoue K., Liu S. and Abe T. (1983). "Numerical simulation of flows in Lake Biwa by means of a three-dimensional mathematical model", *Annuals of Disaster Prevention Res. Inst.*, Kyoto University, No. 26 B-2, pp. 531–542 (in Japanese).

James, W. and Eid, B. (1978). "A three-dimensional model of Hamilton Harbour incorporating spatial distribution of transient surface drag", *J. Civil Eng. Can.*, Vol. 5, pp. 479–488.

Jeftic, L. (1990). "State of the Marine environment in the Mediterranean region", *UNEP Regional Seas Reports and Studies No. 132*.

Johnson, B.H., Kim, K.W., Sheng, Y.P. and Heath, R.E. (1990). "Development of three-dimensional hydrodynamic model of Chesapeake Bay", in *Estuarine and Coastal Modelling*, edited by M.L. Spaulding, ASCE, pp. 162–172.

Johnson, M.G. and Nicholls, K.H. (1988). "Temporal and spatial trends in metal loads to sediments of Lake Simcoe, Ontario", *Water, Air Soil Pollut.*, Vol. 39, pp. 337–354.

Johnson, M.G. and Nicolls, K.H. (1989). "Temporal and spatial trends in metal loads to sediments in Lake Simcoe, Ontario", *J. Great Lakes Res.*, Vol. 15, No. 2, pp. 265–282.

Kawahara, M., Kobayashi, M. and Nakata, K. (1983). "Multiple level finite element analysis and its application to tidal current flow and its application to tidal current flow in Tokyo, Bay", *App. Math. Model.*, Vol. 7, No. 6, pp. 197–211.

Ketchum, B.H. (Editor) (1983). "Estuaries and enclosed seas", in *Ecosystems of the World*, Vol. 26, Elsevier Science Publishers B.V., Amsterdam, The Netherlands.

King, I.P. (1985). "Strategies for finite element modeling of three-dimensional hydrodynamic systems", *Adv. Water Resour.*, Vol. 8, pp. 69–76.

Kinoshita, M. (1989). *A basic study on the three-dimensional analysis methods of lake flows*, Master's thesis, Kyoto University, Japan (in Japanese).

Koutitas, C.G. (1988). *Mathematical Models in Coastal Engineering*, Pentech, London, UK.

Koutitas, C.G. and Gousidou-Koutita, M. (1986). "A comparative study of three mathematical models for wind-generated circulation in coastal areas", *Coastal Eng.*, Vol. 10, pp. 12–138.

Koutitas, C.G. and O'Connor, B. (1980). "Modelling three-dimensional wind-induced flows", *J. Hydr. Div., ASCE*, Vol. 11, pp. 1843–1865.

Krestenitis, Y.N. (1987). "Numerical study of the wind-induced circulation and examination of open-sea boundary conditions — Case study of Thermaikos Gulf", *Technical Report*, Aristotle University, Thessalonika, Greece.

Kurihara, Y. and Holloway, J.L. (1967). "Numerical integration of nine-level global primitive equations model formulated by the box method", *Mon. Weather Rev.*, Vol. 95, pp. 509–530.

Lam, D.C.L., Murthy, C.R. and Simpson, R.B. (1984). *Effluent Transport and Diffusion Models for the Coastal Zone*, Springer-Verlag, New York.

Lardner, R.W. and Cekirge, H.M. (1988). "A new algorithm for three-dimensional tidal and storm surge computations", *Applied Mathematical Modeling*, Vol. 12, pp. 471–481.

Large, W.G. and Pond, S. (1981). "Open ocean momentum flux measurements in model-rate to strong winds", *J. Phys. Oceanogr.*, Vol. 11, pp. 324–336.

Lau, Y.L. and Krishnappan, B.G. (1981). "Ice cover effects on stream flows and mixing", *J. Hydr. Div., ASCE*, Vol. 107, No. 10, pp. 1225–1242.

Lauder, B.E. and Spalding, D.B. (1972). *Lectures in Mathematical Models of Turbulence*, Academic Press, New York, NY.

Leendertse, J.J. (1967). "Aspects of a computational model for long-period water-wave propagation", The RAND Corporation, Memorandum, RM-5294-PR.

Leendertse, J.J. and Liu, S.K. (1975). "A three-dimensional model for estuaries and coastal sea", *Aspects of Computation*, Dept. of Interior, RAND, Santa Monica, Vol. 2, pp. 1–29.

Leendertse, J.J. and Liu, S.K. (1977). "A three-dimensional model for estuaries and coastal seas", Vol. IV, *Turbulent Energy Computation*, The RAND Corporation, R-2187-OWRT.

Leonard, B.P. (1979). "A stable and accurate convective modeling procedure based on quadratic upstream interpolation", *Comput. Methods Appl. Mech. Eng.*, Vol. 19, pp. 59–98.

Leonard, B.P. (1988). "Simple high-accuracy resolution program for convective modelling of discontinuities", *Int. J. Numer. Methods Fluids*, Vol. 8, pp. 1291–1318.

Li, C.Y., Kiser, K.M. and Rumer, R.R. (1975). "Physical model study of circulation patterns in Lake Ontario", *Limnol. Oceanogr.*, Vol. 20, No. 3, pp. 323–337.

Liu, S.-K. and Leendertse, J.J. (1978). "Multidimensional numerical modeling of estuaries and coastal seas", *Adv. Hydrosci.*, Vol. 11, pp. 95–164.

Lynch, D.R. and Werner, F.E. (1987). "Three-dimensional hydrodynamics on finite elements. Part I: Linearized harmonic model", *Int. J. Numer. Methods Fluids*, Vol. 7, pp. 871–909.

Lynch, D.R. and Werner, F.E. (1991). "Three-dimensional hydrodynamics of finite elements", *PartI-Non-linear time-stepping model. Int. J. Num. Meth. in Fluids*, Vol. 12, pp. 507–533.

MacLaren Plansearch (1989). "Water movement study on the Otonabee River and Little Lake", Report for the Peterborough Pollution Control Plan, Toronto, ON.

Malanotte-Rizzoli, P. and Robinson, A.R. (Editors) (1991). "Dynamics of atmospheres and oceans", *Special Issue: The Mediterranean Sea*, Vol. 15, Nos. 3–5, p. 5.

Marsalek, J., Dutka, B.J. and I.K. Tsanis. (1992). "Urban impacts on bacteriological pollution of the St. Clair River in Sarnia: sources and cleanup", *National Water Research Institute*, Contribution 92–138, Burlington, Ontario.

Marsalek, J., Dutka, B.J. and I.K. Tsanis. (1994). "Urban impacts on bacteriological pollution of the St. Clair River in Sarnia, Ontario", Poster BP118, IAWQ Biennial, Budapest, Hungary, July 24–29.

Masch, F.D. (1963). "Mixing and dispersion of wastes by wind and wave action", *Int. J. Air Water Pollut.*, Vol. 7, pp. 697–720.

Matsuno, T. (1966). "Numerical integrations of the primitive equations by a simulated backward difference method", *J. Meteorol. Soc. Japan, Ser. 2*, Vol. 44, No. 1, pp. 76–84.

McCorquodale, J.A., Ibrahim, K. and Hamdy, Y. (1986). "Fate and transport modelling of perchloroethylene in the St. Clair river", *Water Pollut. Res. J. Can.*, Vol. 21, No. 3, pp. 398–410.

Mellor, G.L. (1973). "Analytic prediction of the properties of stratified planetary surface layers", *J. Atmos. Sci.*, Vol. 30, pp. 1061–1069.

Miller A.R. (1972). "Deep convection in the Aegean Sea", in *Processus de formation des eolux oceaniques profondes, Coll. Int. C.N.R.S.*, No. 215.

Miller, M.J. and Blumberg, A. (1985). "Modeling vertical and horizontal diffusivities with the sigma coordinator system", *Monthly Weather Review*, Vol. 113, pp. 1379–1383.

Miller, M.J. and Thorpe, A.J. (1981). "Radiation conditions for the lateral boundaries of limited-area numerical models", *Quart. J. R. Met. Soc.*, Vol. 107, pp. 615–628.

MOE (1974). *Hamilton Harbour Study*, Water Resources Branch, MOE, Toronto, Ontario.

MOE (1989). *Toronto Waterfront RAP, RAND Model Simulation*, Great Lakes section, MOE, Toronto, Ontario.

MOE (1992a). *Collingwood Harbour Remedial Action Plan-Stage 2 Report*, Water Resources Branch, MOE, Toronto, Ontario.

MOE (1992b). *Remedial Action Plan for Hamilton Harbour, Environmental Conditions and Problem Definitions*, Toronto, Canada, RAP stage 1 and 2.

Murakami, K., Morikawa, M. and Horie, T. (1985). "Numerical simulations of storm surges by ADI method", *Technical Note of the Port and Harbour Research Institute*, Ministry of Transport, No. 529, pp. 1–35 (in Japanese).

Murthy, C.R., Simons, T.J. and Lam, D.C.L. (1985). "Dynamic and transport modelling of the Niagara River plume in Lake Ontario", *Rapp. P.-v. Reun. Con. Int. Explor. Mer.*, Vol. 186, pp. 150–164.

Neil, J. (1990). "Lake Simcoe hypolimnion aeration. An assessment for the potential for direct treatment of oxygen depleted hypolimnetic waters in Lake Simcoe", *Technical Report Imp. B.9*. Lake Simcoe Environmental Management Strategy, LSEMS, p. 105.

Nydegger, P. (1979). "Water circulation in lakes, research in situ, using revolving lake models and comparisons with the latest umnosedimen to logical results", in *Hydrodynamics of Lakes*, edited by W.H. Graf, and C.H. Mortimer, Elsevier, Amsterdam, The Netherlands, pp. 161–170.

Ookubo, K., Muramoto, Y., Oonishi, Y. and Kumagai, M. (1984). "Laboratory experiments on thermally induced currents in Lake Biwa", *Bull. of Disaster Prev. Res. Inst.*, Kyoto University, Vol. 34, Part 2, pp. 19–54.

Owen, A. (1980). "A three-dimensional model of the Bristol Channel", *J. Phys. Oceanogr.*, Vol. 10, pp. 1290–1302.

Palmer, M.D. and Dewey, R.J. (1984). "Beach fecal coliforms", *Can. J. of Civ. Eng.*, Vol. 11, pp. 21–224.

Palmer, M.D. and Dewey, R.J. (1987). "Verification of a numerical beach water quality model", *Can. J. Civil Eng.*, Vol. 14, pp. 559–570.

Pearce, B.R. and Cooper, C.K. (1981). "Numerical circulation model for wind-induced flow", *J. Hydraulic Eng., ASCE*, Vol. 107, No. 3, pp. 285–301.

Phillips, O.W. (1981). "The structure of short gravity waves on the ocean surface", in *Spaceborne Synthetic Aperture Radar for Oceanography*, edited by R.C. Beal, P.S. DeLeonibus, and I. Katz, Johns Hopkins University Press, pp. 24–31.

Ramming, H. and Kowalik, Z. (1980). *Numerical Modelling of Marine Hydrodynamics*, Elsevier, Amsterdam, p. 368.

Rasmussen, H. and Badr, H.M. (1979). "Validation of numerical models of the unsteady flow in lakes", *Appl. Math. Model.*, Vol. 3, pp. 410–416.

Reid, R.O. (1957). "Modification of the quadratic bottom-stress law of turbulent channel flow in the presence of surface wind-stress", *Tech. Memo.*, No. 93, Beach Erosion Board, US Army Corps of Engineers, Washington, DC.

Report prepared by a research team from Aristotle University and University of Athens for the Institute of Marine Biology of Crete (in Greek) (1993). "Environmental impact of the sewage disposal in Iraklion Bay", (1993). Report prepared by a research team from Aristotle University and University of Athens for the Institute of Marine Biology of Crete (in Greek).

Roache, P.J. (1972). *Computational Fluid Dynamics*, Hermosa, Albuquerque, New Mexico.

Roberson, J.A. and Croew, C.T. (1990). *Engineering Fluid Mechanics*, Houghton Mifflin, Boston, MA.

Rodi, W. (1978). "Turbulence models and their application in hydraulic — A state of the art review", *Technical Report No. SFB 80/T/127*, Univ. Karlsruhe, Germany.

Roed, L.P. and Cooper, C.K. (1986). "Open boundary conditions in numerical ocean models", *Advances in Physical Oceanography and Numerical Modelling*, edited by James J. O'Brien, D. Reidel, pp. 411–436.

Rumer, R.R. and Hooper, J.A. (1970). "Modeling great lakes circulation", *Proc. Symp. on the Water Environment and Human Needs*, MIT, Cambridge, MA, October 1–2, pp. 212–247.

Ryan, P.J. (1971). "Prediction of the annual cycle of temperature change in a stratified lake or reservoir", *Mathematical Model and User's Manual*, MIT Report, No. 137.

Schlichting, H. (1968). *Boundary Layer Theory*, McGraw-Hill, New York, NY.

Schwab, D.J. (1981). "A two-dimensional lake circulation modelling system", NOAA Technical Memorandum ERL GLERL-38, Ann Arbor, Mi, USA.

Schwab, D.J. (1983). "Numerical simulation of low frequency current fluctuations in Lake Michigan", *J. Phys. Oceanogr.*, Vol. 13, pp. 2213–2224.

Schwab, D.J. (1992). "Hydrodynamic modelling in the Great Lakes from 1950 to 1990 and prospects for the 1990's", in *Chemical Dynamics in Fresh Water Ecosystem*, edited by F.A.B.P. Gobas, and J.A. McCorquodale, Lewis Publishers, Boca Raton, Florida.

Schwab, D.J. and Bedford, K.W. (1994). "Initial implementation of the Great Lakes forecasting system: A real-time system for predicting lake circulation and thermal structure", *Water Pollut. Res. J. Can.*, Vol. 29, No. 2–3, pp. 203–220.

Schwab, D.J., Clites, A.H. and Murthy, C.R. (1989). "The effect of wind on transport and circulation in Lake St. Clair", *J. Geophys. Res.*, Vol. 94, No. C4, pp. 4947–4958.

Shen, H. (1988). *Basic studies on the numerical methods of lake flows*, Master's Thesis, Kyoto University, Japan (in Japanese).

Shen, H. (1991). *Numerical analysis of large-scaled flows and mass transport in lakes*, Ph.D. Dissertation, Kyoto University, Japan.

Shen, H., Iwasa, Y. and Inoue, K. (1990a). "Three-dimensional analysis of flows in Lake Biwa and their characteristics", *Proc. 7th Congr. of APD-IAHR*, Beijing, China, Vol. 3, pp. 239–244.

Shen, H., Iwasa, Y. and Inoue, K. (1990b). "Three-dimensional numerical analysis of flows without using hydrostatic approximation", *Proc. 45th Annual Conf. of JSCE*, Vol. 2, pp. 578–579 (in Japanese).

Shen, H., Tsanis, I.K. and D'Andrea, M. (1995). "A three-dimensional hydro-dynamic/pollutant transport simmulation model for the nearshore areas of Lake Ontario", *J. Great Lakes Res.*, Vol. 21, No. 2, pp. 161–177.

Sheng, Y.P. (1987). "On modeling three-dimensional estuarine and marine hydrodynamics", in *Three-Dimensional Models of Marine and Estuarine Dynamics*, edited by J.C.J. Nihoul, and B.M. Jamart, Elsevier, Amsterdam, The Netherlands, pp. 35–53.

Sheng, Y.P. (1983). "Modeling of three-dimensional coastal currents and sediments dispersion. Vol.1-Model Development and Application", *Technical Report CERC-83-2*, Coastal Engineer Research Center, U.S. Army Engineer Waterways Experiment Station, Vicksburg, MS.

Sheng, Y.P. (1990). "Evolution of three-dimensional curvilinear-grid hydrodynamic model for estuary, lakes and coastal lakes. CH3D", in *Estuarine and Coastal Modelling*, edited by M.L. Spaulding, ASCE, New York, NY, pp. 40–49.

Sheng, Y.P., Choi, J.K. and Kuo, Y. (1990). "Three-dimensional numerical modelling of tidal circulation and sanitary transport in the James River estuary", in *Estuarine and Coastal Modelling*, edited by M.L. Spaulding, ASCE, New York, NY, pp. 209–218.

Simons, D.B. and Senturk, F. (1992). *Sediment Transport Technology — Water and Sediment Dynamics*, Water Resources, Littleton, CO.

Simons, T.J. (1972). "Development of numerical models of Lake Ontario. Part 2", *Proc. 15th Conf. Great Lakes Res. 1972, Internat. Assoc. Great Lakes Res.*, University of Wisconsin-Madison, Wisconsin, pp. 654–669.

Simons, T.J. (1980). "Circulation models of lakes and inland seas", *Can. Bull. Fish. Aquat. Sci.*, Bull. 203, Ottawa, Canada.

Simons, T.J. (1985). "Reliability of circulation models", *J. Phys. Oceanogr.*, Vol. 15, pp. 1191–1204.

Simons, T.J. and Lam, D.C.L. (1986). "Documentation of a two-dimensional X-Y model package for computing lake circulations and pollutant transport", in *Physical Based Modeling of Lake, Reservoirs, and Impoundments*, edited by W.G. Gray, ASCE, New York, NY, pp. 258–308.

Simons, T.J. and Schertzer, W.M. (1989a). "Modelling wind-induced water set-up in Lake St. Clair", *J. Great Lakes*, Vol. 15, No. 3, pp. 452–464.

Simons, T.J. and Schertzer, W.M. (1989b). *The circulation of Lake Ontario during the summer of 1982 and the winter of 1982/1983*. Environmental Canada, Scientific Series, No. 171.

Smith, S.D. and Anderson, R.J. (1984). "Spectra of humidity, temperature, and wind over the sea at Sable Island, Nova Scotia", *J. Geophys. Res.*, Vol. 89, pp. 2020–2040.

Snodgrass, W.J. and Holubeshen, J. (1993). "Hypolimnetic oxygen dynamics in Lake Simcoe, Part3. Model confirmation and prediction of the effects of management", *LSEMS Technical Rep. Imp. B16*.

Spaulding, M.L., Howlett, E., Jayko, K., Anderson, E. and Isaji, T. (1992). "A shell approach to modeling oil spill trajectory and fate and search and rescue operations", in *Estuarine and Coastal Modeling*, edited by M.L. Spaulding, K. Bedford, A. Blumberg, R. Cheng, and C. Swanson, American Society of Civil Engineering, New York.

Spillane, K.T. and Hess, G.D. (1978). "Wind-induced drift in contained bodies of water", *J. Phys. Ocean.*, Vol. 8, pp. 930–935.

Stephens, C.V. (1986). "A three-dimensional model for tides and salinity in the Bristol Channel", *Continental Shelf Res.*, Vol. 6, No. 4, pp. 531–560.

Svensson, U. (1978). "Mathematical model of the seasonal thermocline", *Report No. 1002*, Department of Water Resources Engineering, University of Lund, Sweden.

Swanson, J.C. (1986). *A three-dimensional numerical model system of coastal circulation and water quality*, Ph.D. dissertation, University of Rhode Island, U.S.A.

Tee, K.T. (1987). "Simple models to simulate three-dimensional tidal and residual currents", in *Three-Dimensional Coastal Ocean Models, Coastal and Estuarine Sciences*, edited by N.S. Heaps, Vol. 4, AGU, Washington, DC, pp. 125–147.

Thackston, E.L., Shields, F.D. and Schroeder, P.R. (1987). "Residence time distributions of shallow basins", *J. Environ. Eng. (ASCE)*, Vol. 113, No. 6, pp. 1319–1332.

The Land Sub-Group (1985). "Overview of phosphorus sources, loads and remedial measures", *Technical Report Imp. A*, Lake Simcoe Environmental Management Strategy (LSEMS).

Tomidokoro, G. (1984). *Basic study on the hydraulic characteristics of the wind-driven currents and dispersion in closed shallow water basins*, Ph.D. Dissertation, Kyoto University, Japan (in Japanese).

Tsanis, I.K. (1986). *Characteristics of shear-induced countercurrent flow*, Ph.D. Dissertation. University of Toronto, Canada.

Tsanis, I.K. (1989). "Simulation of wind-induced water currents", *J.Hydr. Eng., ASCE*, Vol. 115, No. 8, pp. 1113–1134.

Tsanis, I.K. and Leutheusser, H.J. (1987). "An example of transient laminar countercurrent flow", *Trans. ASME*, Vol. 109, pp. 262–267.

Tsanis, I.K. and Leutheusser, H.J. (1988). "The structure of turbulent shear-induced countercurrent flow", *J. Fluid Mech.*, Vol. 189, pp. 531–552.

Tsanis, I.K. and Shen, H., (1994a). "North Crete waterfront receiving water modelling study", *Technical Report*, Department of Civil Engineering, McMaster University, Hamilton, ON.

Tsanis, I.K. and Shen, H. (1994b). "Modelling current movements in Cootes paradise", *Technical Report*, Department of Civil Engineering, McMaster University, Hamilton, ON, pp. 285–299.

Tsanis, I.K. and Shen, H. (1994c). "Metropolitan Toronto receiving water model development project, Phase I", Mimico Creek Pilot Study, *Technical Report* submitted to the Metropolitan Toronto and Region Remedal Action Plan Office and the Toronto Waterfront Model Development Steering Committee, McMaster University, Hamilton, ON, Canada.

Tsanis, I.K. and Wu, J. (1991). "Study of water currents in the Canadian Great Lakes", *Atmospheric Environment Service (AES) Report*, March 1991.

Tsanis, I.K. and Wu, J. (1992). "Remedial measures for Sarnia Bay", *National Water Research Institute (NWRI) Report*, Rivers Research Branch, Environment Canada, Burlington, Ontario.

Tsanis, I.K. and Wu, J. (1993). "Modelling current movements in Hamilton Harbour", *Department of Fisheries and Oceans (DEFO) Report*, Phase I & II.

Tsanis, I.K. and Wu, J. (1994). "LMS-an integrated lake modeling system", *Environ. Software,*, Vol. 9, pp. 103–113.

Tsanis, I.K. and Wu, J. (1995). "A nested-grid hydrodynamic/pollutant transport model of nearshore areas in Hamilton Harbour", *Water Quality Res. J. Can.*, Vol. 30, No. 2, pp. 205–229.

Tsanis, I.K., Blaisdell, M. Krestenitis, Y., Chu, V.H. and Engrl, P. (1990). "Physical and mathematical model comparisons for Windermere Basin", *Proc. Int. Assoc. Hydraulic Res. Conf.*, Cambridge, M.I.T., Cambridge, MA.

Tsanis, I.K., Shen, H., Dounas, C. and Eleftheriou, A. (1994b). "Application of POM and IDOR hydrodynamic models in the nearshore areas of North Crete", *Hydrosoft*, Vol. 94, pp. 21–23 September, Porto Carras, Greece.

Tsanis, I.K., Wu, J., Valeo, C. and Boyle, S. (1994a). *Hydrodynamic Modelling of Little Lake, Peterborough*, Environmental Hydraulics Group, Department of Civil Engineering, McMaster University, Hamilton, ON.

Tsanis, I.K., Valeo, C., Wu, J. and Boyle, S. (1996). "Managing contaminated sediments using a hydrodynamic model and a GIS", *Environ. Tech.*, Vol. 17, pp. 877–883.

Tsanis, I.K. and Boyle, S.I. (1997). *A Closely Coupled Hydrodynamic/Pollutant Transport GIS Model*, MEDCOAST 97, Qawra, Malta.

Tsuruya, H., Nakamo, S. and Kato, H. (1985). "Experimental study on wind driven current in a wind-wave tank — Effect of return flow on wind driven current". *The Ocean Surf.*, pp. 425–430.

U.S. Army Corps. of Engineers (1983). "St. Clair River aerial drogue survey-surface velocities and directions", Spring of 1982, Detroit District, September 1983.

Ushijima, S. and Moriya, S. (1988). "Calculation of thermal stratification phenomena in a cylindrical vessel", *Proc. 3rd Int. Sym. Refined Flow Modeling and Turbulence Measurements*, Tokyo, pp. 61–68.

Van Dorn, W.G. (1953). "Wind stress on an artificial pond", *J. Mar. Res.*, Vol. 12, No. 3, pp. 249–276.

Van Zuylen, H.J., Dee, D.P., Mynett, A.E., Rodenuits, G.S., Moll, J.R., Ogink, H.J.M., Most, V.D., Gerritsen, H. and Verboom, G.K. (1994). "Hydroinformatics at delft hydraulics", *J. Hydraulic Res.*, Vol. 32, pp. 83–136.

Vepsa, H., Alasaarela, H. and Sarkkula, J. (1992). "Model development for operational use to help spill combating and sea rescue", in *Estuarine and Coastal Modelling*, edited by M.L. Spaulding, K. Bedford, A. Blumberg, R. Cheng, and C. Swanson, ASCE, New York, NY.

Wang, J. and O'Connor, J. (1975). *Mathematical Modelling of Nearshore Circulation*, R.M. Parsons Lab. for Water Resources and Hydrodynamic T.R. 200 MIT, Cambridge, MA.

Wang, J.D., Blumberg, A.F. and Butler, H.L. (1990). "Transport prediction in partially stratified tidal water", *ASCE Tide Hydraulics Verification Task Committee*, Vol. 116, No. 3, pp. 380–396.

Wu, J. (1969). "Wind stress and surface roughness at air–sea interface", *J. Geophys. Res.*, Vol. 10, No. 34, Part 2, pp. 444–445.

Wu, J. (1975). "Wind-induced drift currents", *J. Fluid Mech.*, Vol. 68, No. Part 1, pp. 49–70.

Wu, J. (1992). "Variation of the heat transfer coefficient with environmental parameters", *J. Phys. Oceanogr.*, Vol. 22, pp. 293–300.

Wu, J. (1993). *An integrated hydrodynamic/pollutant transport model for the nearshore areas of the Great Lakes and their tributaries*, Ph.D. Dissertation, Department of Civil Engineering, McMaster University, Hamilton, ON.

Wu, J. and Tsanis, I.K. (1991). "Quasi 3D hydrodynamic modelling for Lake St. Clair," *11th CSCE Annual Conf.*, May 29–31, Vancouver, BC, pp. 26–35.

Wu, J. and Tsanis, I.K. (1994). "Pollutant transport and residence time in a model basin", *J. Hydraulic Res.*, Vol. 32, No. 4, pp. 583–598.

Wu, J. and Tsanis, I.K. (1995a). "Numerical study of wind-induced currents", *J. Hydraulic Eng. (ASCE)*, Vol. 121, No. 5, pp. 388–395.

Wu, J. and Tsanis, I.K. (1995b). "VHI3D-A vertical/horizontal integrating wind-induced circulation model", *Adv. in Water Res.*, Vol. 18, No. 2, pp. 77–87.

Wu, J., Pu, P., Wang, Q.Q. and Huang, Q. (1987). "Features of current and horizontal dispersion fields at Dapu River mouth in Lake Tai", *Int. Sym. in Lake Tai and Lake Biwa*, Nanjing, China, pp. 21–25.

Yanenko, N.N. (1971). "Method of fractional steps: the solution of problems of mathematical physics in several variables", in edited by M. Holt, Springer-Verlag, New York.

Yen, C.C.J., Keith, W., Bedford, M. and Schwab, D.J. (1992). "Nowcast protocol for the Great Lakes forecasting system", in *Estuarine and Coastal Modelling*, edited by M.L. Spaulding, K. Bedford, A. Blumberg, R. Cheng, and C. Swanson, American Society of Civil Engineering, New York.

SUBJECT INDEX